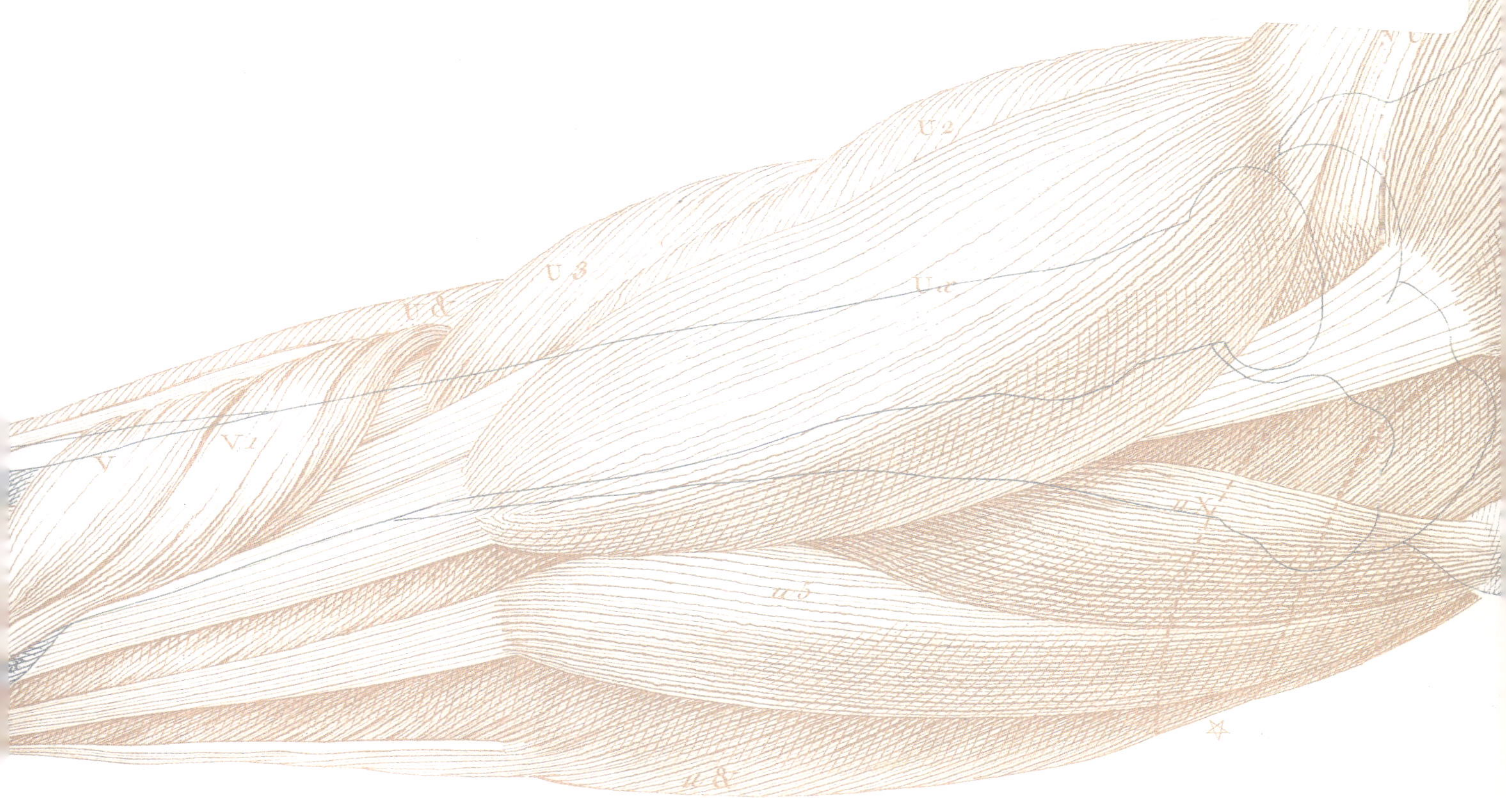

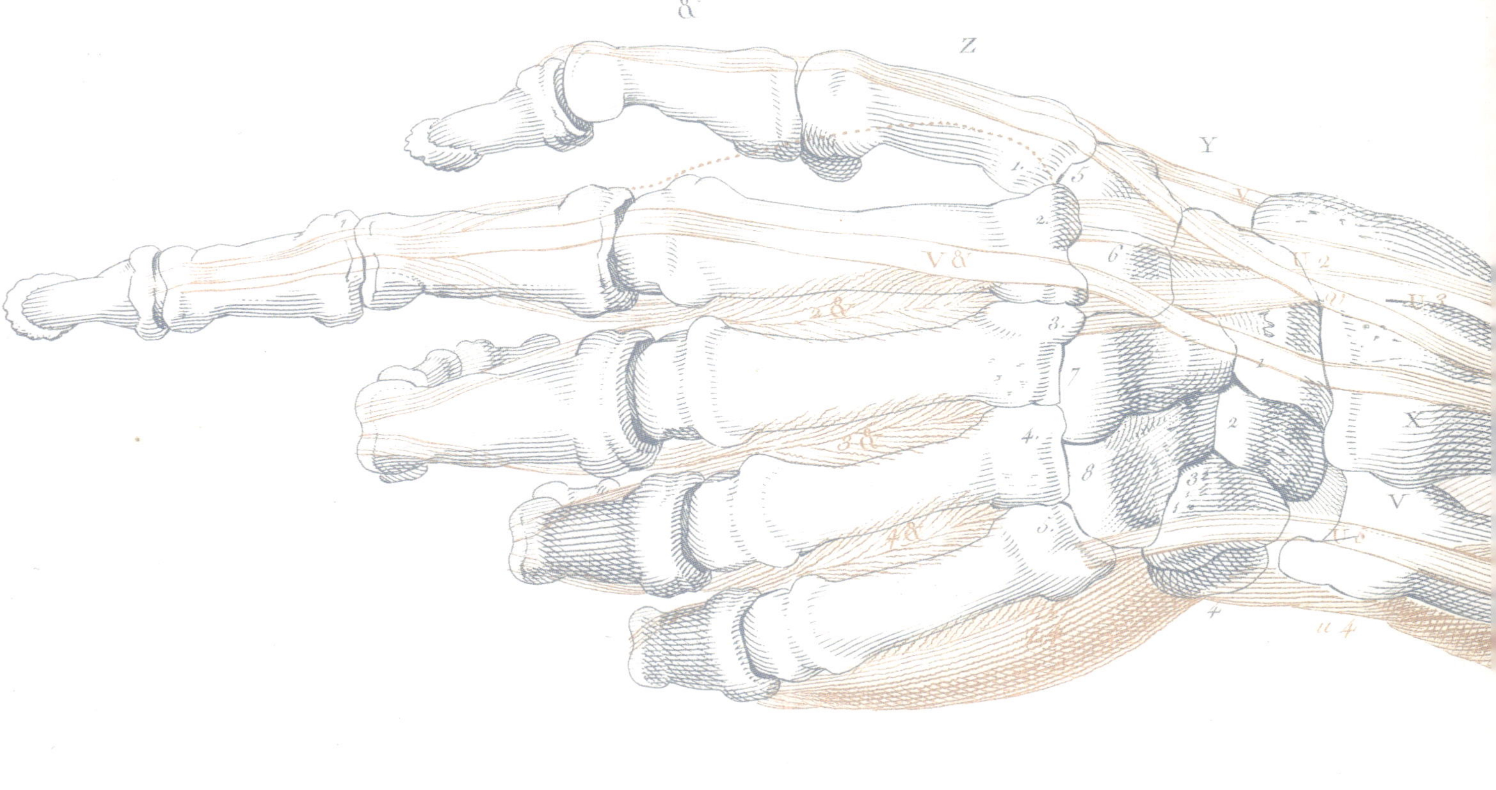

...e Hypothénar et court fléchisseur du Petit Doigt.
...ur du Pouce. ✡ Ce trait pointillé indique la division
... Biceps X s.

1. Os du Pouce.
2. _ De l'Index.
3. _ Du Milieu.
4. _ De l'Annulaire

FLESH AND BONES

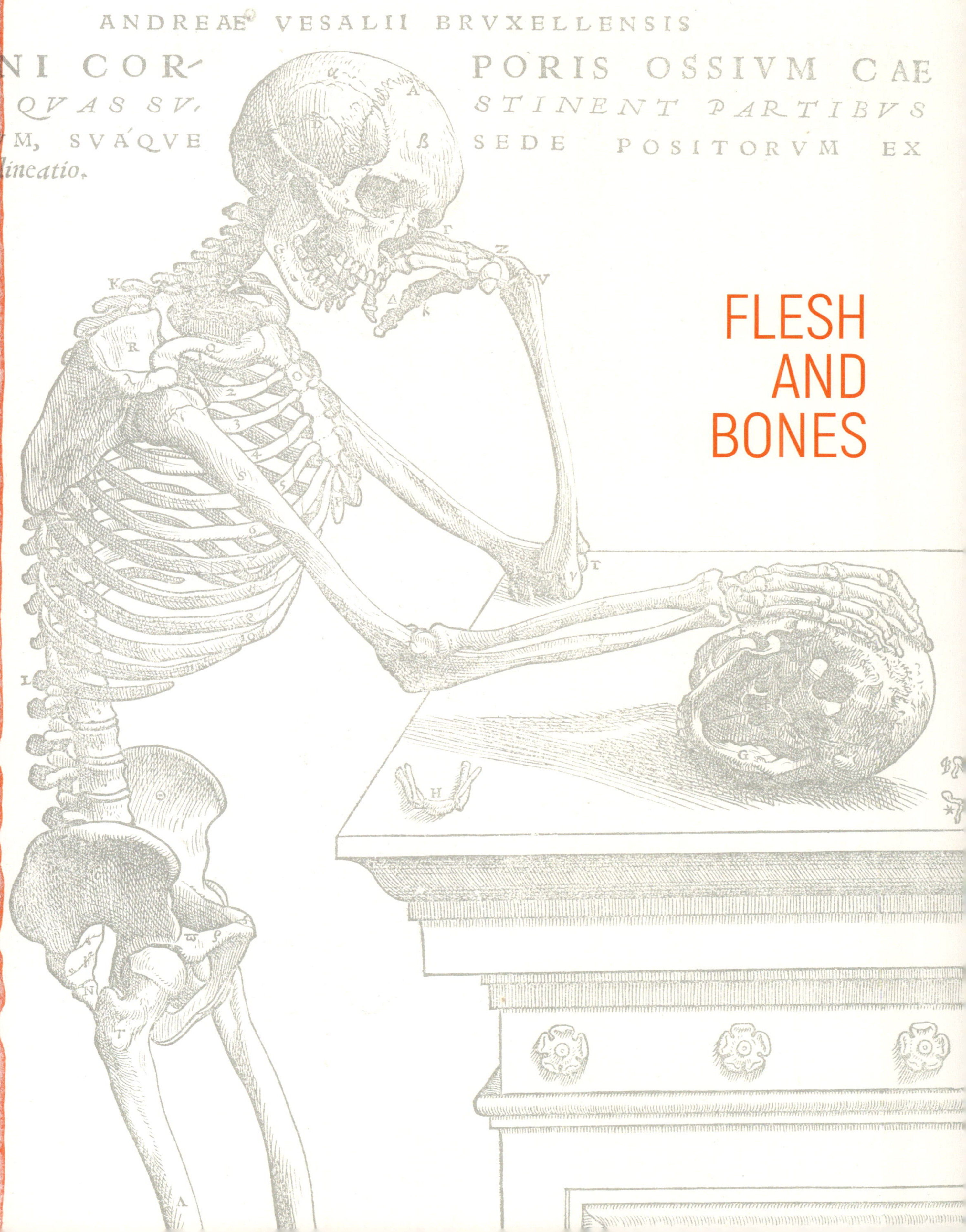

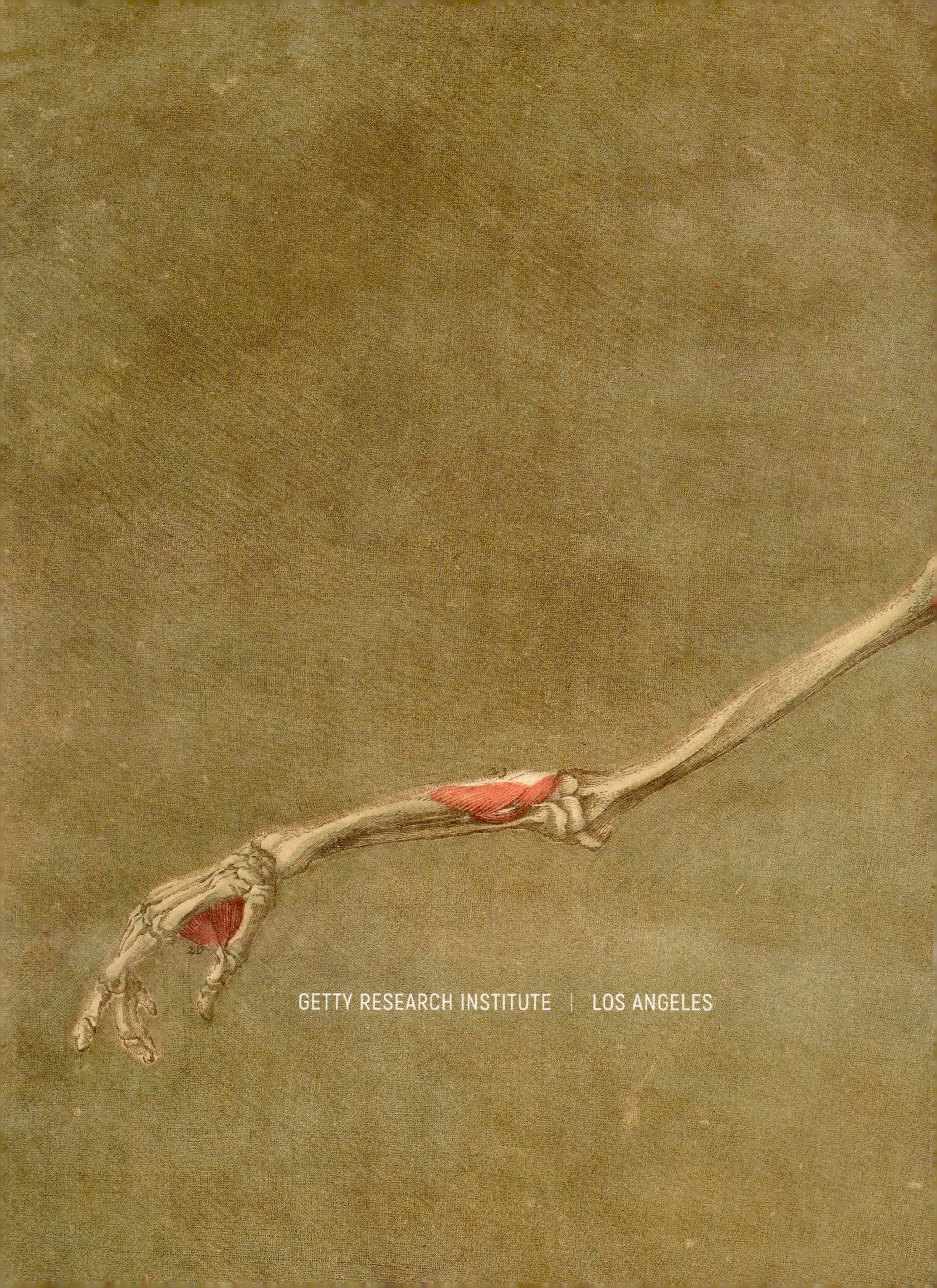

GETTY RESEARCH INSTITUTE | LOS ANGELES

FLESH AND BONES

The Art of Anatomy

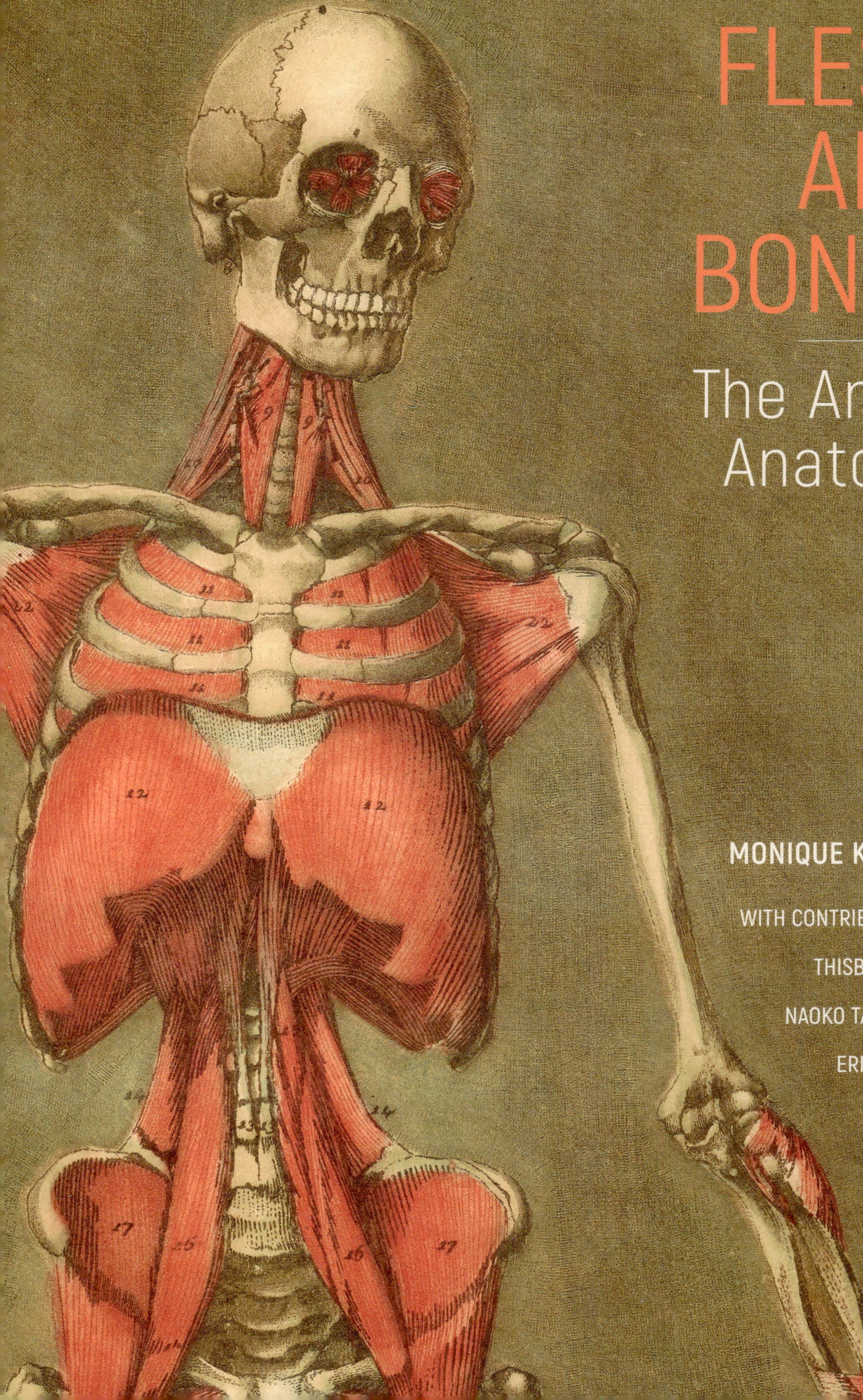

MONIQUE KORNELL

WITH CONTRIBUTIONS BY

THISBE GENSLER

NAOKO TAKAHATAKE

ERIN TRAVERS

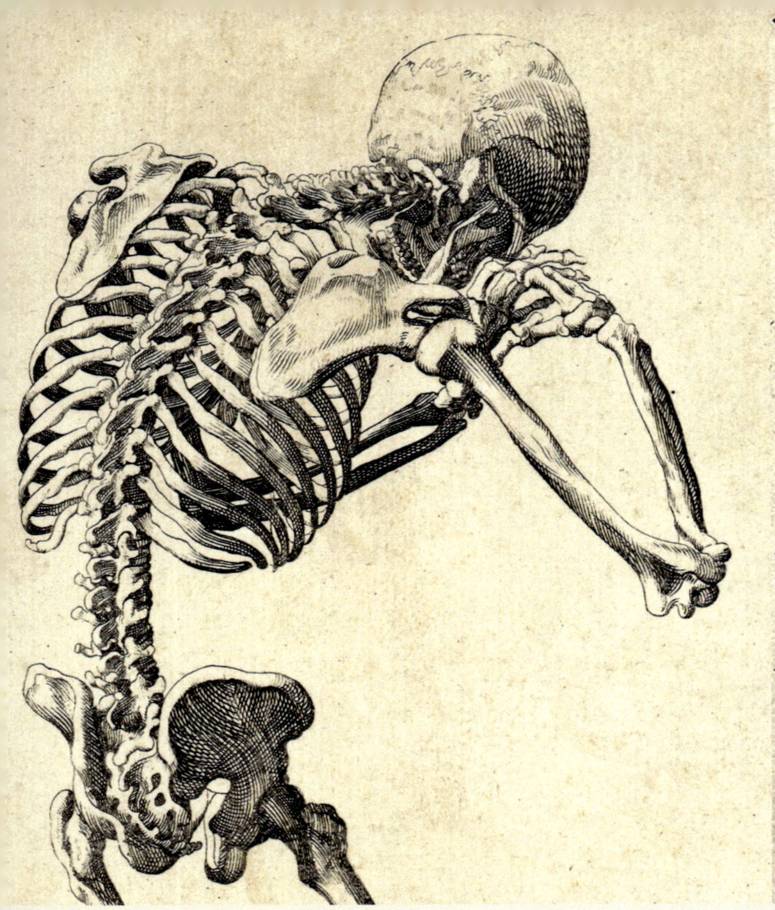

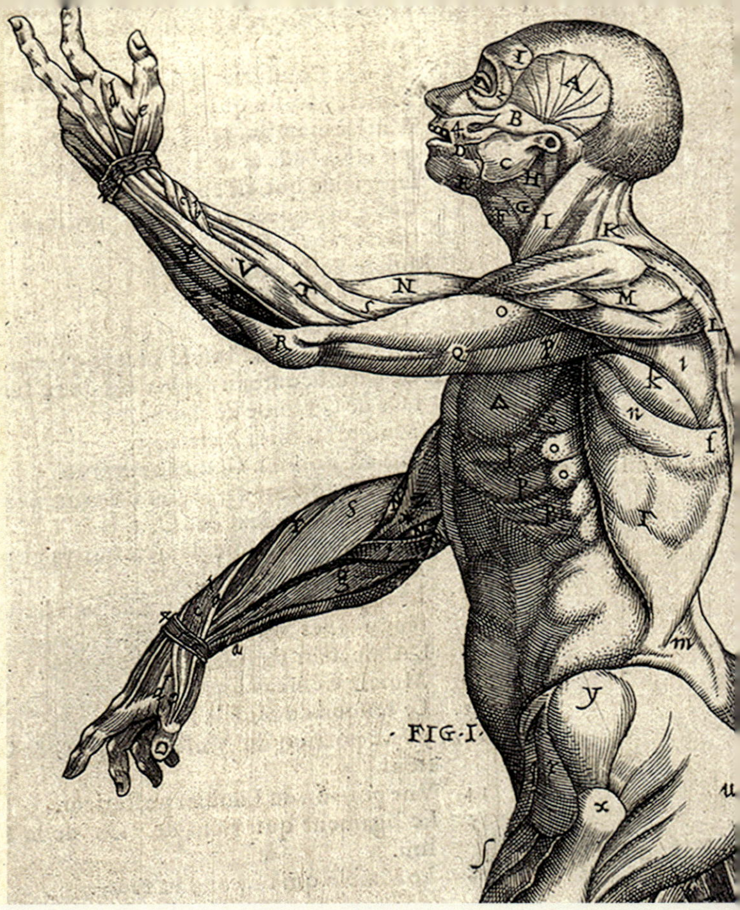

contents

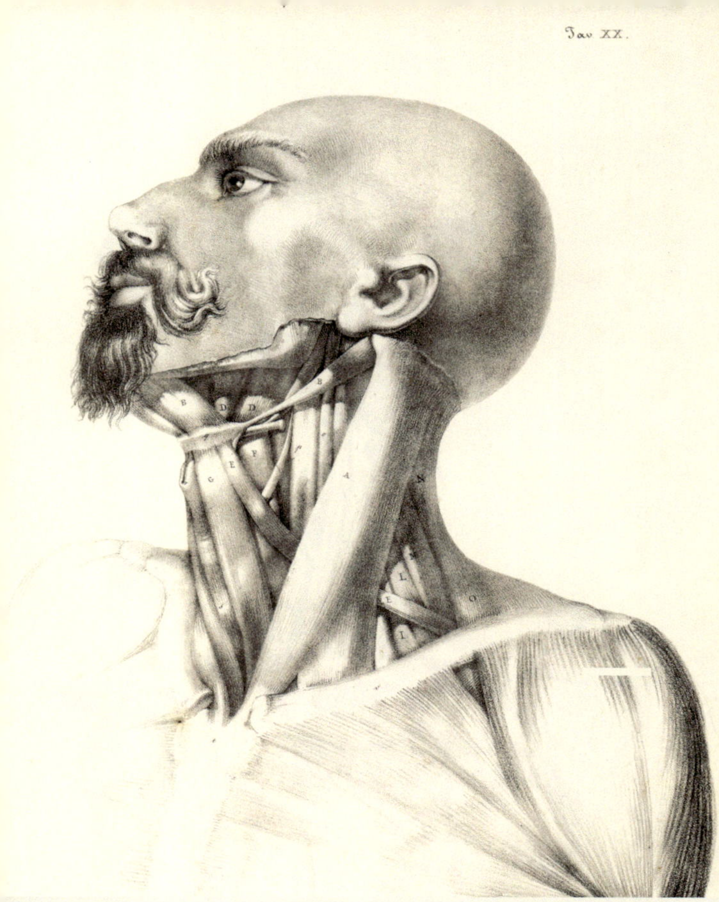
Tav XX.

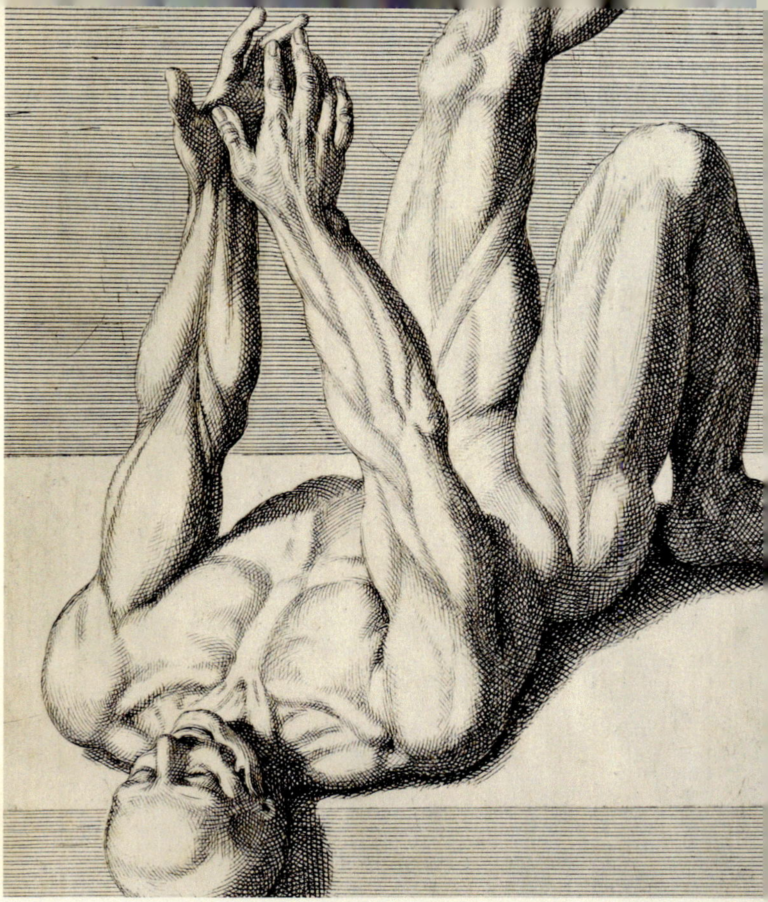

foreword

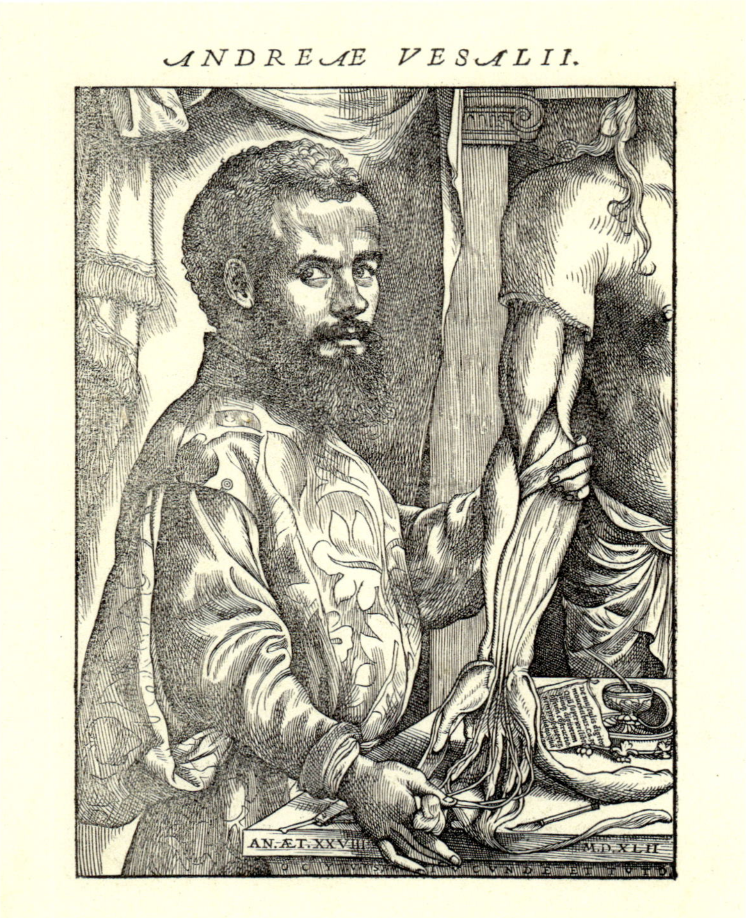

FIG. 0.1
AFTER JAN STEVEN VAN CALCAR (North Netherlandish, ca. 1515–ca. 1546). Portrait of Andreas Vesalius, woodcut. From Andreas Vesalius, *De humani corporis fabrica libri septem* (Basel: J. Oporinus, 1543). Los Angeles, Getty Research Institute, 84-B27611.

Andreas Vesalius proclaimed in his famed anatomy book, *De humani corporis fabrica libri septem* (1543), that the work of the anatomist is defined as *manuum munus:* using the hands to touch what is being studied so that the brain can fully grasp its form and import. In a woodcut portrait in the *Fabrica,* Vesalius is shown studying a flayed right hand to understand its complicated structure and how the ligaments and tendons work (fig. 0.1). The anatomist is tugging a flexor muscle—the muscle of civilization—which allows the hand to grip an object such as a pen, scalpel, or brush. In declaring that tactility has to complement visuality, Vesalius asserts that firsthand experience is just as important as knowledge of the classical medical tradition, and that hands directly exploring the human body could correct the pronouncements of ancient authorities on anatomy, such as Galen. The *Fabrica*'s woodcuts of a normative, idealized male figure—man as God's handiwork—became an instant success in Europe, even though the book was large, weighty, and expensive.

The developing science of investigating the human body using sight and touch, as promoted by Vesalius, occurred concurrently to the European exploration of distant lands. The complex process of naming and renaming places through the continuous creation and modification of atlases and maps found its counterpart in the labeling of illustrated body parts in anatomy books. Humans as a microcosm of the world, in the medieval comprehension of the word, now had to be understood through flesh, bones, nerves, muscles, ligaments, and nodes. That the announcement of this new epistemology came out the same year as Nicolaus Copernicus's *De revolutionibus orbium coelestium* (1543), which revolutionized the understanding of the macrocosm, confirms the momentousness of the changes taking place in Europe during the period.

Artists also engaged in this new exploration of the human body, eagerly taking up the study of anatomy for themselves. A number of anatomy books made specifically for artists were published in subsequent centuries, such as Carlo Cesi's *Cognitione de muscoli del corpo humano per il disegno* (1679) and Jean-Galbert Salvage's *Anatomie du gladiateur combattant* (1812) (cat. nos. 20, 34). They focused on the bones and superficial muscles, often illustrating poses pleasing to the eye. In conveying only the information relevant to the artist, their purpose was to aid the artist in making figures anatomically accurate.

Flesh and Bones: The Art of Anatomy celebrates the fervid exchange between anatomists, surgeons, and artists from the sixteenth through the twenty-first century by offering a historiography of anatomical illustration that directly engages these imaginative artworks. The analytical essays by curator Monique Kornell investigate the birth of anatomical illustration; the creation and use of anatomical books for artists; the utilization of different media to facilitate the rendering and understanding of the body's functions; and the influence of classical body forms on the representation of dissection. The essay by Erin Travers centers on the anatomical collection of Frederik Ruysch, the Dutch physician, botanist, and municipal obstetrician who gained fame for his lifelike preparations made with groundbreaking techniques (cat. no. 6); although primarily meant to instruct medical students, they also amazed their many lay visitors. Thisbe Gensler demonstrates how innovations in the graphic techniques used to visualize the body in three dimensions—such as fugitive sheets, wax models, stereographic photography, and X-rays—transformed how we see and understand the human body.

The inner workings of the human form fascinated both anatomists and artists, and their cooperation in illustrating it became mutually beneficial. The new medium of print facilitated the diffusion of knowledge and the circulation of artworks. Draftsmen prepared their drawings in the same chamber where the anatomy lessons were taking place, and they sometimes lived with anatomists in order to more easily record dissections. Vesalius kept human remains for study in his room, and he wrote of stealing criminals' bodies during his student days. He showed great attention to the production and printing of the woodcut illustrations in his books. Similarly, Bernhard Siegfried Albinus, whose illustrations to the *Tabulae sceleti et musculorum corporis humani* (1747) (cat. no. 15) finally surpassed those of the *Fabrica* in popularity, carefully supervised the engravings of his figures, and when he discovered faults, he had them "rubbed out, and corrected very exactly" (Albinus 1754, xxii). Artists such as Leonardo da Vinci researched human anatomy exhaustively in order to fully represent human proportion and organ function. Michelangelo Buonarroti established a close friendship with Realdo Colombo, Pope Julius III's surgeon and a celebrated anatomist, who one day sent Michelangelo a cadaver to study for his work.

In early anatomical illustration, the body was made sensual, giving it the sense that fragility was part of being mortal. At other times, bits of fabric hid what was irrelevant to the dissections and diagnostics or supplied some modesty. The addition of hair and color personalized the cadavers, intimating that the bodies were once beautiful while also eliciting sympathy. Voyeurism required the viewer to suppress the grotesqueness of the cut-up corpse and the violations committed by the anatomist and to ignore their own complicity in the anxiety-inducing prosection.

In a deliberate retrieval of classical models, artists borrowed marble torsi, similar to one now at the J. Paul Getty Museum (cat. no. 30, fig. I), to stylishly present visceral anatomy in a disemboweled sculpted statue. For the female body, Venus de' Medici offered a beautiful form within which organs and other parts of the anatomy could be illustrated, thereby making the anatomical display lively and pleasing to the eye (cat. nos. 31, 33). Anatomists were keen to point out to artists the differences between a living and a dead body by indicating how muscles could change the appearance of a figure in motion; likewise, artists visually represented changes in surface anatomy through variously posed models or by suspending a skeleton or an excoriated cadaver from a rope, as seen in Cornelis Cort's engraving *Academy of Fine Arts* (1578) (see fig. 3.2).

Printed flap anatomies—manikins in which movable paper flaps reveal different layers of the body—were first introduced in the Renaissance. Through their use, medical practitioners and lay audiences could understand where muscles, nerves, organs, and systems were positioned inside a body; they too could cut apart structures and dissect. Flap anatomies enjoyed great success in subsequent centuries among nonexperts, especially during the Victorian era, when a middle class interested in knowing how the body functions in three dimensions rushed to buy books with male and female anatomical flaps and foldouts rendered in chromolithographs. The viewer's ability to undress and uncover the body was sometimes restricted, as books were often discreetly provided with locks to avoid unchecked exposure to prudish eyes. Anatomical specimens and wax anatomies were carefully cataloged and displayed to visitors, and they became sought-after stops of upper-class patrons during their grand tour. In order to make the visuals more appealing or biologically correct, artists went on to experiment with light, shade, outline, perspective, wax, papier-mâché-like paste, transparencies, and 3-D imaging.

Yet, no matter how much seeing and representing went hand in hand, artists and anatomists were strongly influenced by the entrenched gender ideals of their culture. The anatomical Adam/Apollo was often depicted as a muscular, active, tense, virile, and solid figure, while the anatomical Eve/Venus was illustrated as fleshy, soft, and submissive, and mostly in her reproductive function. Whether artists used paper, copperplate engravings, flaps, ivory, or wax, they typically rendered the female corpse as a reified sexual construct. Characteristic examples show a pregnant woman with long hair, occasionally wearing necklaces, ribbons, or jewelry, and sometimes in an orgasmic state. Good examples of this can be found in La Specola in Florence, for which the visual rhetoric of forensic investigation apparently demanded a supine, voluptuous, pregnant "virgin" gazing afar while seemingly enjoying her own disembowelment (see fig. 8.3).

The prints, flap anatomies, memento mori, wax models, marble torsi, mezzotints, manikins, stereoscopic images, photographs, and neon-filled glass—that is, the abundant and bold material spanning from the representative to the abstract—celebrate the fact that scientific investigation and visual representation require both technical precision and creative flair. Anatomical scrutiny and artistic production were a natural pairing for centuries; their combination often turned into surprising, satisfying, or even deeply unsettling works of art.

—Valeria Finucci, Duke University

preface

The impetus for this exhibition and catalog was the Getty Research Institute's acquisition in 2014 of a spectacular set of three life-size printed figures from late eighteenth-century Bologna by the engraver Antonio Cattani after anatomical sculptures by Ercole Lelli (cat. nos. 36–38). Presenting the muscles of the human body, the figures are based on sculptures of wood and wax produced for the Bolognese anatomy theater and the Istituto delle Scienze, and are representative of the ways in which art and anatomy have been intertwined through the centuries. Although not indicated on the prints, the primary audience for them was artists, as evidenced by a subscription notice that Cattani published in order to raise funds for the project. While anatomy might perhaps seem tangential to a library focusing on the history of art, such as the one at the Getty Research Institute (GRI), it was long considered a foundational subject in the education of an artist.

Understanding the structure of the human body was of vital concern to the practitioners of medicine and art. The will to learn anatomy caused artists to pick up the dissecting knife themselves and to seek out medical contacts for instruction and advice. It also made them part of the market for anatomical illustration. Many of the illustrated books in this exhibition catalog are the result of a close collaboration between an anatomist and a draftsman and printmaker, with these roles sometimes combined to further close the interpretive distance between knowledge and representation. The resulting vocabulary of images shared among anatomists, physicians, surgeons, and artists is the subject of this book. Ranging from the Renaissance to modern day, this vocabulary is explored in a wide range of media drawn primarily from the Special Collections of the GRI, including modes of presenting the anatomy of the human body, such as the animated cadaver, the fictive flaying of antique statues, the physically impressive attempts to illustrate the anatomy of the body at life-size, and the uncanny depths offered by stereoscopic images. Because this catalog is an exploration of the exceptional anatomy collections in the GRI, which are mainly in the western European tradition, it does not offer a global history of the subject.

The GRI possesses both well-known landmarks of anatomical illustration and rare or little-known items. Among the former are an edition of Jacopo Berengario da Carpi's *Isagogae breves* (1523) (cat. no. 8), the earliest anatomy book in the library; Andreas Vesalius's *De humani corporis fabrica libri septem* (1543) (cat. no. 30), a transformative model for luxury anatomy atlases for centuries; and Govard Bidloo's *Anatomia humani corporis* (1685) (cat. no. 40), which introduced a new style of realist imagery. Rarities include Cattani's life-size figures, an anatomy book by John Walker (1787) (cat. no. 25) that was designed to fit in an artist's pocket, and the first edition of Carlo Cesi's book on muscles and bones (1679) (cat. no. 20). The last two are part of the GRI's impressive collection of anatomy books produced specifically for artists. The anatomical holdings of the GRI are complemented and contextualized by works from the J. Paul Getty Museum and other collections in the Los Angeles area, including those of the University of California, Los Angeles; the Los Angeles County Museum of Art; the Hammer Museum; and the Huntington Library.

—Monique Kornell, curator, *Flesh and Bones: The Art of Anatomy*

acknowledgments

The impressive breadth and depth of the holdings of the Getty Research Institute (GRI) and the GRI's encompassing vision are eminently suited for a multidisciplinary exhibition such as *Flesh and Bones: The Art of Anatomy*. I am grateful for the GRI's interest in this project and for the support of Mary Miller, director; Gail Feigenbaum and Kathleen Salomon, associate directors; and Marcia Reed, associate director and head curator, who proposed the exhibition. Christa Aube, exhibitions coordinator, played a key role in its organization, and I must acknowledge her guidance and unflagging enthusiasm. I have consistently benefited from looking at works with Naoko Takahatake, curator of prints and drawings, who served as the curatorial contact for the exhibition.

An earlier idea for a multimedia exhibition on art and anatomy was discussed with Anne-Lise Desmas, senior curator and head of sculpture and decorative arts at the J. Paul Getty Museum, and Louis Marchesano, former curator of prints and drawings at the GRI, now senior curator of prints, drawings, and photographs at the Philadelphia Museum of Art. Many of the books and prints in this exhibition were presented during a meeting of the American Association for the History of Medicine in 2018, which I had the pleasure of organizing with David Brafman, rare books curator at the GRI.

Warm thanks go to the lenders to the exhibition: the University of California, Los Angeles (UCLA); the Hammer Museum at UCLA; the Los Angeles County Museum of Art (LACMA); the Huntington Library, Art Museum, and Botanical Gardens; the J. Paul Getty Museum; a private collector; and the artist Tavares Strachan (Marian Goodman Gallery), all of whom are additionally acknowledged for accommodating the shifting schedule of the exhibition, which was delayed by the pandemic. I am particularly grateful for the assistance and early support of Russell Johnson, special collections librarian of UCLA's Biomedical Library, who believes that the collection in his care at the Louise M. Darling Biomedical Library of UCLA should be seen and used. Joel Klein, Molina Curator for the History of Medicine and Allied Sciences at the Huntington Library, kindly alerted me to the life-size mezzotint by Jacques Fabien Gautier Dagoty in the library's collection.

I am indebted to Anne Varick Lauder, Dániel Margócsy, Vivian Nutton, Jeffrey Spier, and an anonymous reader for their helpful comments on the text. I have also greatly enjoyed stimulating conversations with catalog contributors Thisbe Gensler and Erin Travers. The incisive observations of Andrew Perchuk, deputy director of the GRI, helped guide the format of the exhibition and catalog.

The catalog greatly benefited from the anatomical observations of Prof. Lyndon Da Cruz of Moorfields Eye Hospital in London; Dr. Michael Eisenberg of Pueblo Medical Imaging in Las Vegas; Prof. Dr. Jörg-Elard Otten at the Universitätsklinikum Freiburg; Dr. Elena Stark and Dr. Naveen El Farra of the David Geffen School of Medicine at UCLA; and Francis Wells of the Royal Papworth Hospital in Cambridge, United Kingdom.

I express my appreciation to everyone who responded to my queries and requests, particularly during a period of restricted access: Giorgio Marini, Istituto Centrale per la Grafica, Rome; Benedetta Basevi, Museo della Città, Bologna; Charlotte Beck, Woodward Library, University of British Columbia; Mélanie Bernuz, Musée Granet, Aix-en-Provence; Alexandra K. Carter, Thomas Fisher Rare Book Library, University of Toronto; Jean-Gérald

Castex, Musée du Louvre, Paris; Jack Eckert, Francis A. Countway Library of Medicine, Harvard University; Devin Fitzgerald, UCLA Library Special Collections; Vicki Gambill, The Broad, Los Angeles; Wim Hupperetz, Allard Pierson Museum, University of Amsterdam; Teresa Johnson, formerly at the Biomedical Library, UCLA; Serena Keller of Gilhofer & Ranschburg Rare Books & Manuscripts; Clara Maldini, Biblioteca comunale dell'Archiginnasio, Bologna; Britt Salvesen, LACMA; William Schupbach, Wellcome Collection, London; Patrizia Tomba, Studio Putti, Istituto Ortopedico Rizzoli, Bologna; Morgan Webb, Philadelphia Museum of Art; Susan Wheeler, Cushing/Whitney Medical Library, Yale University; and Matthew Wood, Royal College of Physicians, London. I also wish to express my gratitude to Carolyn Latham, Tatton Park, Knutsford, United Kingdom, for facilitating the viewing of the portrait of Realdo Colombo.

This project has been aided and supported in many ways by the able and knowledgeable staff of the GRI and of the J. Paul Getty Museum. The work of the excellent team at Getty Publications and GRI Publications is acknowledged, and particular thanks are due to Lauren Edson, senior editor; Michelle Deemer, senior production coordinator; and Kurt Hauser, senior designer. My thanks also to exhibition designers Erin Hauer and Alan Konishi; Jennifer Park, associate exhibitions coordinator; and Lisa Forman, associate conservator. It is my great pleasure to thank the GRI's helpful and accommodating library staff, who are too many to name here. I am grateful for the assistance and comments of Pietro Rigolo, Idurre Alonso, and Alexa Sekyra. Annie Rana swiftly arranged for high-quality scans of the few books that had escaped the GRI's excellent digitizing campaign, and this proved especially valuable during the library's closure.

Many people have generously offered their assistance and shared their insights with me: Sarah Bane, Judith Barr, Emily Beeny, Nicole Budrovich, Lony Castro, Sara Cole, Mazie Harris, Leon Fine, Laura Frahm, David Franklin, Jack Hartnell, Stephen Joffe, Jorrit Kelder, Martin Kemp, Jeanette Kohl, Sachiko Kusukawa, Gideon Manning, Nino Nanobashvili, and Maria Cristina White-da Cruz, with special thanks to Giulia da Cruz for photographing the tomb of Giuseppe del Medico in the Church of San Rocco in Rome.

—Monique Kornell

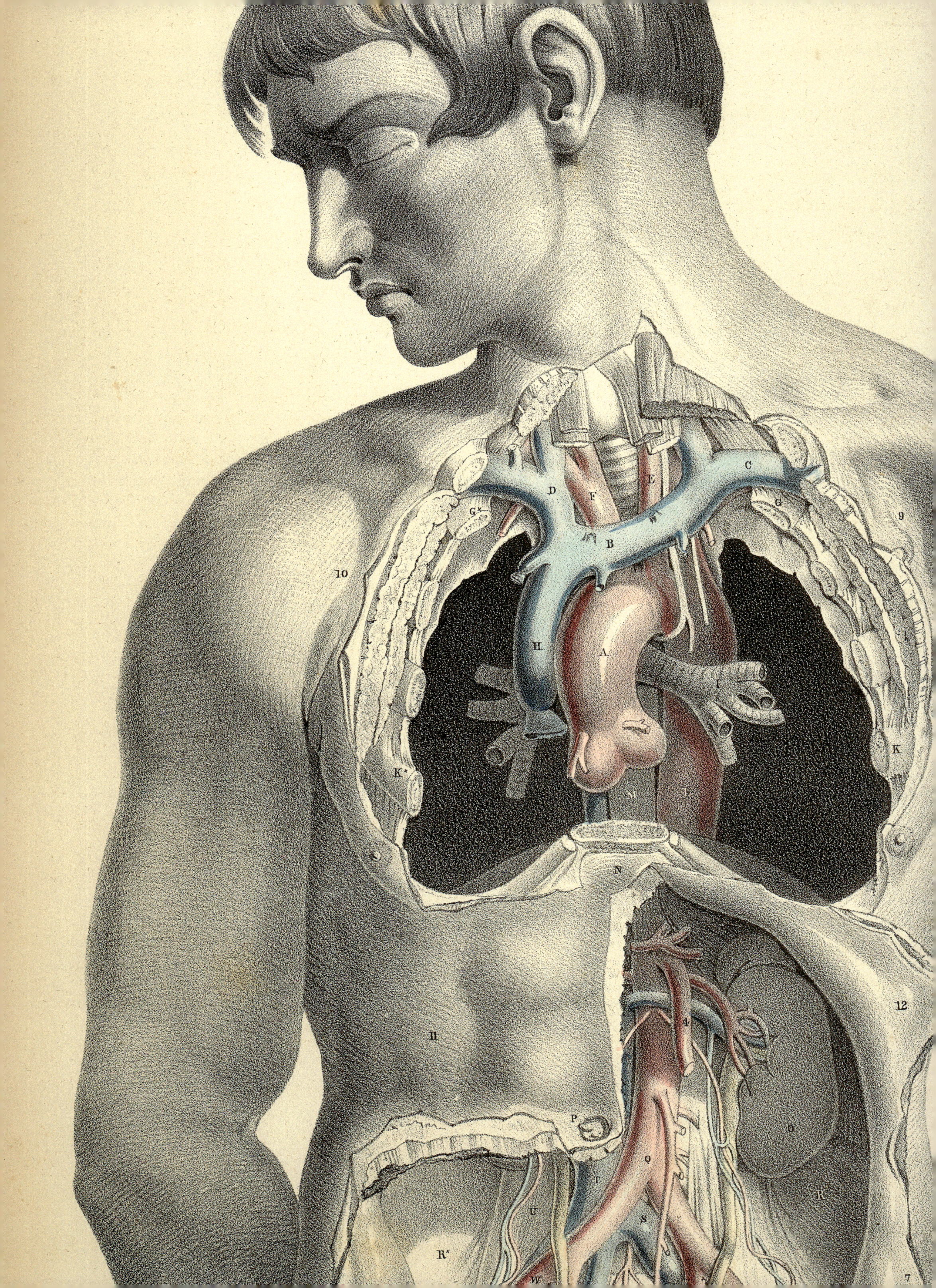

D
F
E
C
B
G
9
10
H
A
I
K
M
N
12
4
P
O
Q
T
U
S

chapter one

THE ILLUSTRATION OF ANATOMY

Monique Kornell

While composing *The Anatomy of the Bones, Muscles, and Joints,* published in 1793, the Scottish surgeon and anatomist John Bell (1763–1820) increasingly felt the need to provide a "system of drawings" for it.[1] The following year, he published a separate volume of anatomical illustrations, which he had drawn and etched himself. Bell stated his belief in their power to make clear what could only be laboriously understood through description alone and declared that a book of anatomy without illustrations seemed "no better than a book of geography without its maps," or "like teaching mathematics without diagrams, or solving Euclid's problems without the help of figures or lines."[2]

Two hundred and fifty years earlier, the Flemish anatomist Andreas Vesalius (1514–64), in the dedication of his *De humani corporis fabrica libri septem* (1543) to Emperor Charles V, had referred to mathematical and geometric diagrams as an example of the power of an image to "aid the understanding of these things and place a subject before the eyes more precisely than the most explicit language" (fig. 1.1; cat. no. 30).[3] The German physician and botanist Leonhart Fuchs (1501–66), in the contemporaneously published herbal *De historia stirpium* (1542), likewise ranked images before words in their ability to convey form.[4] Both Fuchs and Vesalius were defending their use of illustrations in response to traditional beliefs that ancient authors had discouraged the use of pictures because images could not replace empirical knowledge.[5] A confidence in the efficacy of images was also shared by the Paduan anatomist Hieronymus Fabricius ab' Acquapendente (ca. 1533–1619), who toward the end of a long career published illustrated books of human and comparative anatomy and had

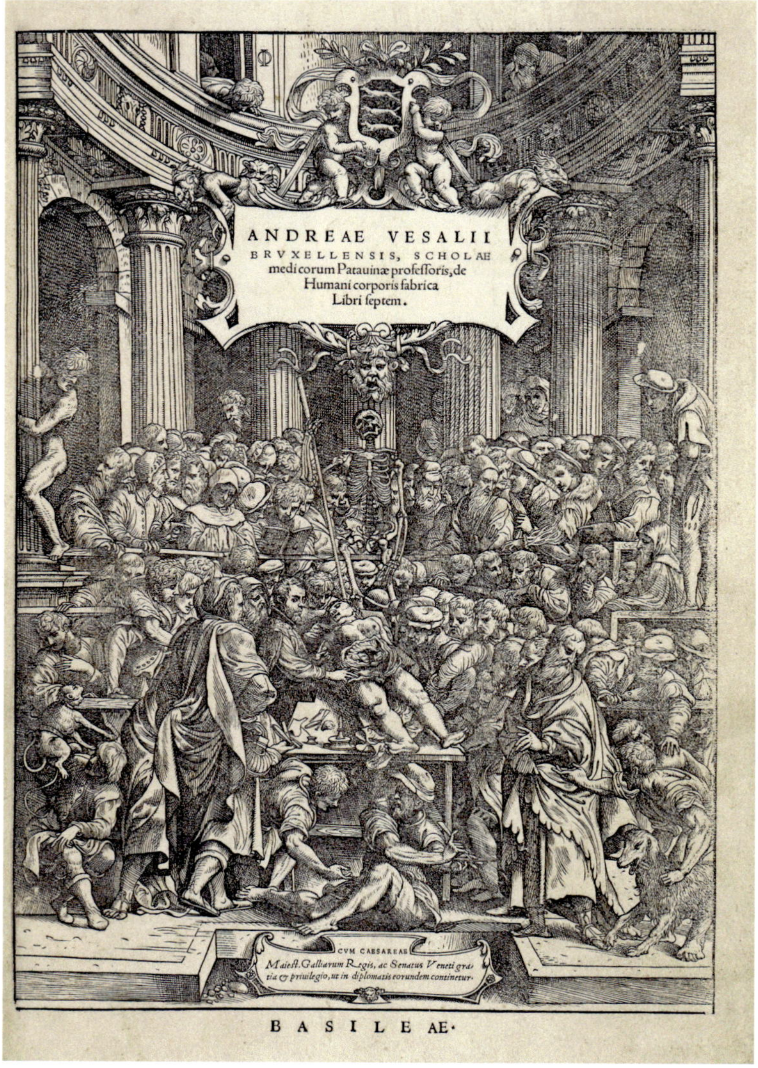

FIG. 1.1
AFTER DOMENICO CAMPAGNOLA
(Italian, 1500–1564) or Jan Steven van Calcar (North Netherlandish, ca. 1515–ca. 1546).
Title page showing Andreas Vesalius performing a public dissection, woodcut.
From Andreas Vesalius, *De humani corporis fabrica libri septem* (Basel: J. Oporinus, 1543).
Los Angeles, Getty Research Institute, 84-B27611.

plans for a larger illustrated work.[6] At the conclusion of the text to his *De venarum ostiolis* (1603), Fabricius refers his readers to the illustrations for the number of valves in the veins and where they may be found, since these and "all other matters will become better known from an actual inspection of the Plates than from any written account."[7]

In this age of the omnipresent image, the idea of a long, technical description in words alone of the complicated, multilayered structure that is the human body strikes one as labored and unusual. But in fact, many early anatomy texts relied on words rather than images to explicate the body.[8] Illustrations were an added expense and effort, yet one that sixteenth-century authors and printers were increasingly willing to undertake and that they marketed in their titles. Early examples are Giovanni Battista Canani [Canano]'s *Musculorum humani corporis picturata dissectio* (ca. 1542), a book on the muscles of the arm with engravings after Girolamo da Carpi (ca. 1501–56), and Thomas Geminus's *Compendiosa totius anatomie delineatio, aere exarate* (1545), with engravings after the woodcut illustrations in Vesalius's *Fabrica* and its succinct compendium, the *Epitome* (both published in 1543).[9] The presence of illustrations was also promoted in the extended titles of Charles Estienne and Estienne de la Rivière's *De dissectione* (1545), Jacques Guillemeau's *Tables anatomiques* (1586), and *Vivae imagines*, published by Christophe Plantin in 1566 (cat. nos. 1, 9, 31).

A lingering distrust of illustrations was expressed in the early nineteenth century by the Scottish anatomist John Barclay (1758–1826), who, while conceding their usefulness in the absence of cadavers, was of the opinion that when substituted for dissection they "serve only to mislead, to diffuse error, and to perpetuate it."[10] Robert Knox (1791–1862)—a student of Barclay's who had been an extremely popular lecturer on anatomy in Edinburgh before his career was derailed by his involvement in the William Burke and William Hare murder and body-theft scandal in the late 1820s—shared this belief, though he gained a more nuanced view through his lecturing experience, as he records in the preface to his translation of Antonio Scarpa's *Tabulae neurologicae* (1794): "I knew nothing at the time of the difficult duties of the Teacher of Anatomy, how he must adapt his mode of instruction to the capacity of the individual student, how some acquire knowledge most readily by description and demonstration, whilst to others a single glance at a diagram or drawing, compared at the same moment with nature, will at once convey the truth."[11] Scarpa's life-size illustrations of the nerves of the heart (cat. no. 52, fig. Q) were reduced with the aim of providing Knox's students with handy and accurate visual aids "for the Dissecting Table and for the Anatomical Classroom."[12] Demonstrating the continued relevance of anatomical illustration as a guide in the dissection lab, students at the University of California, Los Angeles, can access reference illustrations on screens set up by the cadaver.[13]

Aside from their efficacy in imparting information, various other reasons were given for the use of employing anatomical illustrations. Although Vesalius was at pains to point out that his illustrations, although accurate, were in no way meant to replace a dissection, he also readily described them as helpful for those interested readers who were not able to attend a dissection or those who could not bear to.[14] In his *Tabulae anatomicae sex* (1538), illustrations were characterized as helpful memory aids. In the preface to Estienne and La Rivière's *De dissectione,* illustrations are offered to the reader not as replacements but as stopgaps until they are able to consult a cadaver (cat. no. 1). Seeing an actual dissection did not necessarily guarantee comprehension, though. Canani's book of about 1542 was prepared with illustrations at the instigation of the dedicatee, the Ferrarese nobleman Bartolomeo Nigrisoli, so that those "who cannot recognize the parts of the body by personal inspection during dissection might gain some knowledge of them at least through pictures and thus become more reliable in their medical consultations for the health of those in their care."[15]

The practical reasons the Swiss physician Felix Platter (1536–1614) gives for adding etched illustrations, mostly adaptations from Vesalius, to his *De corporis humani structura et usu libri III* of 1583, are that it is difficult to understand the description of the parts of the body without seeing them and that opportunities to dissect are limited.[16] Dissections hosted by medical schools usually occurred in the winter for the better preservation of the corpse and were often crowded affairs, as seen on the title page of Vesalius's *Fabrica* (see fig. 1.1). It was not unusual for students to supplement their studies with private dissections, which allowed better access to a cadaver. Platter's claim of having dissected more than fifty-five bodies is given some weight by accounts in his remarkable diary of participating in grave robbing and private dissections while a student in Montpellier.[17] In it, he describes the danger, physical labor, and inconveniences involved in stealing bodies, something that Vesalius also conveys in his account of the trouble he went to when, as a student in Louvain, he smuggled parts of a dried cadaver from a gibbet outside the city walls.[18] The efforts of the artist Aert Mytens (1556–1602) to retrieve a body from a gallows outside Brussels, though farcical in Karel Van Mander's account, still carried the threat of official reprisal. On the first try, an accomplice, spooked by the movement of the corpse as it was being cut loose, ran away, causing Mytens to give chase. While a second attempt was successful, Mytens's father caught wind of his son's activities and soon sought out a local official to smooth the matter over.[19]

Vesalius, with the *Fabrica,* was the first author to explore fully the potential of a book to describe with word and image what was an unavoidably ephemeral subject—the dissected human body. The *Fabrica* changed the way anatomical illustration was used and served as a model of a luxury publication for ambitious anatomists in the centuries to follow. Its illustrations were disseminated through countless iterations of copies, only displaced in popularity by the appearance of Bernhard Siegfried Albinus's *Tabulae sceleti et musculorum corporis humani* in 1747 (figs. 1.2, 1.3, 1.8, 1.9; cat. nos. 15, 22).[20] None of the illustrated books that preceded the *Fabrica* had integrated text and image to the same extent. In terms of the number, quality, and size of the illustrations, it set new standards. Influential precedents were the *Fasciculus medicinae,* first published in Venice in 1491 under the name of Johannes de Ketham, and Jacopo Berengario da Carpi's commentary of 1521 on the Italian anatomist Mondino de' Luzzi (ca. 1275–1326), and Berengario's handy textbook, the *Isagogae breves* (1523) (cat. no. 8). Had a court case not delayed its publication until 1545, Estienne and La Rivière's *De dissectione* would likely have

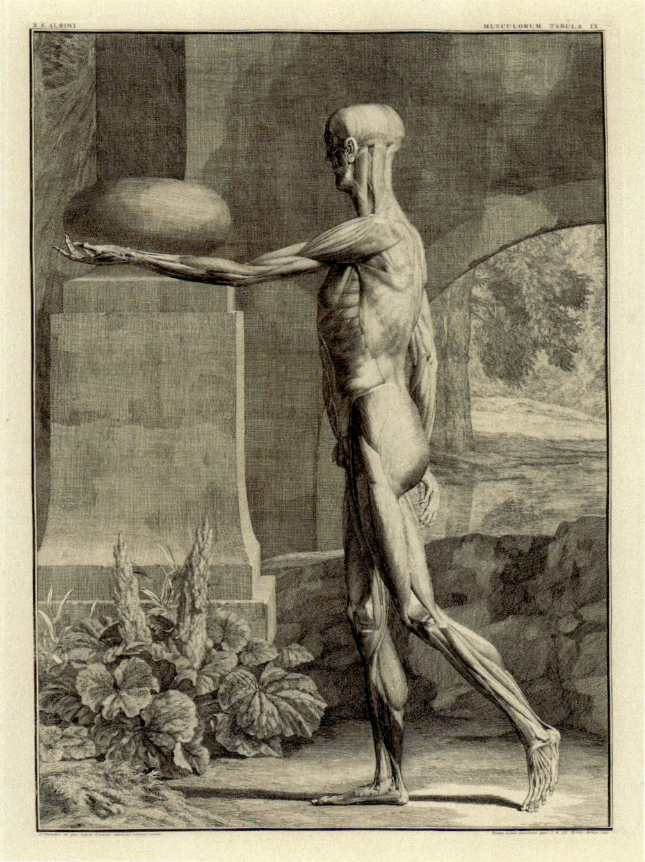

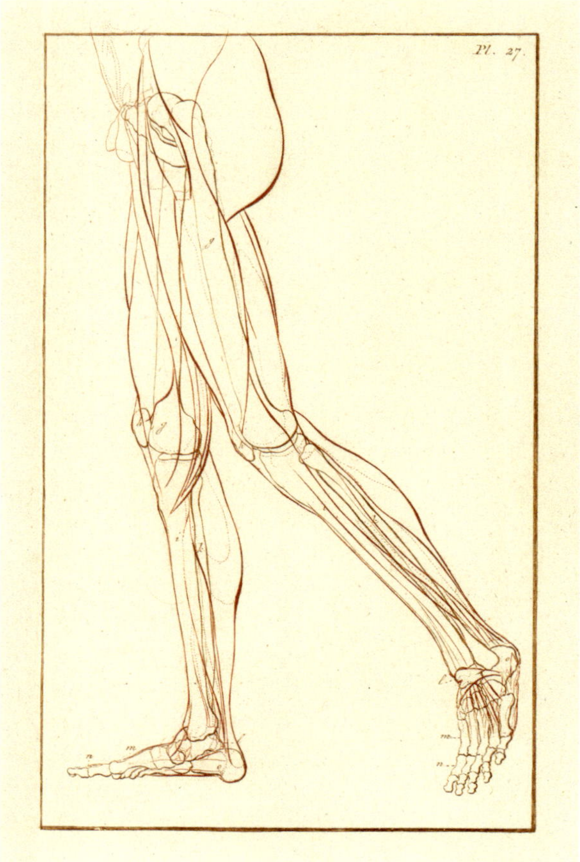

FIG. 1.2
JAN WANDELAAR
(Dutch, 1690–1759).
Ninth muscle figure, 1743,
etching and engraving,
platemark: 56.1 × 40 cm;
sheet: 76 × 54 cm.
From Bernhard Siegfried
Albinus, *Tabulae sceleti
et musculorum corporis
humani* (Leiden: Johannes
and Herman Verbeek,
1747), unbound sheet.
Los Angeles, Library
Special Collections for
Medicine and the Sciences,
Louise M. Darling
Biomedical Library, UCLA,
BIOMED ** WZ 260 A337t
1747 plates.

FIG. 1.3
AFTER JAN WANDELAAR
(Dutch, 1690–1759).
Walking legs in profile,
transparent anatomy view,
etching, printed from two
plates in sepia and red.
From Cornelis Ploos van
Amstel, *Aanleiding tot de
kennis der anatomie, in
de tekenkunst, betreklyk tot
het menschbeeld*
(Amsterdam: J. Yntema,
1783), pl. 27.
Los Angeles, Getty
Research Institute,
86-B21477.

ushered in the new style of luxurious anatomy book: copiously illustrated and in folio size.

What Berengario (ca. 1460–ca. 1530), La Rivière (d. 1569), and Vesalius all shared was their interest in visual representation. Berengario, who continually made revisions to his woodcut illustrations between editions, collected paintings and drawings and owned a headless, cuirassed ancient statue.[21] In an accompanying letter published in the *Fabrica,* Vesalius—who sent his printer Johannes Oporinus the proofs of his illustrations as well as the blocks in order to set a standard for their print quality—expresses his delight in the variation of line and the elegance of the shading of the woodcut illustrations. Vesalius also drew, and during lectures he would use this skill to elucidate anatomical structures for the benefit of his students—at times sketching directly onto the dissecting table.[22] Some of these lecture drawings were the basis for the first three been a discouraging factor for books with original illustrations in the second half of the sixteenth century. The anatomist Realdo Colombo (ca. 1515–59) had been at work on illustrations in the late 1540s, but his book appeared without them in 1559 (cat. no. 16). By this time, copies after Vesalius's illustrations had been published in books by Colombo's student Juan Valverde de Amusco (ca. 1525–ca. 1588) (1556; cat. no. 31), and by Geminus (1510–62). The majority of Bartolomeo Eustachi's (ca. 1500/1510–1574) plates, prepared by 1552, only appeared in 1714. They were first published by the papal physician Giovanni Maria Lancisi (1654–1720) and then republished using the same plates in 1741 by the physician and surgeon Gaetano Petrioli (fig. 1.4). The illustrations by Pietro da Cortona (1597–1669) prepared in the early seventeenth century were likewise only published in the eighteenth century (cat. no. 14).

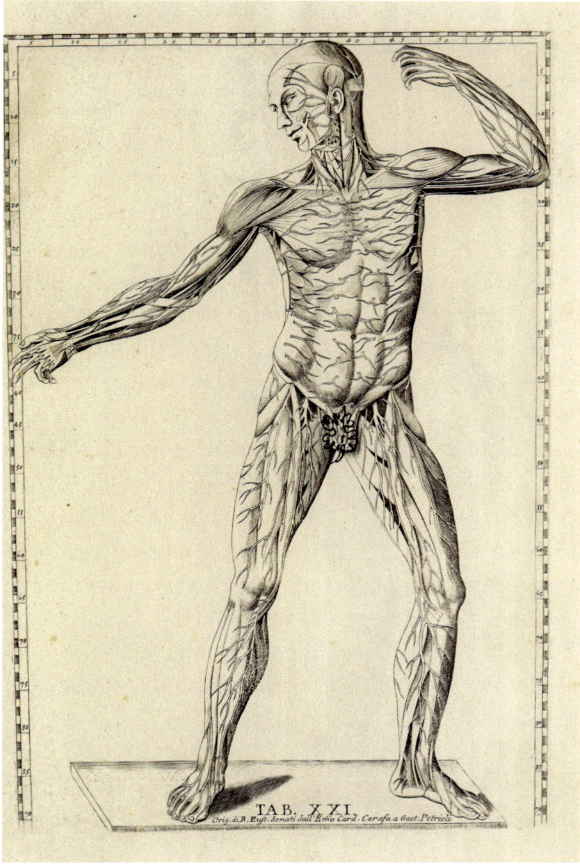

FIG. 1.4
ATTRIBUTED TO GIULIO DE MUSI
(Italian, active 1550–55).
Nerve and muscle figure, ca. 1552, engraving, with fig leaf added in pen and ink.
From Gaetano Petrioli, *Riflessioni anatomiche sulle note di Monsignor Gio: Maria Lancisi fatte sopra le tavole del celebre Bartolomeo Eustachio* (Rome: Giovanni Zempel, 1740 [1741]), pl. 21.
London, Wellcome Collection, EPB/D/40614.

of his *Tabulae anatomicae sex,* published in 1538.[23] The French surgeon La Rivière prepared the dissections and illustrations after them for the *De dissectione,* on which he collaborated with Estienne (cat. no. 1).

The impact of Vesalius's *Fabrica* was all the greater because of the many delayed or unrealized projects of both anatomists and artists from the sixteenth and early seventeenth centuries. One can only wonder how much different the history of anatomy and anatomical illustration would have been if the extensive studies of Leonardo da Vinci (1452–1519) (see fig. 8.1) had been published or even widely circulated, or how the education of artists would have been affected if the planned anatomy books by Michelangelo Buonarroti (1475–1564), Alessandro Allori (see fig. 6.6), and perhaps Peter Paul Rubens (see fig. 2.7), among other artists, had seen publication. The success of the *Fabrica* and its numerous copies may have

A Collaborative Endeavor

Making a visual record of soft body parts before decay set in necessitated a close collaboration between anatomist and artist. An artist would have to be physically close at hand if, as William Hunter (1718–83) describes in the preface to *The Anatomy of the Human Gravid Uterus* (1774) (cat. no. 42), the anatomist "will not allow the artist to paint from memory or imagination but only from immediate observation." While in Padua, Vesalius worked side by side with an artist, likely the northern artist Jan Steven van Calcar (ca. 1515–ca. 1546), for the series of muscle figures in the second book of the *Fabrica.* Vesalius describes suspending cadavers by a rope from a pulley attached to a beam and arranging them in poses to be drawn.[24] Some fifty years later in Padua in 1593, the Swiss glass painter and draftsman Josias Murer II (1564–1630), who was then living in the house of the anatomist Giulio Casseri (1561–1616), drew a private dissection carried out by Casseri for the benefit of his German students (cat. nos. 11, 12).[25] What is perhaps an artist at work is seen at the lower left of a dissection scene on the title page to Colombo's *De re anatomica libri XV* (1559) (cat. no. 16).[26]

While working on colored life-size anatomical images sometime between 1605 and 1613, the German artist Christoph Gertner (ca. 1575/80–after 1623) is described as working right next to the cadavers dissected by the physician Henning Arnisaeus (1570–1636).[27]

Some of the collaborations between artist and anatomist extended over several years. Albinus (1697–1770) worked for over three decades with the artist and engraver Jan Wandelaar (1690–1759), whose work he closely oversaw and who in later years resided in the anatomist's house. Paolo Mascagni worked for fifteen years, until his death in 1815, mainly with the engraver and wax modeler Antonio Serantoni (1780–1837) on life-size copperplates for the *Anatomia per uso degli studiosi di scultura et pittura* (1816) and for the posthumously published *Anatomia universa* (1823–31) (cat. no. 45).

Sometimes collaboration was defined by an effort to control and correct the artists' works, not only to eliminate error but also to rein in artistry and avoid possible distortion. In botanical illustration, Fuchs had his artistic team avoid shading and other artistic effects.[28] Both William Cheselden (1688–1752) and Albinus introduced mechanical means through which their artists viewed the body, Cheselden employing a camera obscura and Albinus a fixed eyepiece and grids of rope (cat. nos. 33, 15).[29] But whereas Cheselden and Hunter's aim was the exact representation of the specific body (cat. no. 42), Albinus's was the perfect amalgam of many.

Anatomist Artists

Leonardo da Vinci fused observation, understanding, and artistic ability to an exceptional degree, as he himself recognized.[30] The seventeenth-century Dutch anatomist Nicolaas Hoboken (1632–78) was of the opinion that anatomists should illustrate their own works, avoiding the intermediary of the artist, as he did for his book on uterine anatomy, and there are many others who did so, some with considerable talent.[31] Providing the most direct interpretations were those anatomists, such as Volcher Coiter (1534–76) and John Bell, who were responsible for the etchings after their own drawings, and there is evidence that William Cowper (1666/67–1710) did the same (cat. no. 13).[32] The surgeon La Rivière's role as both dissector and illustrator for *De dissectione* was paralleled three hundred years later by the surgeon Henry Vandyke Carter (1831–97), who provided the illustrations for Henry Gray's enduring best seller, *Anatomy, Descriptive and Surgical* (1858) (fig. 1.5).[33] The German physician Ludwig Choulant considered Antonio Scarpa's life-size delineation of the nerves of the heart to be Scarpa's "anatomic masterpiece"[34] (cat. no. 52; fig. Q), and Cowper's figures of muscles won him great praise from his contemporaries. Jean-Galbert Salvage's *Anatomie du gladiateur combattant* (1812) and Joseph Maclise's *Surgical Anatomy* (1851) are two examples where medical training and artistic ability are combined to truly impressive effect (cat. nos. 34, 53).

"Impossible to improve upon"

As with naturalist illustration, such as herbals, the history of anatomical illustration is one of great invention but also one of continually reused images. In his foreword to Edward Mitchell's set of engravings of human and comparative osteology after Albinus, Cheselden, George Stubbs, and Jean-Joseph Sue the Elder, John Barclay asks rhetorically why he did not advise Mitchell "to copy from nature than from the engravings of first-rate artists."[35] The reason he gives is that error was less likely to be introduced into the engravings if they were based on accurate images that were the result of a lengthy and costly process by anatomists such as Sue and Albinus, who had "watched incessantly over the artists whom they employed" and who were "less intent on pecuniary emolument than lasting reputation."[36] The avoidance of error in copying may not have always worked in practice, and John Bell was closer to the mark when he decried "the careless copying from book to book."[37] There was, however, a sense from early on of capitalizing on the excellence of previously published

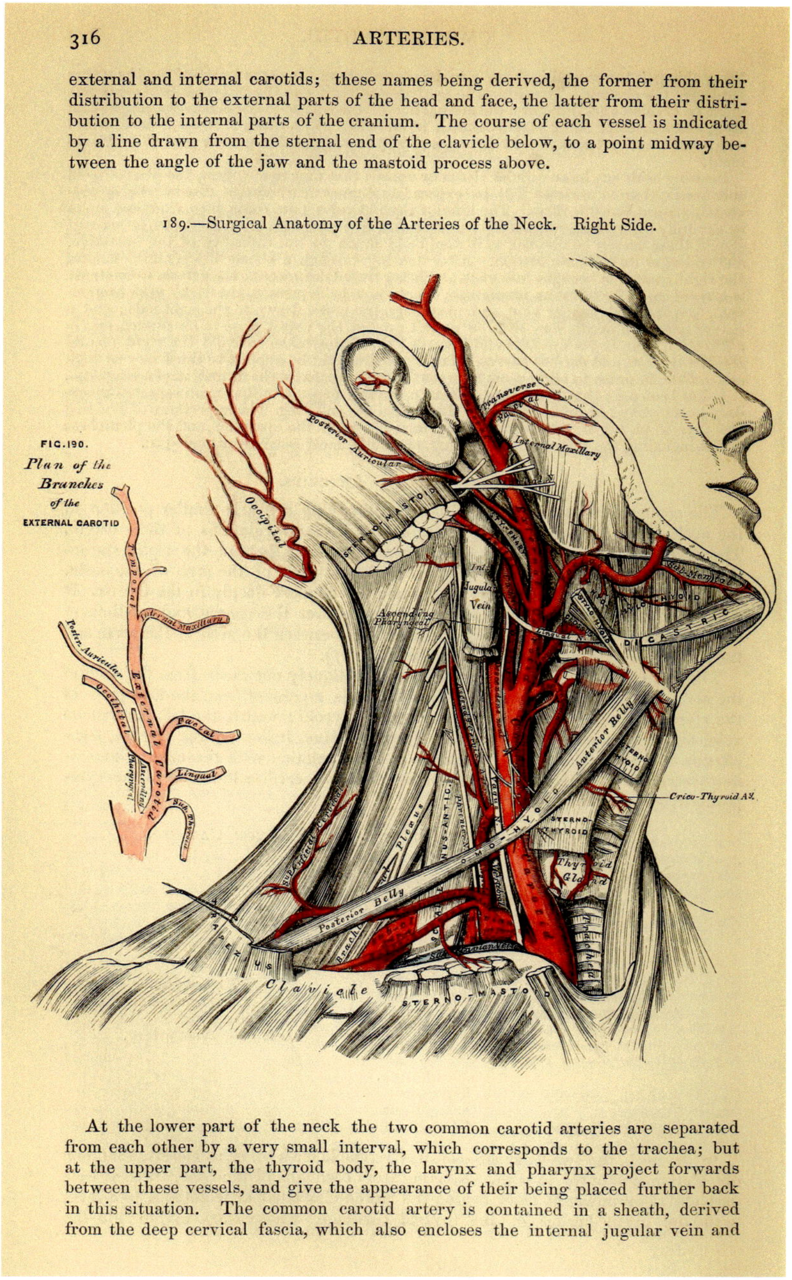

316 ARTERIES.

external and internal carotids; these names being derived, the former from their distribution to the external parts of the head and face, the latter from their distribution to the internal parts of the cranium. The course of each vessel is indicated by a line drawn from the sternal end of the clavicle below, to a point midway between the angle of the jaw and the mastoid process above.

189.—Surgical Anatomy of the Arteries of the Neck. Right Side.

At the lower part of the neck the two common carotid arteries are separated from each other by a very small interval, which corresponds to the trachea; but at the upper part, the thyroid body, the larynx and pharynx project forwards between these vessels, and give the appearance of their being placed further back in this situation. The common carotid artery is contained in a sheath, derived from the deep cervical fascia, which also encloses the internal jugular vein and

FIG. 1.5
HENRY VANDYKE CARTER (English, 1831–97).
Surgical Anatomy of the Arteries of the Neck. Right Side; Plan of the Branches of the External Carotid, hand-colored wood engraving.
From Henry Gray, *Anatomy, Descriptive and Surgical* (London: John W. Parker and Son, 1858), 316, figs. 189, 190.
Boston, Francis A. Countway Library of Medicine, 005100436.

examples as well as avoiding the effort and the cost involved in new illustration.

Juan Valverde de Amusco, in his own much-copied anatomy book first published in Rome in 1556, stated that the Vesalian woodcuts were so well done that it would seem "envious or malignant" not to use them, and that it would be easier to indicate where he concurred or disagreed with Vesalius (cat. no. 31).[38] The same mixture of admiration and opportunity for dialogue with Vesalius were the reasons given by Platter for his adaptations of Vesalius's illustrations.[39] In 1668, the artist and engraver François Tortebat (1616–1718) published *Abregé d'anatomie,* an anatomy book for artists dedicated to the Académie Royale de Peinture et de Sculpture, Paris, with prints after Vesalius's *Fabrica* and *Epitome.* In the address to the reader, anonymously penned by the artist and art critic Roger de Piles (1635–1709), de Piles voiced the opinion that the Vesalian illustrations were "most correct and impossible to improve upon."[40] The figures, though reversed, are otherwise quite faithful to

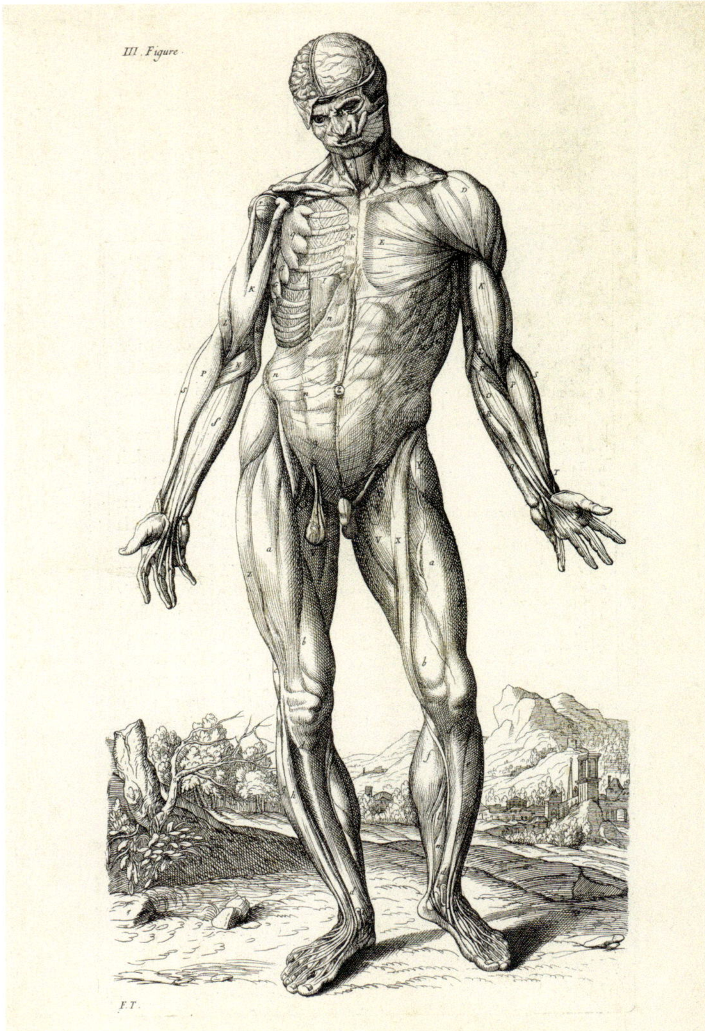

FIG. 1.6
FRANÇOIS TORTEBAT (French, 1616–1718), after Jan Steven van Calcar (North Netherlandish, ca. 1515–ca. 1546).
Third muscle figure, etching and engraving.
From François Tortebat and Roger de Piles, *Abregé d'anatomie, accommodé aux arts de peinture et de sculpture* (Paris: Tortebat, 1668).
Los Angeles, Getty Research Institute, 92-B12688.

the originals; however, to make the publication appear seamless, those from the *Fabrica* were enlarged to match the dimensions of the *Epitome* figures.[41] Furthermore, as Edouard Turner observes, selected backgrounds from the *Fabrica* were added to the *Epitome* figures.[42] In copying the first *Epitome* figure, Tortebat eliminated the abbreviated landscape and the surrounding text that hemmed it in, and inserted the background from the thirteenth muscle figure of the *Fabrica* (figs. 1.6, 1.7). He also omitted the two eyes between the feet of the original figure, in keeping with the promise to the reader that only those elements of anatomy of relevance to the artist would be included in the book. In the *Abregé d'anatomie,* the backgrounds of the *Fabrica*'s tenth and eleventh plates appear behind the male and female nudes of the *Epitome*—the first demonstration that the *Fabrica* landscape settings are contiguous.[43]

In general, though, the elaborate backgrounds in the skeleton and muscle figures of Vesalius's *Fabrica* and Albinus's *Tabulae* were often reduced or left out altogether. Some exceptions are luxury copies, such as the edition of Vesalius's complete works published by Herman Boerhaave (1668–1738) and his former student Albinus in Leiden in 1725 with plates by Wandelaar, and the London editions of Albinus's *Tabulae,* published by John and Paul Knapton in 1749.[44] Arnauld Éloi Gautier

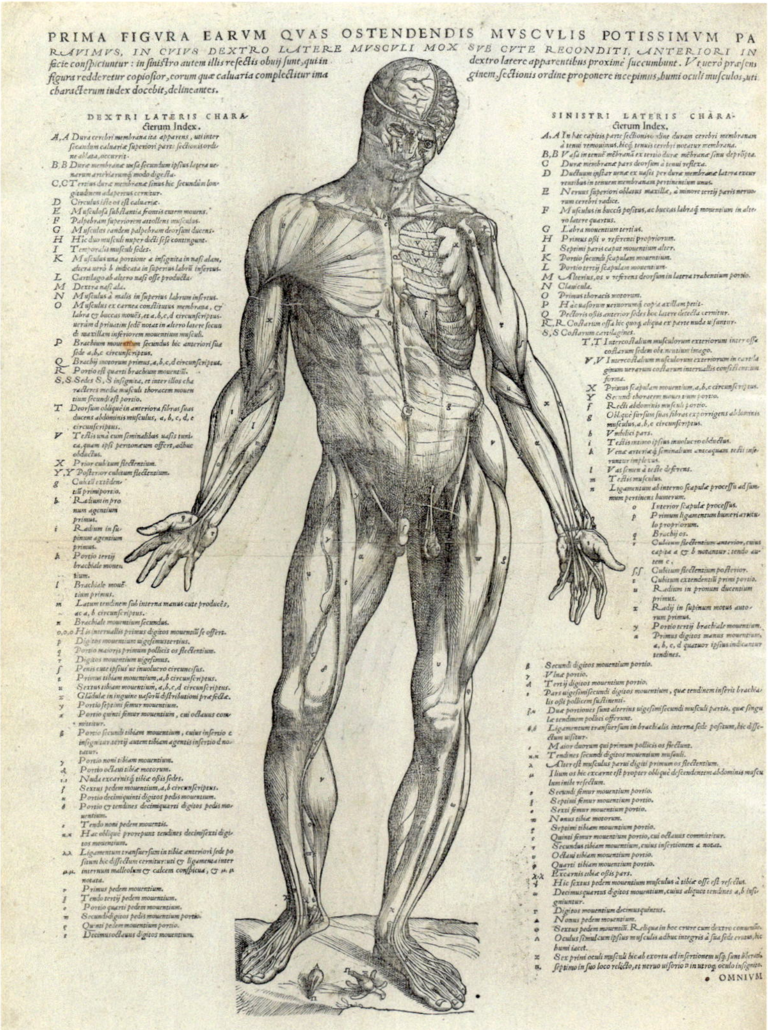

FIG. 1.7
AFTER JAN STEVEN VAN CALCAR
(North Netherlandish, ca. 1515–ca. 1546).
First muscle figure, woodcut.
From Andreas Vesalius, *De humani corporis fabrica librorum epitome* (Basel: J. Oporinus, 1543).
London, Wellcome Collection, EPB/F/6565.

Dagoty (1741–before 1780), the son of Jacques Fabien Gautier Dagoty (1716–85) (cat. no. 41), placed Albinus at the pinnacle of his history of anatomical illustration prefixed to his *Cours complet d'anatomie* (1773), yet he still removed Wandelaar's elaborate backgrounds for his color mezzotint copies.[45] In the copy after Albinus's fourth muscle figure, the red muscles in Gautier Dagoty's mezzotint are in vivid contrast to the subdued green-tinted floor and wall that replaces the grazing rhinoceros and vegetation of the original (figs. 1.8, 1.9). In this figure of deep dissection, the added color has the benefit of bringing to prominence muscles such as those of the eye sockets or the hands, both less apparent in the original.

Like Vesalius before him, Albinus complained about the quality of copies after his illustrations, referring to the Knapton editions.[46] Cornelis Ploos van Amstel (1726–98)—a Dutch timber merchant, collector, printmaker, and codirector of the Stadstekenacademie in Amsterdam—however, had the approval of Albinus before he died in 1770 for the use of the illustrations in his anatomy book for artists, *Aanleiding tot de kennis der anatomie* (1783), where they appear much altered.[47] In the final plate, the outlines of Albinus's écorché and skeleton figures have been fragmented and recombined to create a transparent anatomy of muscle over bone. Color has also been added, with the muscles rendered in a dark red-brown

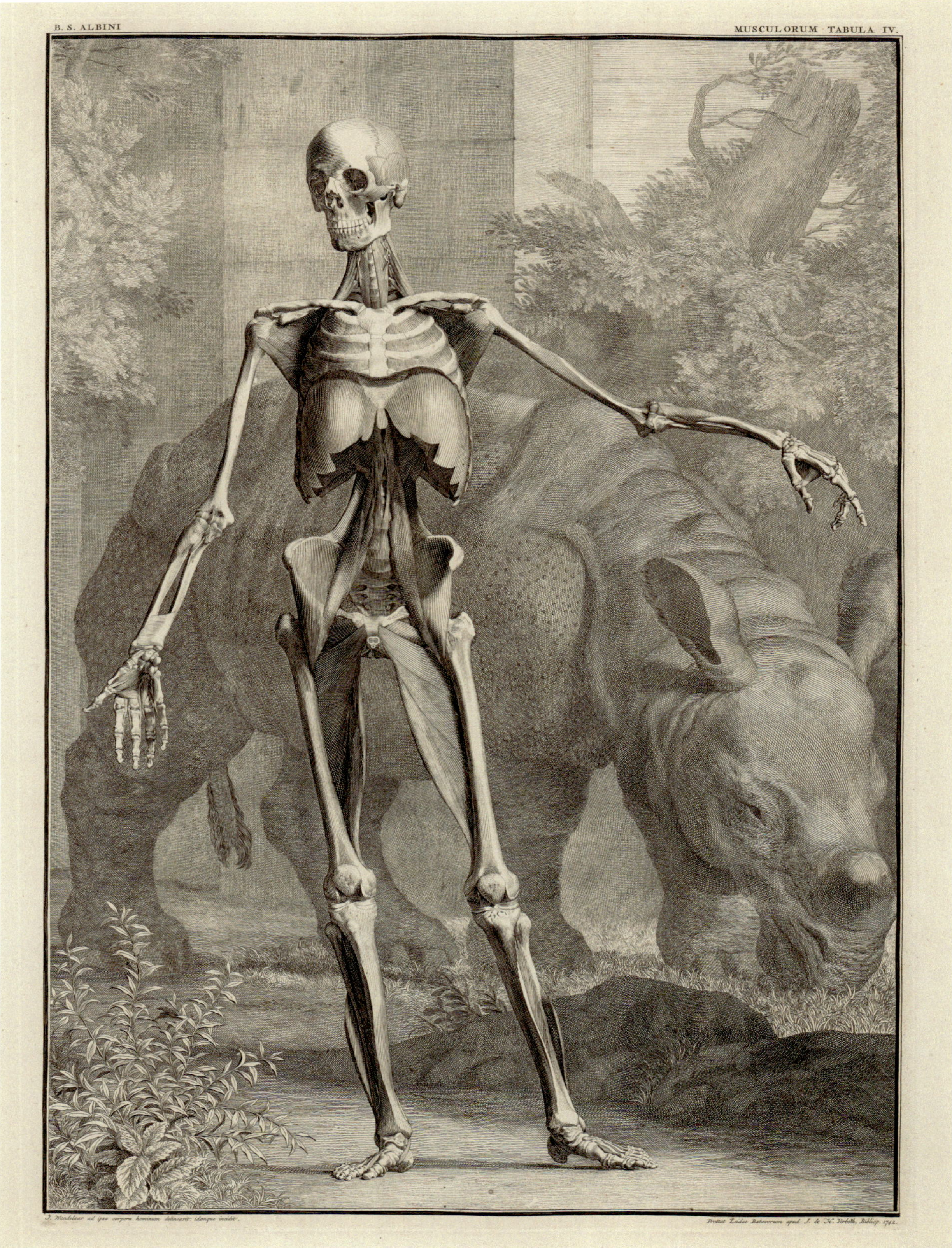
B. S. ALBINI
MUSCULORUM TABULA IV.
J. Wandelaar ad ipsa corpora hominum delineavit: idemque incidit.
Prostat Lugduni Batavorum apud J. & H. Verbeek, Bibliop. 1742.

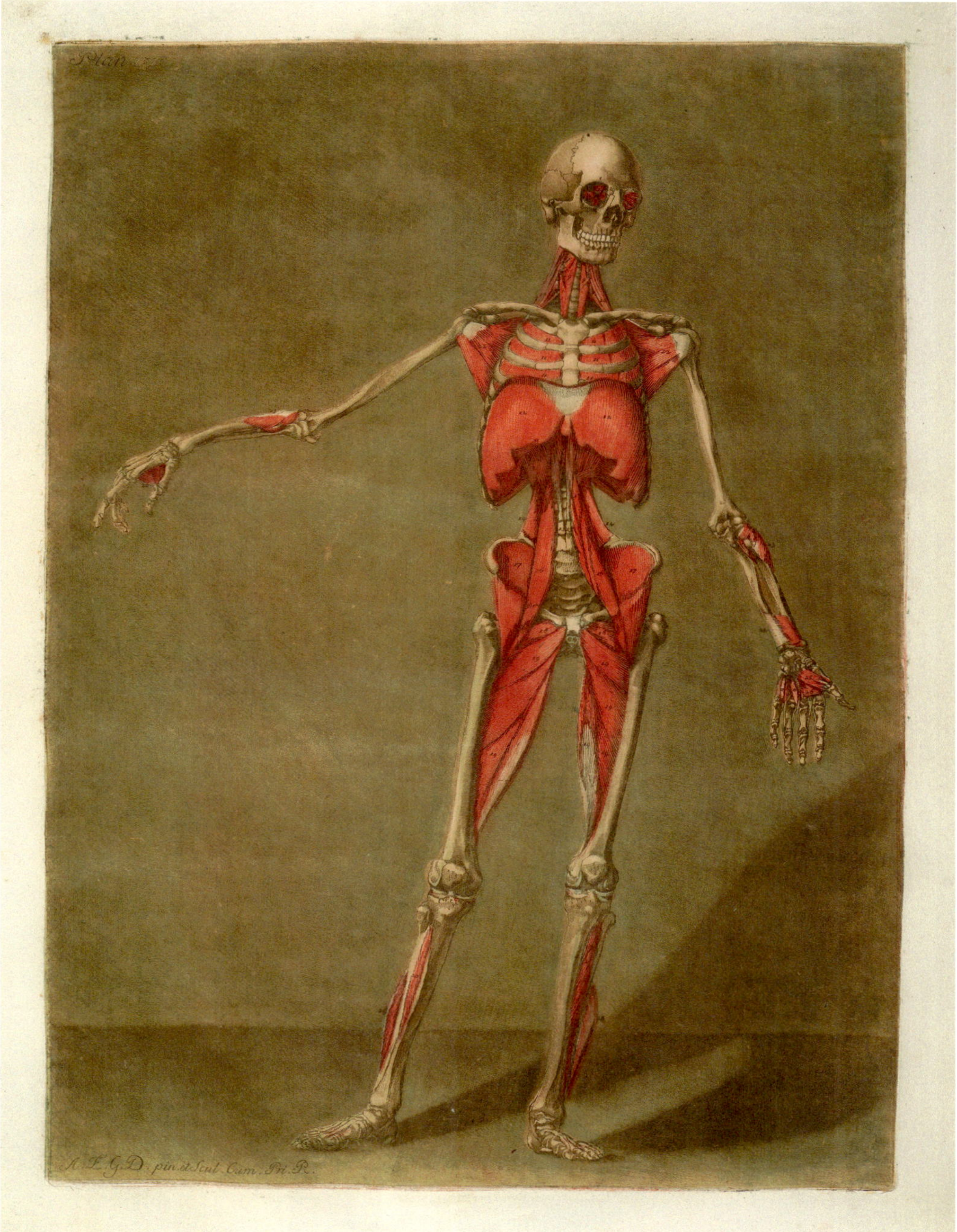

FIG. 1.8
JAN WANDELAAR
(Dutch, 1690–1759).
Fourth muscle figure, 1742, etching and engraving, platemark: 56.4 × 40.7 cm; sheet: 76.7 × 54.4 cm.
From Bernhard Siegfried Albinus, *Tabulae sceleti et musculorum corporis humani* (Leiden: Johannes and Herman Verbeek, 1747), unbound sheet.
Los Angeles, Library Special Collections for Medicine and the Sciences, Louise M. Darling Biomedical Library, UCLA, BIOMED ** WZ 260 A337t 1747 plates.

FIG. 1.9
ARNAULD ÉLOI GAUTIER DAGOTY
(French, 1741–before 1780), after Jan Wandelaar (Dutch, 1690–1759).
Sixth muscle figure, color mezzotint.
From Arnauld Éloi Gautier Dagoty, *Cours complet d'anatomie: Peint et gravé en couleurs naturelles* (Nancy: Jean-Baptiste-Hyacinthe Leclerc, 1773).
Los Angeles, Getty Research Institute, 42–2.

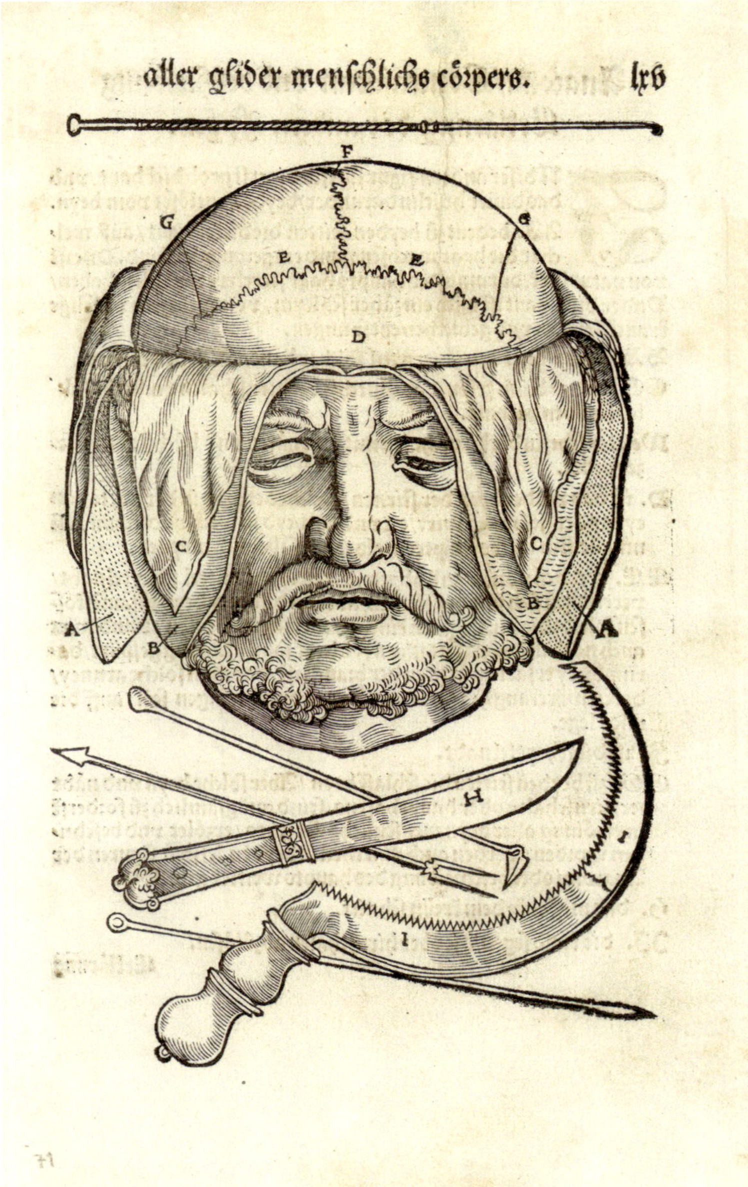

ink to distinguish them from the lighter-colored bones (see figs. 1.2, 1.3).

The German author Walther Hermann Ryff (d. 1548), a serial plagiarizer in many subjects, also borrowed anatomical images. Figure 1.10 is one of a series of illustrations of the dissection of the head that Ryff adapted from Johannes Dryander's *Anatomia capitis humani* (1536).[48] Another compendium of copied illustrations is found in Thomas Bartholin's gathering of his father's writings and those of others, although in the case of the images taken from Casseri's publications, there was a personal connection, since Caspar Bartholin the Elder had been a friend of Casseri's in Padua (cat. no. 11). The lithographs in Jules Galet's *Le corps de l'homme* (1835–41), a survey more ambitious in scope than Bartholin's but similar in its magpie nature, were not always exact in its copies (cat. no. 52)—just one of innumerable examples that demonstrate that John Barclay was far too optimistic about the avoidance of error through copying.

FIG. 1.10
ATTRIBUTED TO HANS BALDUNG
(German, 1484/85–1545).
Dissection of the head, exposing the skull, woodcut.
From Walther Hermann Ryff, *Des aller fürtrefflichsten, höchsten vnnd adelichsten Gschöpffs aller Creaturen* (Strasbourg: Balchassar Beck, 1541), fol. 65r.
Los Angeles, Getty Research Institute, 85-B959.

Notes

1 Bell 1793, advertisement, n.p.

2 Bell 1794, iii.

3 Vesalius 1543a, 4r; and Vesalius 2014, 1:7–8. Kusukawa 2012, 192–94. For a historical survey of images and objectivity, see Daston and Galison 1992. For Renaissance attitudes toward medical images, see Kusukawa 2012 and Pantin 2014. Unless otherwise noted, all translations are mine.

4 See Kusukawa 2012, 112–13.

5 Vesalius 1543a, 4r. See Carlino 1999b, 19; Kusukawa 2012, 20–21; 233; and Pantin 2014, 16.

6 Rippa Bonati and Pardo-Tomás 2004; and, in particular, Kemp 2004; and De Caro 2018.

7 Fabricius ab' Acquapendente 1933, 56 (translation), 75 (facsimile).

8 Nutton 2001, 76–77; For evidence that it was the text of the *Fabrica* that most engaged its earliest readers, rather than the images, see Margócsy, Somos, and Joffe 2018; and Margócsy 2019, 317–18.

9 Canani 1925; Lind 1975, 307–12; and Geminus 1545. This also occurs in the titles of printed illustrated herbals of the period.

10 Barclay 1827, 106 (see also 146–47). For the similar reservations of Alexander Monro (primus) (1697–1767), see Kemp 1993, 103. In his introduction to Edward Mitchell's engravings of the skeleton, Barclay considers illustrations "merely as auxiliaries" and never replacements for nature (Barclay 1819–20, 7).

11 Scarpa 1832, 4. Knox's translation first appeared sometime before the second edition of 1829. For the Burke and Hare case, see Richardson, R. 1987.

12 Scarpa 1832, 3.

13 I thank Prof. Elena Stark for this information.

14 Vesalius 1543a, 4r; and Vesalius 2014, 1:8.

15 Lind 1975, 309. For facsimile edition, see Canani 1925. On Canani, see Van Glabbeek and Biesbrouck 2020, with further references.

16 Platter 1583, bk. 3, address to reader. For a discussion of Platter and anatomy, see Herrlinger 1970, 129–31; and Kusukawa 2012, 241–47. For his collection of natural history drawings, see Egmond 2017.

17 Platter 1583, dedication, [2]r; and Platter 1961, 88–90, 92–93, 110.

18 Vesalius 1543a, 161–62; and Vesalius 2015, 227. See O'Malley 1964, 64.

19 Van Mander 1994–99, 1:313, trans. Jacqueline Pennial-Boer and Charles Ford. For a fine imposed on the Florentine physician Ostilio Giunta for body theft in 1563, see Kornell 1993, 56, 217–18. For a fourteenth-century example of prosecution after grave robbing, see Park 1994, 7.

20 For a listing of copies after Vesalius, see Cushing 1962. On the reception history of the *Fabrica* illustrations, see Margócsy 2019; and for consideration of the text and illustrations by the *Fabrica*'s owners, see the recent and invaluable census of copies of the first and second editions of the *Fabrica* by Margócsy, Somos, and Joffe 2018. For copies after Albinus, see Punt 1983, 128.

21 Cellini 1901, 55–56; and Vasari 1568, pt. 3, 1:83. For the statue, excavated in Bologna in 1514 and mounted by Berengario on a revolving base, see Putti 1937, 40n4. For the suggestion that Berengario's illustrations show the influence of contemporary prints and drawings, see Kornell 1989b, 846–47; and Laurenza 2012, 19.

22 Heseler 1959, 137. On Vesalius's lecture drawings, see O'Malley 1958, 13n22; and, most recently, Shotwell 2018.

23 For Vesalius's references to his own drawings in the *Fabrica*, see Saunders and O'Malley 1950, 29; and for the *Epitome*, see Vesalius 2015, 232.

24 Vesalius 1543a, 266 [268]; and Vesalius 2014, 2:538. For an overview of artist collaborations in the making of scientific prints, see Dackerman 2011a.

25 Casseri names him as the German artist Josepho Murero (Casseri 1600–1601, pt. 2, *De auris auditus organi historia anatomica*, bk. 1, ch. 13, 79; and Choulant 1945, 223). For another private anatomy by Casseri for his northern students, see Klestinec 2010, 41–42.

26 For the stipulation of 1491 that drawings be made of dissections of bodies of the poor in the Ca' Granda hospital, Milan, see Azzolini 2006, 162. I thank Francis Wells for drawing my attention to this document.

27 Conring 1687, 179. See Kornell, "'As Large as Nature'" this volume, p. 49.

28 Fuchs 1542, dedication.

29 Kemp 1993, 107. For these and later examples, see Daston and Galison 1992, 100–103; For examples of "mechanical objectivity" and exclusion of pictorial effect in the nineteenth and twentieth centuries, see Galison 1998, 327–34.

30 Keele and Pedretti 1978–80, 1:W 19070v, no. 113r, 362.

31 Hoboken 1675, 264–65.

32 On Coiter's illustrations, see Herrlinger 1970, 127–29; and on Cowper and etching, see Kornell 2019, 493.

33 On Carter's contributions, see Richardson, R. 2008.

34 Choulant 1945, 299.

35 Barclay 1819–20, 5. The book does include, however, figures after human bones and animal skeletons in Barclay's own collection.

36 Barclay 1819–20, 5.

37 Bell 1794, vii.

38 Valverde 1556, address to reader.

39 Platter 1583, bk. 3, address to reader. See Kusukawa 2012, 241–42.

40 Tortebat and de Piles 1668, address to reader, [iv] v. For editions of the *Abregé*, see Cushing 1962, 144–47; Bridson and White 1990; and Loire 1992, 447–50. For de Piles's acknowledgment of his authorship, see Dufresnoy 1668, 91; and de Piles 1708, 153.

41 Something that Vesalius himself did not do when he reused the second skeleton of the *Fabrica* in the *Epitome*.

42 Turner 1878, 178.

43 For modern rediscoveries of this, see Cushing 1962, 87, figs. 58, 59, with further references; and Cavanagh 1983.

44 Russell 1987, 2, nos. 5, 6.

45 Gautier Dagoty, A. É. 1773, "Plan de l'ouvrage."

46 See Vesalius's letter to his publisher Johannes Oporinus printed in the *Fabrica* (Vesalius 1543a) and *The China Root Epistle* of 1546 (Vesalius 2015, 232; and Albinus 1753, preface.).

47 Ploos van Amstel 1783, iii.

48 On Ryff, see R. Sadan in Dackerman 2011b, 64–67, no. 10; and Marr 2014.

3.

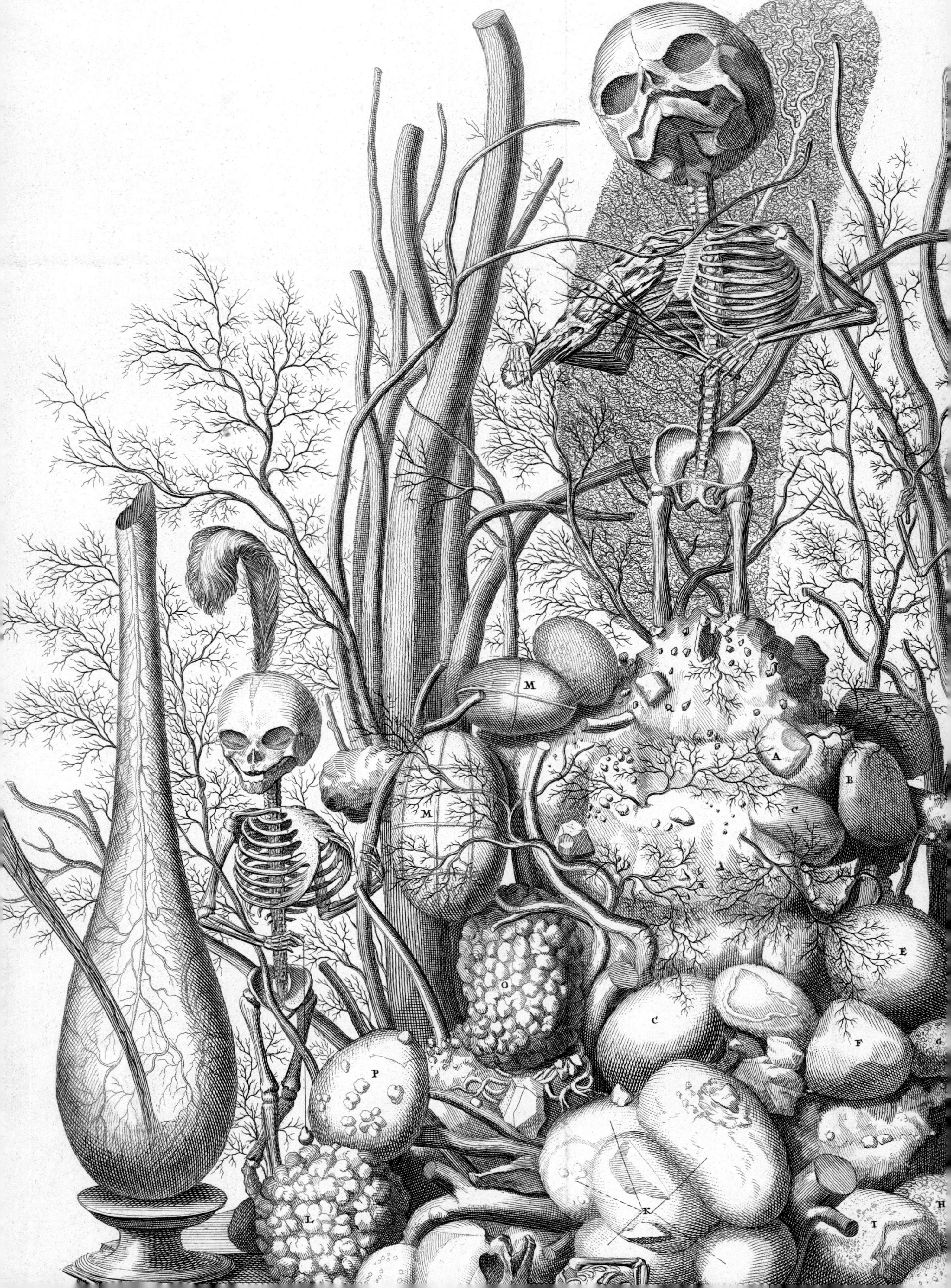

chapter two

THE LIVING DEAD: ANIMATED ANATOMY

Monique Kornell

Striding across a rolling landscape, a pair of dissected legs walk toward the viewer, with a distant port city and sailing ships in the background (cat. no. 12). An uncanny vision that is amusing in its incongruousness, these bounding legs appear to be so full of life that nothing, not even their current anatomized state, will stop them from gamboling across the countryside.[1] This is how, in the early seventeenth century, the Italian artist Odoardo Fialetti (1573–1626/27) depicted for the anatomist Giulio Casseri the muscles of the upper leg and their attachments. The energy of these walking legs is shared in several of Fialetti's figures who twist and turn and raise dissected flaps to display their own anatomy. With one arm akimbo, a figure of deep dissection jauntily raises one of his last remaining muscles with a delicate grasp, hardly inconvenienced by the loss of most of his body (fig. 2.1).

Casseri's wandering dissected legs are an extreme example of what was an enduring theme of early anatomical illustration: the animated cadaver and skeleton, usually depicted in an outdoor setting.[2] Such figures were popularized by Jacopo Berengario da Carpi's best-selling illustrated textbook, the *Isagogae breves* (cat. no. 8), first published in 1522, and in turn adopted by Andreas Vesalius in 1543 for the *Fabrica* (fig. 2.2). Rather than the series of obviously dead figures forlornly propped up on trees or stony chairs found at the beginning of the second book of Charles Estienne and Estienne de la Rivière's *De dissectione* (1545) (fig. 2.3), it was the muscle men, elegant in bearing, and the contemplative skeletons of the *Fabrica* (fig. 2.8; cat. no. 2) that were the subject of innumerable copies, thus establishing a convention that survived, although with decreasing frequency, into the nineteenth century.

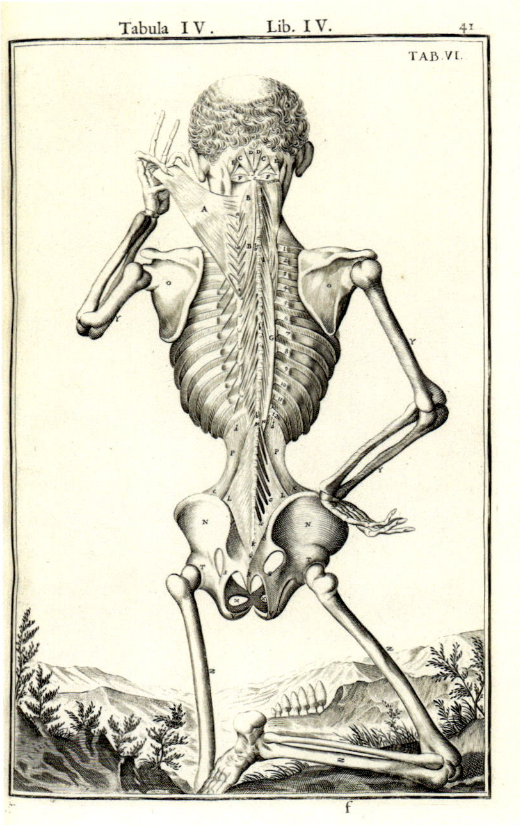

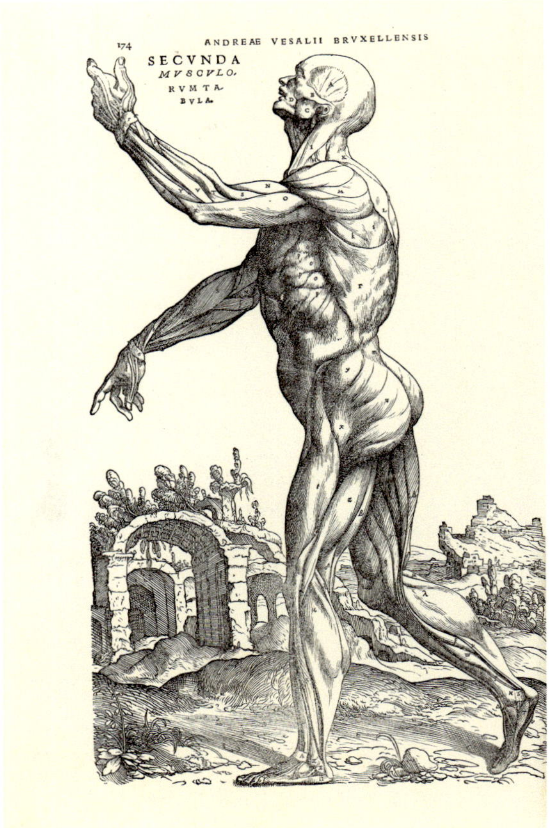

FIG. 2.1
FRANCESCO VALESIO
(Italian, active 1598–1624),
after Odoardo Fialetti (Italian, 1573–1626/27).
Deep dissection, showing muscles that move the head, engraving completed by 1616, and first published in Giulio Casseri, *Tabulae anatomicae* (Venice: Evangelista Deuchinus, 1627).
From Adriaan van den Spiegel, *Opera quae extant, omnia* (Amsterdam: Joan Blaeu, 1645), bk. 4, p. 41, pl. 6.
Los Angeles, Getty Research Institute, 84-B2833.

FIG. 2.2
AFTER JAN STEVEN VAN CALCAR
(North Netherlandish, ca. 1515–ca. 1546).
Second muscle figure, woodcut.
From Andreas Vesalius, *De humani corporis fabrica libri septem* (Basel: J. Oporinus, 1543), bk. 2, p. 174.
Los Angeles, Getty Research Institute, 84-B27611.

FIG. 2.3
AFTER ESTIENNE DE LA RIVIÈRE
(French, d. 1569).
Abdominal dissection demonstrating the layers of the peritoneum, woodcut.
From Charles Estienne and Estienne de la Rivière, *De dissectione partium corporis humani libri tres* (Paris: Simon de Colines, 1545), bk. 2, p. 168.
Los Angeles, Getty Research Institute, 84-B31171.

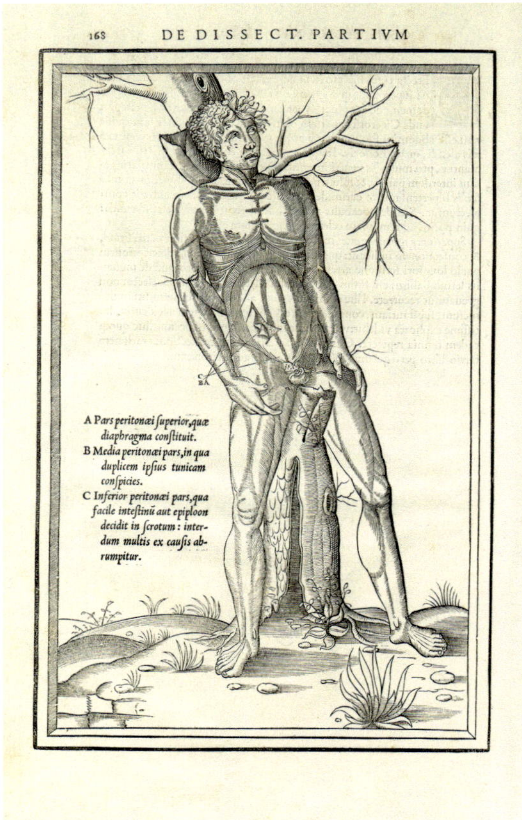

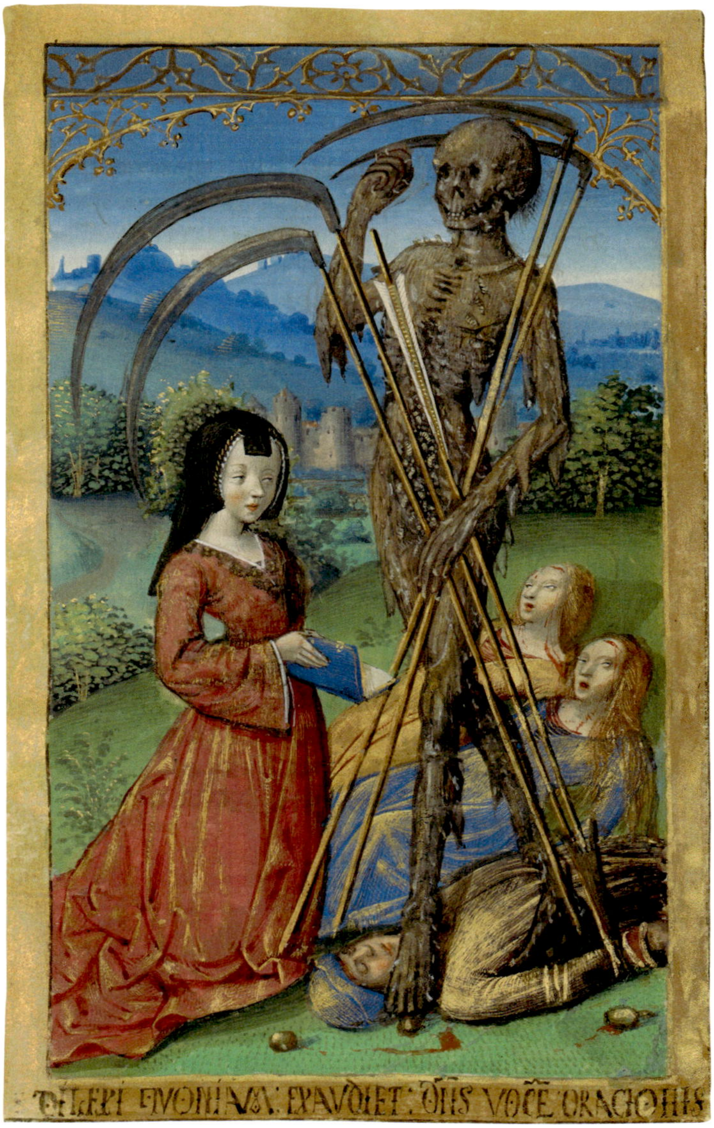

The lifeless cadaver was more often to be found in depictions of the act of dissection, as in the *Fasciculus medicinae* (1491), published under the name of Johannes de Ketham (cat. no. 56), or on the title pages of the *Fabrica* and other books (cat. nos. 16, 19; see fig. 1.1).[3]

The conceit of the animated cadaver inhabiting the land of the living was one already well known in the tradition of the Dance of Death, where the cavorting dead prance about unhindered by their decomposing state. However, unlike memento mori figures who regularly intrude upon the daily lives of often unsuspecting mortals (fig. 2.4; cat. no. 4),[4] active cadavers and skeletons in early anatomical illustration do not regularly interact directly with the living world; instead, views of towns, buildings, and the occasional boat drifting on a calm body of water are seen at a distance (cat. nos. 1, 12).[5] In an illustration to Samuel van Hoogstraten's art treatise of 1678 (fig. 2.5), a boat full of passengers passes by a pair of écorchés on the shore. These muscle men remain separate from the living world but are nevertheless anchored to the here and now. While the activities of the cadavers and skeletons of the Dance of Death serve to remind us of the fleetingness of life and the inescapability of death, the aim of the animated corpse in early illustrated anatomy books has a different purpose, which is to imitate and recapture life and, in doing so, reveal its inner workings.

FIG. 2.4
MASTER OF THE CHRONIQUE SCANDALEUSE (French, active ca. 1493–1510).
Denise Poncher before a vision of Death, tempera colors, ink, and gold on parchment.
From Master of the Chronique scandaleuse, *Poncher Hours*, ca. 1500, MS 109, fol. 156.
Los Angeles, J. Paul Getty Museum, 2011.40.156.

In anatomical illustration, the beauty of the body, the wonder of its structure, the delight in discovering its hidden layers, can never be completely separated from the horror of death and the sense of imminent decay. Whatever empathetic response a body being cut and denuded, layer by layer, might evoke, the living corpses in early anatomical illustration, if sometimes melancholic, are usually shown without expressions of pain. They endure their progressively diminishing state with the same equanimity that is found in early images of "wound figures" for early surgeons,[6] one that is not far removed from the serene expressions given images of martyred saints, who are typically shown stoically enduring their torture, often with eyes cast to heaven (see fig. 6.5), or else calmly continuing to exist while exhibiting emblems of their martyrdom. This can be seen in a late fifteenth-century altarpiece of Saint Stephen, who was stoned to death; two of the stones remain improbably balanced on his head, from which blood drips (fig. 2.6).[7] The irreal calmness of the animated dissected cadaver negates any potential concern on the viewer's part for the suffering of the sentient or for the taboo of vivisection on human subjects.

Unlike saints, dissected cadavers were unlikely to claim the consolation of heavenly reward, particularly given that a main source for anatomies in the early modern period was executed criminals, as is hinted at by the noose slung over the shoulder of the "rope man" that first appeared in Berengario's commentary on Mondino de' Luzzi (1521) and then reappeared in editions of his *Isagogae breves* (cat. no. 8, fig. A).[8] In a humorous dialogue in verse by Bernardino Partenio added to the Bologna 1523 edition of *Isagogae breves,* the dissected cadaver of an executed thief named Harpagus descends to the underworld in order to plea to Pluto for the return of his missing body parts, which medical students had absconded with.[9] Pluto has to lend Harpagus a tongue so that he may speak. In telling his tale to Pluto, he gives the reader the sequence of a Renaissance dissection.

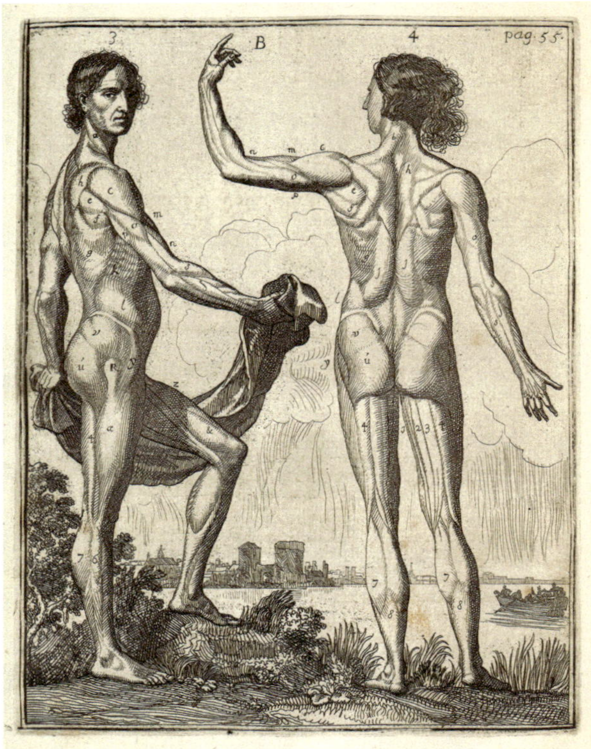

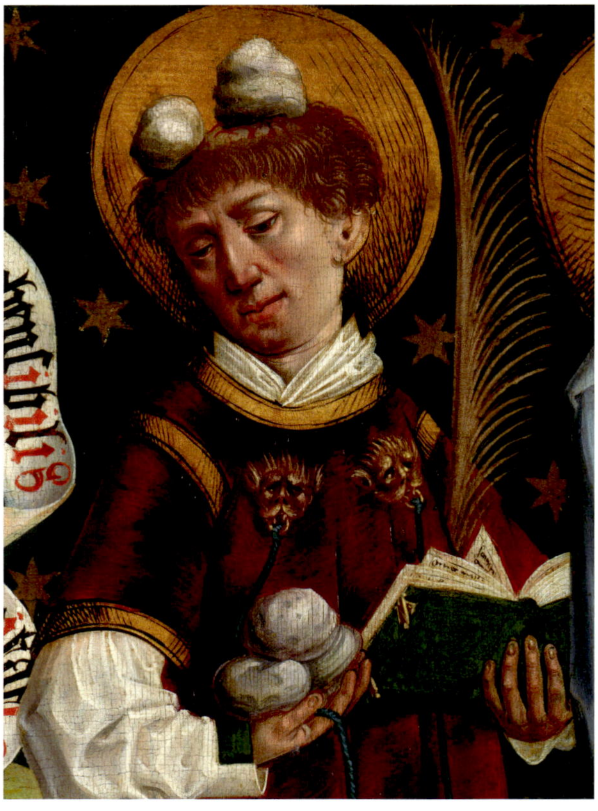

Motion is one of several factors that denotes liveliness in anatomical illustration.[10] In Vesalius's second muscle figure, the raised heel of the right foot, the toes pushing off the ground, and the eloquent hand gestures demand the viewer consider that this figure still has the use of his muscles, however bare they may be of skin (see fig. 2.2).[11] The striding pose and arrangement of arms allow for a simultaneous view of the muscles on the inside and the outside of the legs and arms. Unlike Fialetti's figures for Casseri who remain animated to the end, the Vesalian muscle men of book two falter and require support as they are progressively divested of their muscles, and it is this debilitation, as pointed out by Glenn Harcourt, that reinforces the conceit that the muscle figures are alive.[12] Also from the mid-sixteenth century, but not published in their entirety until the eighteenth, are the vivacious figures engraved for Bartolomeo Eustachi, such as one in an archer-like pose used to display the muscles and nerves (see fig. 1.4). But the figures of Vesalius, Eustachi, and later Bernhard Siegfried Albinus (cat. no. 15) all seem sedate and collected compared to the incredible dynamism of the écorchés by Peter Paul Rubens, who rush past each other in the sheet of anatomical studies in the J. Paul Getty Museum; figure 2.7 is one of a series of anatomical drawings kept as a group by Rubens.[13] The same impulse to show the muscles in an active pose is seen in the écorché gladiators in the anatomy books for artists by Carlo Cesi (1679),[14] Genga and Lancisi (1691), and in Salvage (1812) (cat. nos. 20, 32, 34).

FIG. 2.5
SAMUEL VAN HOOGSTRATEN
(Dutch, 1627–78).
Écorché figures in profile and anterior view, etching. From Samuel van Hoogstraten, *Inleyding tot de Hooge Schoole der Schilderkonst, anders de Zichtbaere Werelt* (Rotterdam: François van Hoogstraten, 1678), pl. B.
Los Angeles, Getty Research Institute, 86-B12032.

FIG. 2.6
PETER HEMMEL VON ANDLAU WORKSHOP
(German, ca. 1420/25–after 1501).
Saint Stephen, with rocks of his martyrdom, detail from *The Trinity with the Virgin, Saints John the Evangelist, Stephen and Lawrence and a Donor,* 1479, oil on panel, 79.8 × 140.2 cm.
Los Angeles, J. Paul Getty Museum, 2012.22.

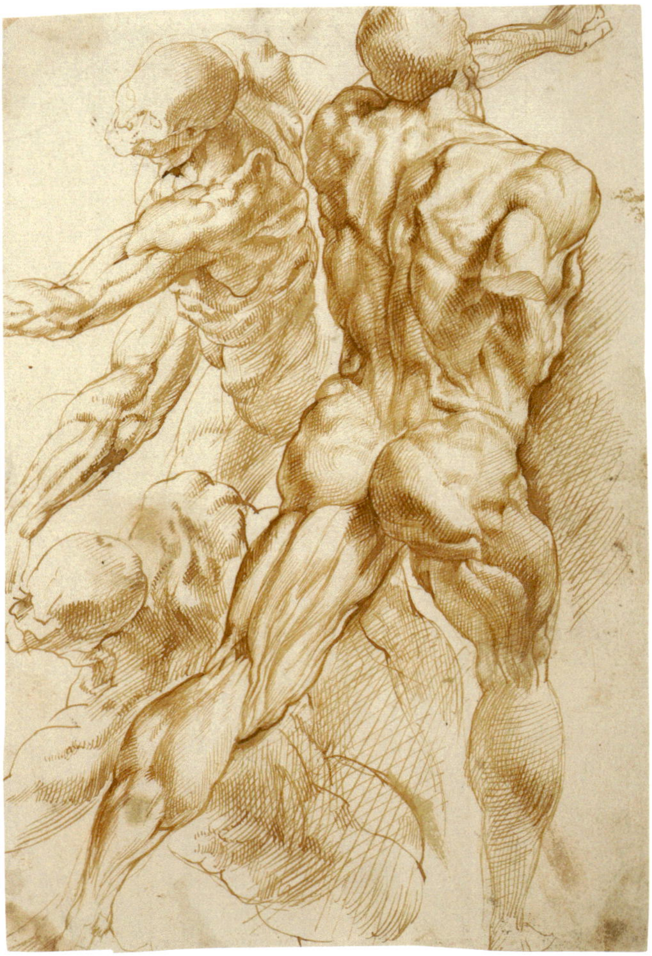

FIG. 2.7
PETER PAUL RUBENS
(Flemish, 1577–1640).
Anatomical studies, ca. 1600–1605, pen and brown ink, 27.9 × 18.7 cm.
Los Angeles, J. Paul Getty Museum, 88.GA.86.

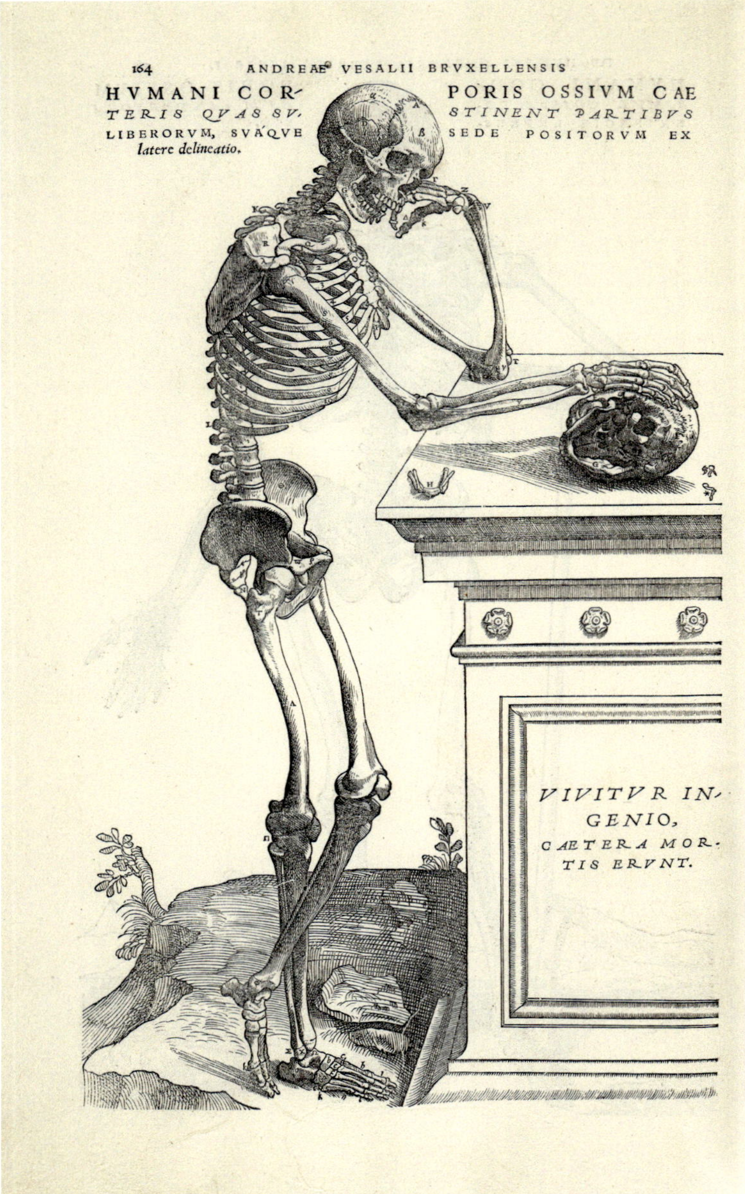

FIG. 2.8
AFTER JAN STEVEN VAN CALCAR
(North Netherlandish, ca. 1515–ca. 1546).
Profile skeleton, woodcut.
From Andreas Vesalius, *De humani corporis fabrica libri septem* (Basel: J. Oporinus, 1543), bk. 1, p. 164.
Los Angeles, Getty Research Institute, 84-B27611.

Exuberant animated skeletons featured in the decorations for aristocratic funerals of the sixteenth and seventeenth centuries (cat. no. 3) are descendants of the Dance of Death and the small posable skeletons of the ancient world, like the one that amused Trimalchio's dinner guests in Petronius's *Satyricon*.[15] The artists who designed them could call on their early training, which included drawing from posed skeletons. This is an activity seen in Cornelis Cort's engraving after Stradanus of an ideal academy, in which the education of the artist along with the practice of art in every media is represented (see fig. 3.2). At the lower left, youths are drawing a skeleton suspended by a rope and set in a pose of great torsion. It is a visualization of the method suggested by the sculptor and goldsmith Benvenuto Cellini (1500–1571), who in a treatise dated to the mid-1560s described a course of study that started with the purposefully undaunting exercise of drawing a single bone—which, according to Cellini, would seem to the timorous student as nothing more than a small stick—and culminated with the drawing of an entire skeleton set in a variety of animated poses, "as if he was a living man."[16] The suspended skeleton that the Carracci kept "for the benefit of all" in their Bolognese academy proved far too animated for one ambitious artist. Pietro Faccini (1575/76–1602) was in the habit of returning secretly at night to draw it. When the brothers Annibale and Agostino Carracci became aware of this, they waited in hiding and manipulated the skeleton's ropes, inclining it toward Faccini, which caused him to flee in terror.[17]

In addition to the promise of motility, a sense of consciousness imparts life to what we know cannot be alive. The emotiveness of the *Fabrica* skeletons who despair, lament, and contemplate grant them an inner life (fig. 2.8; cat. no. 2). The skeleton viewed in profile, its skull supported on its hand in a familiar gesture of thought, is remarkable for simultaneously evoking a consideration of the ephemeral nature of life while uniting the author and the reader in considering the physical structure of the skull.[18] This is reinforced by the inscription

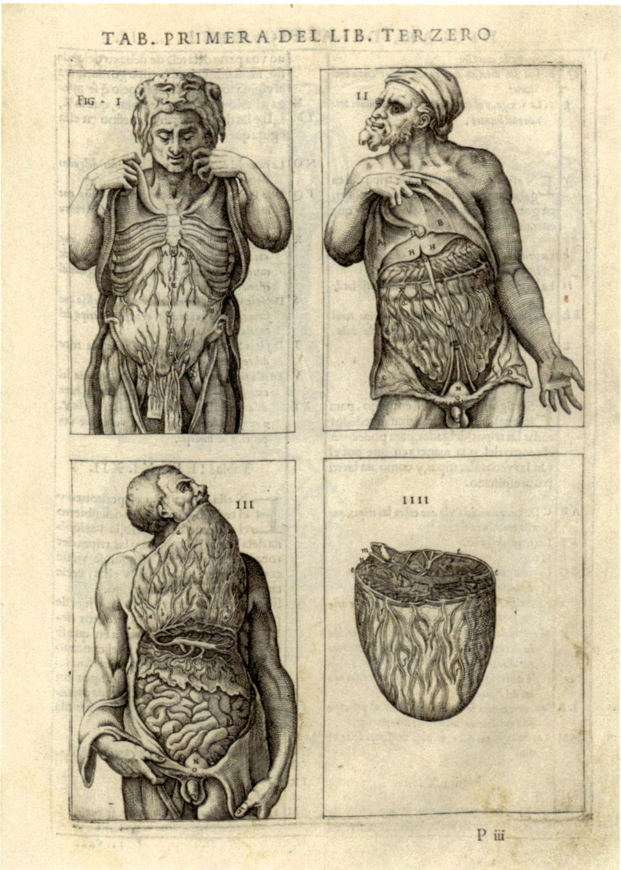

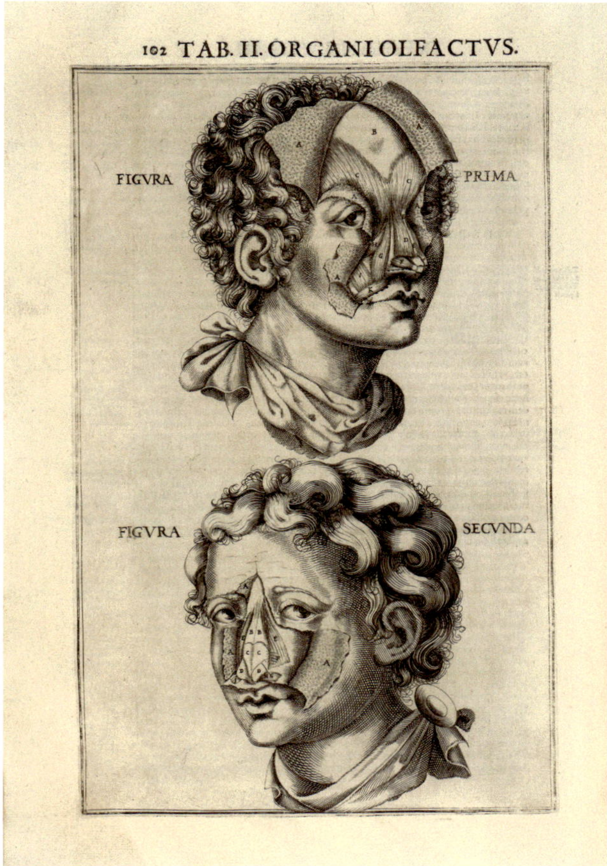

FIG. 2.9
NICOLAS BEATRIZET (French, 1507 or 1515–ca. 1565), after Gaspar Becerra (Spanish, ca. 1520–ca. 1570), after Jan Steven van Calcar (North Netherlandish, ca. 1515–ca. 1546).
Abdominal dissections and the omentum, engraving.
From Juan Valverde de Amusco, *Historia de la composicion del cuerpo humano* (Rome: Antonio Salamanca and Antonio Lafrerij, 1556), bk. 3, pl. 1.
Bethesda, National Library of Medicine, 2294023R.

FIG. 2.10
Nose dissections, engraving.
From Giulio Casseri, *Pentaestheseion, hoc est de quinque sensibus liber* (Venice: Nicolo Missirini, 1609), bk. 3, p. 102, pl. 2.
San Marino, California, Huntington Library, 631661.

on the plinth that declares, "One lives on by genius, the rest will belong to death." The self-displaying cadavers that are seen in Berengario (1523) (cat. no. 8), Valverde (1556) (fig. 2.9), Casseri (1627), and Gaetano Petrioli (1741) could be said to be aware of their dissected states and, by extension, the reader, who benefits from the flaps they lift away from their bodies to grant a better view of the structures beneath.[19] What they all have in common is a reluctance to engage directly with the viewer as they either look off into the distance or look back at their own bodies. This, however, is not the case with a selection of illustrations in Casseri's book on the senses, *Pentaestheseion, hoc est de quinque sensibus liber* (1609). In an illustration of the muscles of the eye appearing in the section devoted to the sense of sight, there is a frisson in the connecting gaze of an eye, loose in its socket, rotated to look at the viewer (cat. no. 11). The same effect is seen in the youthful faces whose eyes look out to us somewhat resignedly on either side of their anatomized noses in an illustration in the section concerning the sense of smell (fig. 2.10). Another rare example is Van Hoogstraten's écorché in profile, whose head is turned to look directly at the viewer, catching our eye with a calm sense of self that bridges the gap between his physical state and that of our own (see fig. 2.5).

The retention of hair and a living face was another means of eliminating the sense of death and otherness. Setting an early and influential example, all the figures in Berengario's *Isagogae breves* retain their facial features and hair (cat. no. 8). The "rope man" echoes not only the pose of Michelangelo Buonarroti's *David* but also his resolute expression (cat. no. 8, fig. A). The locks on the two heads from Casseri's *Pentaestheseion* (see fig. 2.10) have been charmingly detailed with small curls for the male head and larger waves for the female head below, both interspersed with individuated curly strands. In the *Elemens de pourtraiture*, a drawing manual by the painter and engraver Jean de Saint-Igny, an écorché is

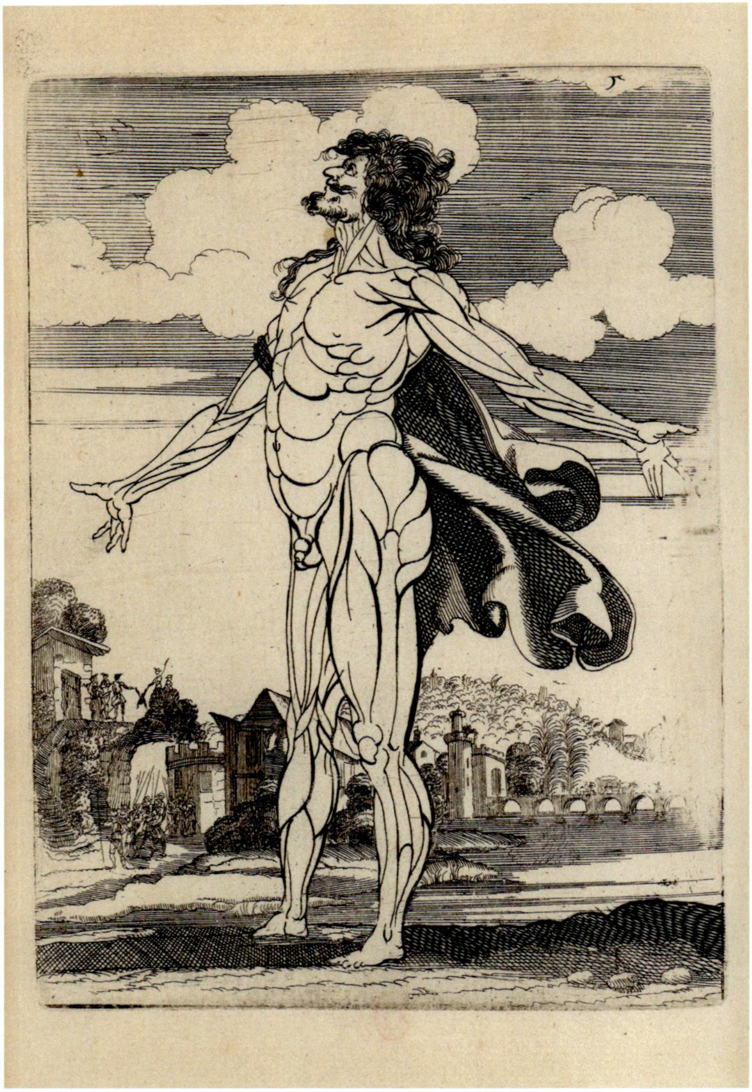

FIG. 2.11
JEAN DE SAINT-IGNY
(French, ca. 1660–after 1649).
Écorché in profile, etching.
From Jean de Saint-Igny, *Elemens de pourtraiture; ou, La metode de representer & pourtraire toutes les parties du corps humain* (Paris: François Langlois dit Chartres, [1630]), part 4, pl. 5.
Paris, Bibliothèque de l'Institut National d'Histoire de l'Art, 12 RES 2100 (1-2).

presented with a luxuriant wig and a Van Dyke beard (fig. 2.11).[20] Elegantly posed with outstretched arms and a turned-out leg before a contemporary populated townscape, he is an anatomical counterpart to the fashionable figures in *Le jardin de la noblesse françoise,* published the previous year in 1629 after drawings by Saint-Igny and Abraham Bosse (1602–76).[21] The anatomized women in the mezzotints of Jacques Fabien Gautier Dagoty appear elegantly coiffed (cat. no. 41). Below an image of a dissected woman who had died in labor, but who is nevertheless shown with rosy cheeks and carefully arranged hair, Gautier Dagoty remarked, "These figures have been given an air of life to remove a more disagreeable aspect."[22] They are akin to the wax

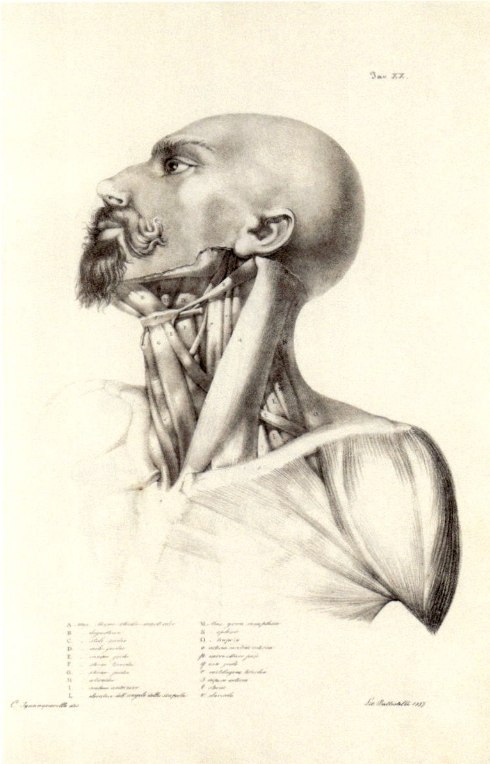

anatomical Venuses of the eighteenth century, who have lengthy tresses, rosy lips, and sometimes necklaces of pearls. The attention to these attributes makes these figures alluring vanitas symbols (see fig. 8.3).[23] In an illustration from Costantino Squanquerillo's anatomy book for artists of 1841 (fig. 2.12), the scalpel's progress was halted just above the jaw—at a similar point seen in Antonio Scarpa's *Tabulae neurologicae* (1794) and again in Henry Gray's *Anatomy, Descriptive and Surgical* (1858) (see fig. 1.5)—thereby sparing the face and leaving its splendid facial hair intact.[24] A delight in hairstyles is also displayed in three of Squanquerillo's écorché figures, who sport the coiffures of dandies of the period.

The paradox of using a fantastical mode to present the factual was rejected emphatically by Govard Bidloo (1649–1713), who had his artist Gérard de Lairesse draw inanimate cadavers and dissected parts in a still-life fashion, for his *Anatomia humani corporis* (1685), which Bidloo describes on the title page and in the address to the reader as "ad vivum" (cat. no. 40).[25] William

FIG. 2.12
PACIFICO BATTISTELLI (Italian, active late 1830s–1840), after Costantino Squanquerillo (Italian, active late 1830s–1840). *The Superficial Muscles of the Neck*, 1837, lithograph. From Costantino Squanquerillo, *Trattato di anatomia pittorica* (Rome: Tipografia delle Belle Arti, 1841), pl. 20. Los Angeles, Getty Research Institute, 89-B25581.

Hunter declares his allegiance to Bidloo's example of representing the actual rather than the imagined in the preface to the *Gravid Uterus* of 1774 (cat. no. 42). In the same spirit, Charles Nicholas Jenty describes the mezzotints of a dissection of a woman at full term (after drawings by Jan van Rymsdyk [active 1750–90]) in his 1758 obstetric atlas as "not done at Random, or from Fancy, as some have been."[26] For the illustrations to his book of anatomy, the Scottish anatomist and surgeon John Bell drew and etched the figures himself, showing cadavers on the dissecting table and including the ropes and props of the dissecting room. He had little patience for the "imagination of the painter," pointing to the "ludicrous contrast" between the sturdy, active figures demonstrating their own anatomy and their dissected state.[27] As the portrayal of the entire figure was progressively abandoned in favor of depictions of specific areas of the body, as seen in Gray's *Anatomy*, the lively corpse ultimately faded from view.

Notes

1 On the element of play in anatomical illustration, see Sappol 2006, 17–19.

2 On this subject, see Sawday 1995, 112–29; Cazort 1996, 27–29; Caldwell 2006; Kornell 2007, 209–14; and Cuir 2009. Cuir sees the presentation of the animated cadaver as an expression of the early modern concern with the functioning body (Cuir 2009, 56, 60, 66–75). For late fifteenth-century examples of a so-called Zodiac man with dissected abdomen in an outdoor setting, see *Der "teutsche Kalender": "Meister Almansor spricht,"* printed in Augsburg by J. Blaubiren circa 1483 (London, Wellcome Collection) and *Hours of Louis Quarré*, Oxford, Bodleian Library, MS Douce 311, fol. 1v, after 1488.

3 For Hans Wechtlin's woodcut fugitive sheet of an eyewitness's recording of a dissection dated 1517, see Carlino 1999b, 57, fig. 33; 82. See Carlino 1999a, ch. 1, for title pages to early anatomy books.

4 For examples in early Renaissance manuscript illumination, see Morrison 2017; and idem, 88–89, for an analysis of figure 4.

5 Boats are also seen in the background of illustrations in the *Fabrica* (Vesalius 1543a, 178), in Dryander's *Anatomia Mundini* (1541, fol. 5r), and in the *Tabulae anatomicae* of Pietro da Cortona (Petrioli 1741, pl. 14).

6 Karp 1985, no. 4b; and Carlino 1999a, 14, 17, fig. 6.

7 In Marco d'Agrate's statue of 1562 in Milan's cathedral, Saint Bartholomew's skin is loosely draped around his flayed living body. For examples of the flayed Saint Bartholomew shown not only alive but also preaching, see Mittman and Sciacca 2017, 147–49, 159–63; and Gregory 2018, 794, 796, fig. 8.

8 Park 1994, 23. On criminals as cadaver sources, see also Carlino 1999a, 92–98, 214–19.

9 "Plutonis & Harpagi dissecti dialogus," in Berengario 1523, fols. 73r–74r; and Roth 1889.

10 For considerations of lifelikeness, see Jacobs 2005; and Syson et al. 2018.

11 The raised hand may reflect the oratorical gesture in Titian's *The Allocution of Marquis del Vasto to His Troops* (1540–41) in Museo Nacional del Prado, Madrid (P000417). See Panofsky 1969, 74–75.

12 Vesalius 1543a, bk. 2, tables 7, 8, 14; and Harcourt 1987, 48.

13 See Jaffé 1987; Heinen and Jaffé 2005; Laurenza 2012, 41–45; and Woollett 2021, p. 7.

14 Cesi 1679, pl. 16. Illustrated in Kornell 1996, 63, fig. 34.

15 See, for example, one in bronze in the J. Paul Getty Museum, 78.AB.307. For examples of skeleton drawings for funerals, see Gaeta Bertelà and Petrioli Tofani 1969, nos. 32, 43, 82–83, 120.

16 Cellini 1979, 1935–36, 1939.

17 Malvasia 1678, 1:565.

18 For an exploration of these themes, see San Juan 2012.

19 Valverde 1556, bk. 3, pl. 1. For Casseri, see Walters 2009, 1:138–39; and Petrioli 1741, pls. 4, 6, 9, 14, 17, 27. I thank Jack Hartnell for drawing my attention to a medieval example in the scroll of the English surgeon John Ardenne dated to circa 1430, in the National Library of Sweden, Stockholm, MS X 118: https://exhibitions.lib.cam.ac.uk/vesalius/artifacts/saggital-section-john-ardene/.

20 Saint-Igny [1630]. Dated by the privilege granted on 18 October 1630.

21 Bosse and Saint-Igny 1629.

22 Gautier Dagoty, J. F. 1773, pl. 6.

23 Jordanova 1989, 50. For a late fifteenth-century illumination with Dance of Death figures sporting fashionable headdresses, see Morrison 2017, 90–91, fig. 4.

24 Scarpa 1794, pls. 1-3. Agnolo Bronzino's *St. Bartholomew* (ca. 1555–56) in the Galleria dell'Accademia Nazionale di San Luca, Rome (inv. 423), provides an earlier example of selective flaying, retaining the skin on the fingers and toes as well as his bearded face (Kemp and Wallace 2000, 73, no. 66).

25 Fend 2019.

26 Jenty 1758, 8.

27 Bell 1794, vi.

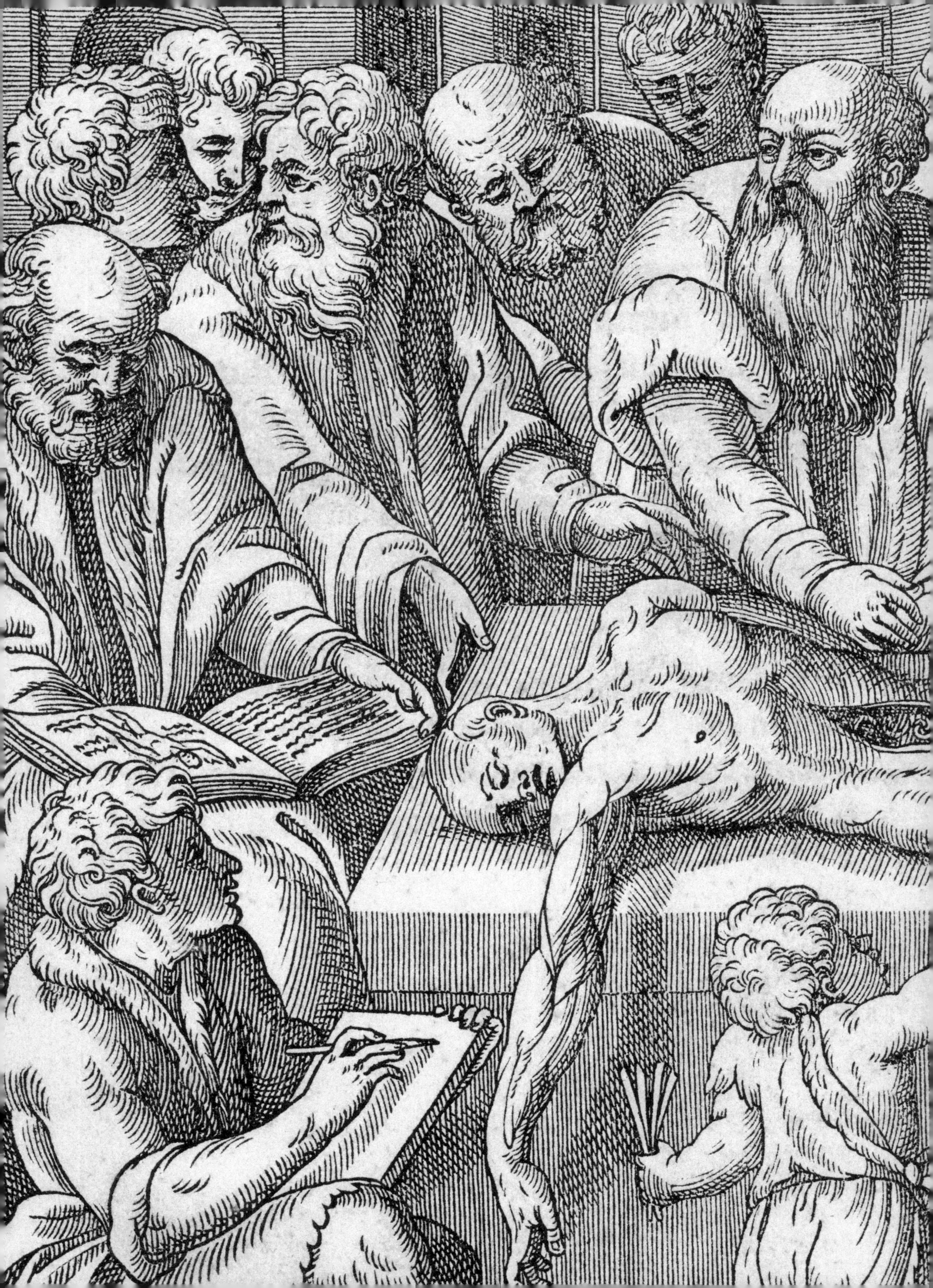

chapter three

ARTISTS AND ANATOMY BOOKS

Monique Kornell

In his *China Root Epistle* (1546), Andreas Vesalius complains of the ill humor of engravers and painters who often made him feel "more hapless than the corpses who had fallen to my lot for dissection."[1] Having overseen and personally financed the production of more than two hundred woodcuts for his *Fabrica* and *Epitome* (both 1543) in a span of less than four years (cat. no. 30), it is not surprising that the anatomist would have wearied of some aspects of his collaboration with his artistic team. Overall, in the *Fabrica,* Vesalius is supportive of artists and solicitous of their interest in anatomy, and he displays evidence that he has been looking critically at their works. He advises them to note the changing appearance of the muscle in action, and in his explanation of the second muscle figure, Vesalius brings them to task over their frequent misrepresentation of the brachialis muscle in the arm (see fig. 2.2, "N").[2] The first two muscle figures in book 2, which he likens to the "square-built" figures of learned artists and sculptors, leave out the membranous tissues and muscle striations depicted in the face and neck of the third figure, which Vesalius observes "are rather disturbing to the painter, the sculptor, and the modeler of statuary (whose studies it also seems right to benefit)."[3] An earlier if briefer recognition of artists is made by the anatomist Jacopo Berengario da Carpi, who signals that his woodcut illustration of the anterior view of the muscles below the skin was useful to physicians and to artists in the depiction of limbs (fig. 3.1).[4]

The sixteenth century, a period in which great emphasis was placed on the mastery of the nude figure in emulation of the antique and of Michelangelo Buonarroti, saw an intense desire by artists to acquire a knowledge of anatomy, particularly in Italy. Artists carried out dissections, made casts from flayed bodies, modeled écorché statues, and circulated anatomical drawings, and some owned books on anatomy. Like medical students, they occasionally resorted to disturbing graves to obtain cadavers, as the Florentine painter Rosso Fiorentino (1494–1540) did in the late 1520s while a guest of his friend Leonardo Tornabuoni, the bishop of Sansepolcro.[5] For dissection opportunities, artists also took advantage of their medical contacts. Michelangelo's biographer, Ascanio Condivi (1525–74), records that Realdo Colombo (cat. no. 16), Michelangelo's friend and physician, had conferred with the artist on anatomy and sent him the body of a "young, most handsome Moor" to dissect.[6] The relationship between artists and the medical community would become formalized in the late sixteenth century, when anatomy became a fundamental part of the curriculum of the newly established formal art academies, and when physicians, surgeons, and anatomists were often appointed as lecturers on the

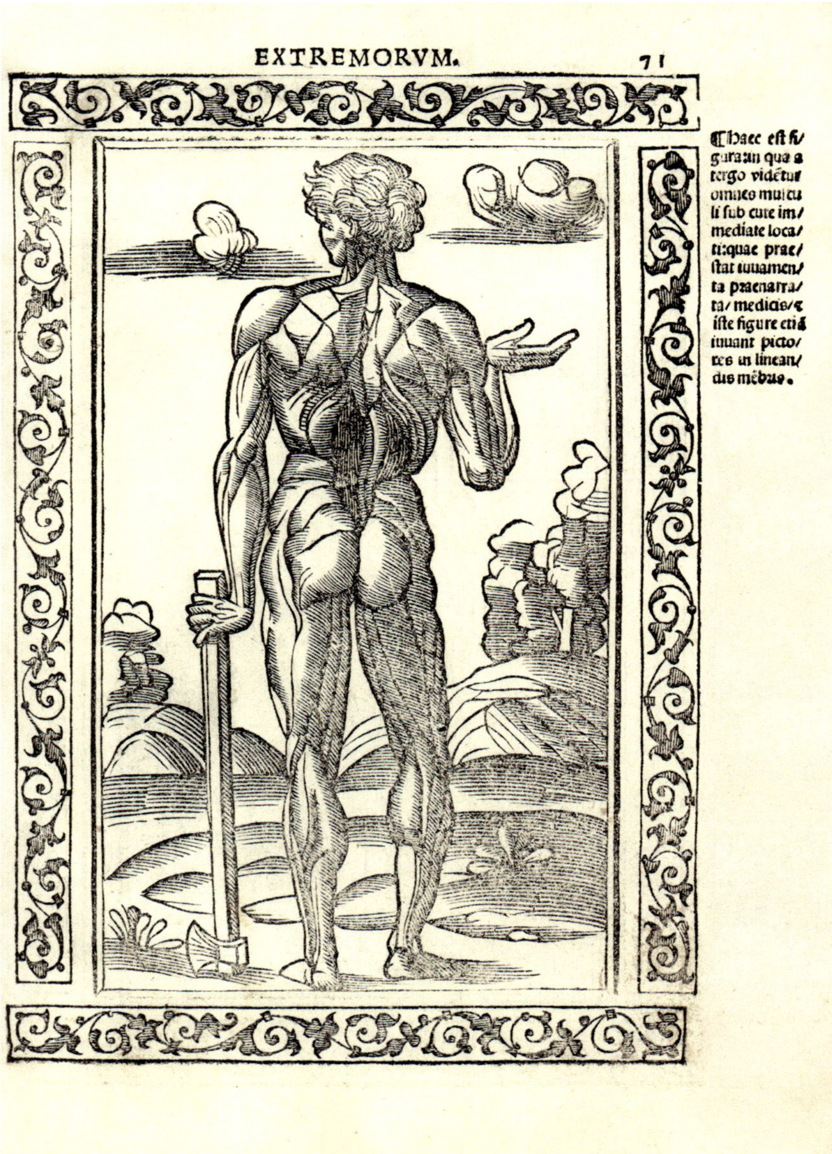

FIG. 3.1
View of the anterior muscles, woodcut.
From Jacopo Berengario da Carpi, *Isagogae breves* (Bologna: B. Hectoris, 1523), fol. 71r.
Los Angeles, Getty Research Institute, 84-B28214.

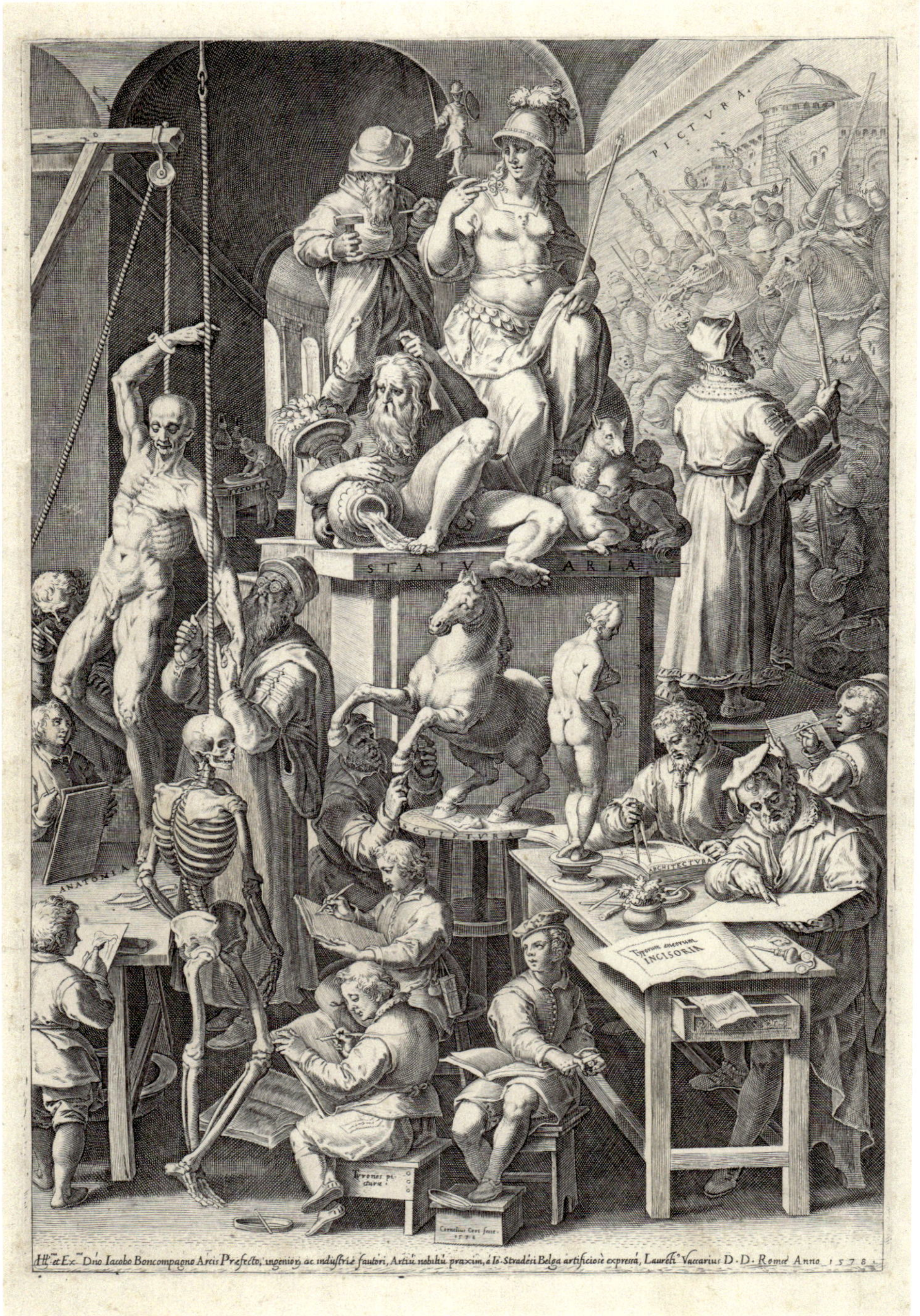

FIG. 3.2
CORNELIS CORT
(North Netherlandish, 1533–before 1578), after Stradanus (Flemish, active in Italy, 1523–1605). *Academy of Fine Arts,* 1578, engraving, plate mark: 43 × 29.5 cm; sheet: 46 × 31 cm. Los Angeles, Getty Research Institute, 1410-207.

subject. Anatomy would remain an important element of an artist's education until the waning of the academic system in the twentieth century.

Some indication of how anatomy was taught and how central its study had become in the education of artists by the late sixteenth century is found in early scenes of art academies. In a sixteenth-century engraving by Cornelis Cort (1533–before 1578) after Stradanus (1523–1605) of an ideal academy, in which art is pursued in all its media, young apprentices draw from a posed skeleton and an écorché suspended from ropes (fig. 3.2).[7] The latter is in the process of being flayed; a bespectacled man and his assistant work together at the task in a race against putrefaction. Pietro Francesco Alberti's print dated to

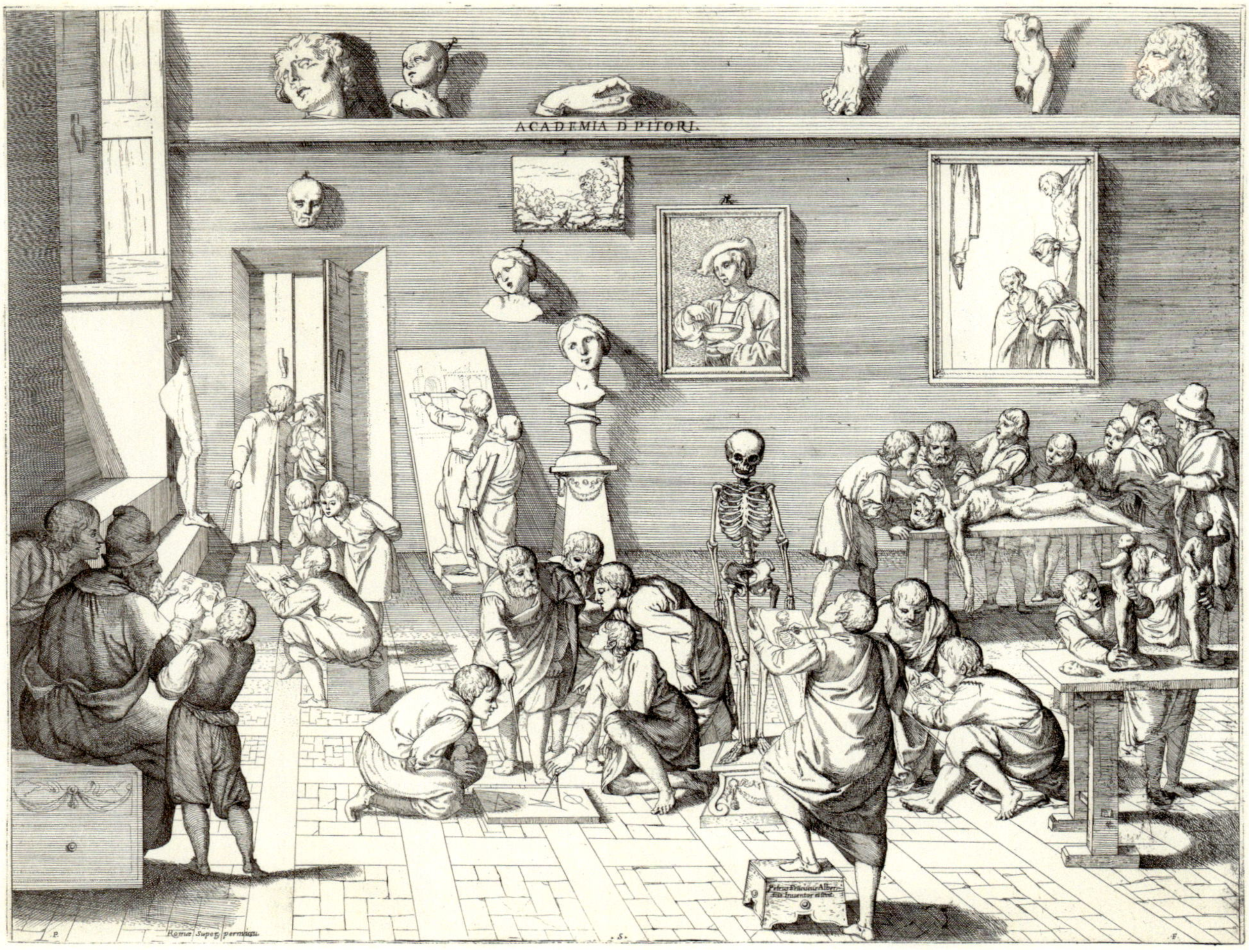

FIG. 3.3
PIETRO FRANCESCO ALBERTI
(Italian, 1584–1638).
Academia d'pitori,
ca. 1600, etching, image: 41 × ca. 53 cm, sheet 43 × 55 cm.
Los Angeles, Getty Research Institute, 2007.PR.29.

the turn of the seventeenth century of a Roman academy is focused on the education of the young artist. Here again, an articulated skeleton is the subject for young draftsmen, while in the background eager youths look on as a cadaver is dissected (fig. 3.3).[8]

According to Condivi, Michelangelo had planned a book of anatomy for artists, one that would deal with the appearance of the body in motion.[9] Michelangelo's was one of several reported but unfinished anatomy books by sixteenth-century Italian artists.[10] Paradoxically, the first successful illustrated publications dealing with anatomy for artists occurred outside of Italy at the end of the century in Seville, as part of a book by Juan de Arfe y Villafañe (1585–87) (cat. no. 17), and in drawing manuals with texts like those of Heinrich Lautensack (1564) (fig. 3.4), Philips Galle (1589),[11] and Jean Cousin the Younger (1595) (cat. no. 18). Anatomical exemplars would continue to appear in later drawing manuals (see fig. 2.11).[12] Whether or not Peter Paul Rubens intended his anatomy drawings for publication either as a book or as part of a drawing manual, they were the subject of copies by his students and were disseminated as engravings after his death (see fig. 2.7).[13] His renditions of energetic écorché figures indicate that like Michelangelo, Rubens was interested in the anatomy of the moving body.

A new stand-alone genre of anatomy books for artists—one that exists to this day—first appeared in the seventeenth century (cat. nos. 2, 19, 20; see fig. 1.6).[14] Initially produced by artists, several examples were later authored by anatomists or surgeons, many of whom,

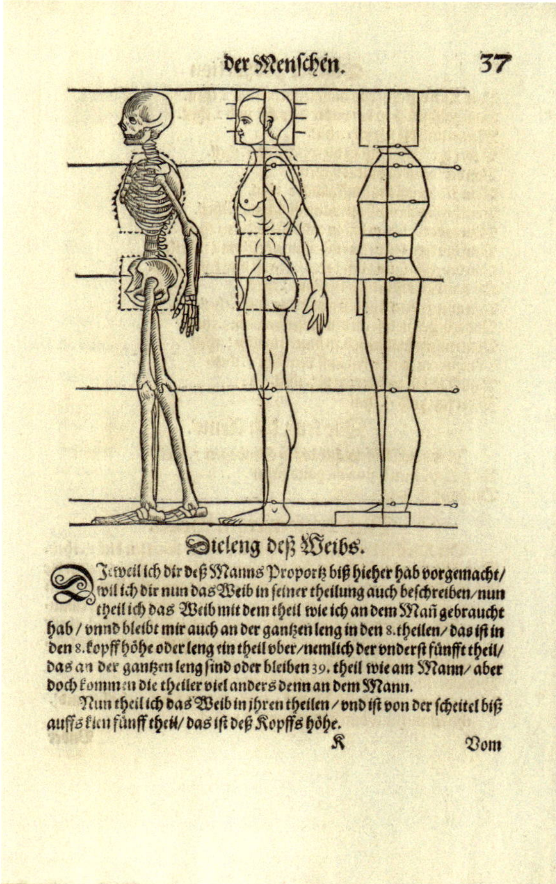

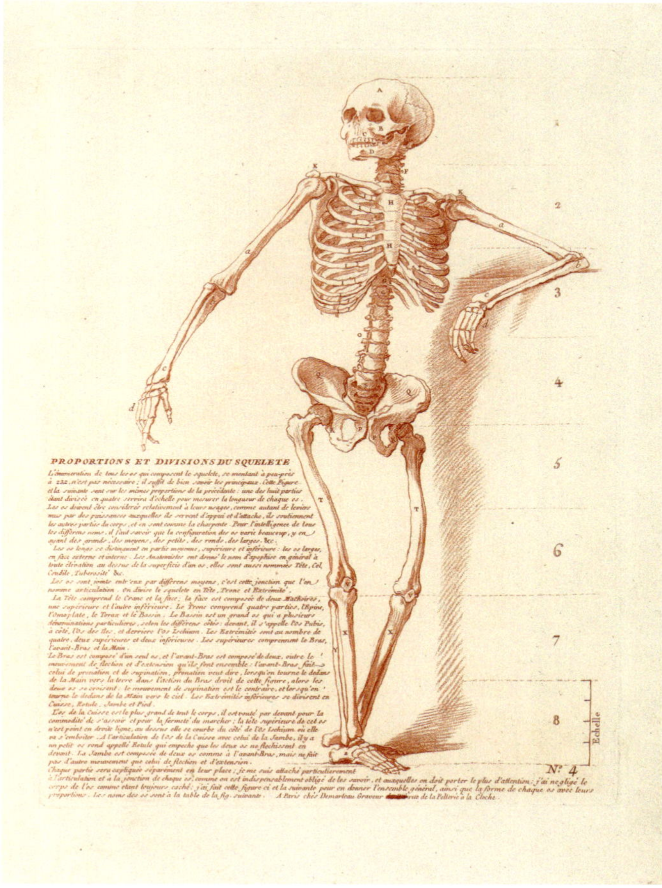

FIG. 3.4
HEINRICH LAUTENSACK (German, 1522–68). Proportional studies, woodcut.
From Heinrich Lautensack, *Dess Circkelss und Richtscheyts, auch der Perspectiva, und Proportion der Menschen und Rosse, kurtze, doch gründtliche Underweisung, dess rechten gebrauchs,* 2nd ed. (Frankfurt: Egenolff Emmel, in verlegung Simonis Schambergers, 1618), fol. 37r.
Los Angeles, Getty Research Institute, 87-B1738.

FIG. 3.5
GILLES DEMARTEAU (French, 1722–76), after Charles Monnet (French, 1732–after 1808).
Skeleton, with proportional measures, crayon-manner engraving.
From Charles Monnet, *Études d'anatomie à l'usage des peintres* (Paris: Demarteau, 1774, pl. 4.
Los Angeles, Getty Research Institute, 84-B28065.

such as Bernardino Genga (1636–95) and Giuseppe del Medico (1772–1825), were connected to academies (cat. nos. 32, 27).[15] Anatomy books tailored for artists generally concentrated on the bones and muscles and avoided the viscera, nerves, and veins. They also frequently promised simplicity and a brevity of approach. This pledge is ably fulfilled in *L'anatomie nécessaire pour l'usage du dessein* (1741) (cat. no. 21), by the French sculptor and draftsman Edme Bouchardon (1698–1762); the names and functions of the bones and muscles are summarized on just a single page. Proportional studies were occasionally included, such as those found in the late eighteenth-century book of the French artist Charles Monnet (fig. 3.5; cat. no. 23), for which there are precedents in earlier anatomical drawings[16] and drawing manuals (see fig. 3.4; cat. nos. 5, 18).

An indication of the resources that artists relied on before the appearance of these specialized anatomy books for artists is found in the *Dialogos de la pintura* (1633), by Vicente Carducho (ca. 1576–1638), an Italian-born artist who spent most of his life in Spain. Carducho advises artists to consult the books of Vesalius and Juan Valverde de Amusco, as well as the anatomical drawings of the Italian sculptor Prospero Antichi il Bresciano (ca. 1555/65–92) and the Italian painter Romulo Cincinnato (ca. 1502–93).[17] An inventory of Carducho's possessions records that he owned a copy of Valverde and a bronze anatomical model by Bresciano, who, according to Giovanni Baglione (1571–1644), excelled at this type of figure, both large and small.[18] The Florentine artist Alessandro Allori (1535–1607) referred to both Vesalius's *Fabrica,* a copy of which he owned, and Valverde in his dialogue on the rules of design.[19] It is to Vesalius that the French painter Nicolas Poussin (1594–1665) first turned during a period of intense artistic study after his arrival in Rome in 1634, followed by an examination of cadavers and skeletons with the French surgeon Nicolas Larche (1602–65).[20] Artists continued to acquire general anatomy books even after the appearance of anatomy books for artists. The English architect Nicholas Hawksmoor (1661–1736) obtained Govard Bidloo's *Anatomia* a few years after its publication in 1685 (cat. no. 40).[21] In addition to copies of François Tortebat and Roger de Piles's *Abregé d'anatomie* (1668), Bouchardon owned illustrated editions of works by Vesalius and the French surgeon Ambroise Paré (1510–90), as well as books on the anatomy of the horse, for which he had a passionate interest.[22]

An episode of a continued preference for Vesalius's *Fabrica* is found in correspondence between the director of the Académie de France in Rome, Matthieu de La Teulière (d. 1702), and the royal superintendent of buildings, Édouard Colbert, marquis de Villacerf (1628–99). In 1690, La Teulière, in what seems to have been an economizing measure, dispensed with the services of the surgeon Bernardino Genga, who had long been the lecturer on anatomy at the academy. By citing the risk of plague, La Teulière blocked Genga from visiting the academy because of a recent suspected case at the hospital of Santo Spirito, to which Genga was attached. Pointing out to Colbert that Genga lectured in Italian, a language in which most of the current pensioners did not have fluency, La Teulière reports that he himself had been teaching anatomy, using the figures of "the great Vesalius."[23] He considers the visual example of the woodcuts more useful to painters and sculptors than Genga's lectures and describes walking room to room with one of the two copies of the *Fabrica* to hand, correcting the artists as they worked. This could not have been easy, given the weightiness of the *Fabrica.* A more portable resource would have been Tortebat and de Piles's *Abregé d'anatomie* with engravings after Vesalius (see fig. 1.6). La Teulière was aware of Tortebat's engravings but was of the opinion that they did not have "the beauty of the originals."[24] That he also eschews the use of Genga and Giovanni Maria Lancisi's book—which was published two years before, in 1691, for use by the academy, and illustrated with engravings after Charles Errard (ca. 1606/9–89), its previous director (cat. no. 32)—speaks of La Teulière's known estrangement from Errard and from Genga.

A Shared Market

A recognition of artists' interest in anatomy is seen in their inclusion in the marketing of anatomy books that also targeted medical practitioners. An early example is a book issued by the Lyonnaise publisher Clément Baudin in 1560 with a series of engravings after the abdominal and chest dissections that first appeared in the fifth and sixth books of the *Fabrica.* The translation of the title is "Description and demonstration of the inner parts of man and woman, in twelve tables, taken from nature, according to the true anatomy of André Wesal [Vesalius], philosopher and doctor of medicine. A work useful and necessary not only to physicians and surgeons, but also to

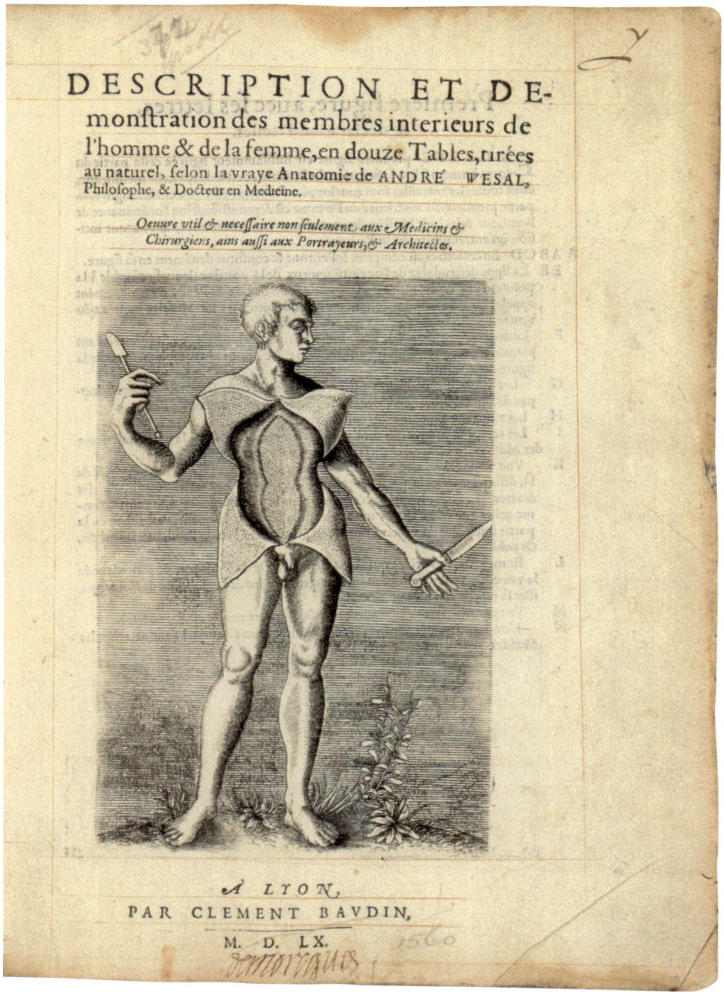

FIG. 3.6
Dissected figure holding surgical implements, etching.
Title page from Andreas Vesalius, *Description et demonstration des membres interieurs de l'homme & de la femme, en douze tables, tirées au naturel, selon la vraye anatomie de André Wesal* (Lyon: Clément Baudin, 1560).
London, Christie's, 9 December 2020, lot 99.

artists [*portrayeurs*] and architects" (fig. 3.6).[25] Just how useful these plates showing the location and form of the internal organs were for artists and architects is debatable, but the aim of capturing as wide an audience as possible for them is clear. The book is a reedition of one published in German the previous year, the extended title of which describes the figures as "not only for physicians and surgeons, but also for those who have a love of art."[26] The youth on the title page of both editions is copied from a woodcut in Joannes Dryander's *Anatomia Mundini* (1541, fol. 4r), which in turn was inspired by Berengario's standing abdominal figures (cat. no. 8), but instead of supporting his own dissected flaps, his hands hold a knife and a surgical implement.

Volcher Coiter, an anatomist who provided his own etchings of human and comparative anatomy for his works, had also planned a book aimed at a varied market that included artists. This was to be on the muscles of the human body for "the surgical profession and those who apply themselves to matters concerning natural phenomena and the arts of painting and sculpture."[27] In his preface to the 1575 edition of his notes on the lectures of Gabriele Falloppio, he cites a lack of suitable bodies as a factor causing him to abandon the book.

The same cross-marketing was at work for books produced mainly for artists. Jacob van der Gracht's anatomy book of 1634 is promoted on the title page as "useful for painters, sculptors, etchers, as well as Surgeons" (cat. no. 19).[28] In the title of the second edition of 1740 of engraved plates of anatomy and proportion by Crisóstomo Martínez (ca. 1638–ca. 1694), first published in the 1690s, a similarly wide net for a prospective audience was cast.[29] *Anatomy Improv'd and Illustrated with Regard to the Uses Thereof in Designing*, John Senex's English edition of 1723 of Genga and Lancisi's anatomy book for artists of 1691, is dedicated to the physician and noted collector Richard Mead (1673–1754), and the list of subscribers published in the book includes surgeons and physicians, among them the Scottish anatomist, physician, and man-midwife James Douglas (1675–1742), who was engaged at that time on two ultimately unpublished illustrated books of anatomy.[30] The full title of the *Compendium anatomicum* (1743), by the English printmaker John Tinney (ca. 1707–61), describes it as "A Work of very great Service to Painters, Statuaries, and all Professors of Drawing and Design; as well as a proper Introduction to the Study of Anatomy for the Use of young Surgeons."[31] Jacques Gamelin's *Nouveau recueil d'ostéologie et de myologie* (1779)—an ambitious but odd mélange of osteology, myology, nudes, and genre scenes—was composed with both "anatomists and draftsmen" in mind, according to its prospectus; but, failing to find broad appeal, it was a financial failure that appeared in a single edition (cat. no. 24).

The *Tabulae anatomicae*, first published in 1741 with prints after Pietro da Cortona (cat. no. 14), appeared in a second edition published in Rome in 1788 with the added declaration that the work was indispensable for surgeons and artists.[32] It is likely that the editor, Francisco Petraglia, was capitalizing on Cortona's fame as an artist, since the focus of this work was on the nerves at different dissected levels, only selected plates of which would have been of immediate use for artists. The engraved vignette on the title page shows an anatomy school in the ancient world, with a dissection taking place in an anteroom to a barrel-vaulted space (fig. 3.7). On the right, as in the early academy scenes of Cort and Alberti (see figs. 3.2, 3.3), youths draw from an articulated skeleton, and on a pedestal in the background is what appears to be an écorché statue. The scene of human dissection and of artists studying anatomy reinforces the appeal of anatomy to the surgeons and artists addressed in the title, although the incidence of these activities in the ancient world would become points of contention in the nineteenth century (see Kornell, "Surface Anatomy," this volume, p. 43). With both the vignette and the title, Petraglia recognized the joint interest of medical practitioners and artists in the anatomy of the human body, a marketing practice that continued into the next century as well.[33]

The veristic depiction of the human figure is no longer a primary requirement for artistic practice, and while anatomy is still often taught in art schools, it is not as pervasive nor does a student receive as thorough a grounding as,

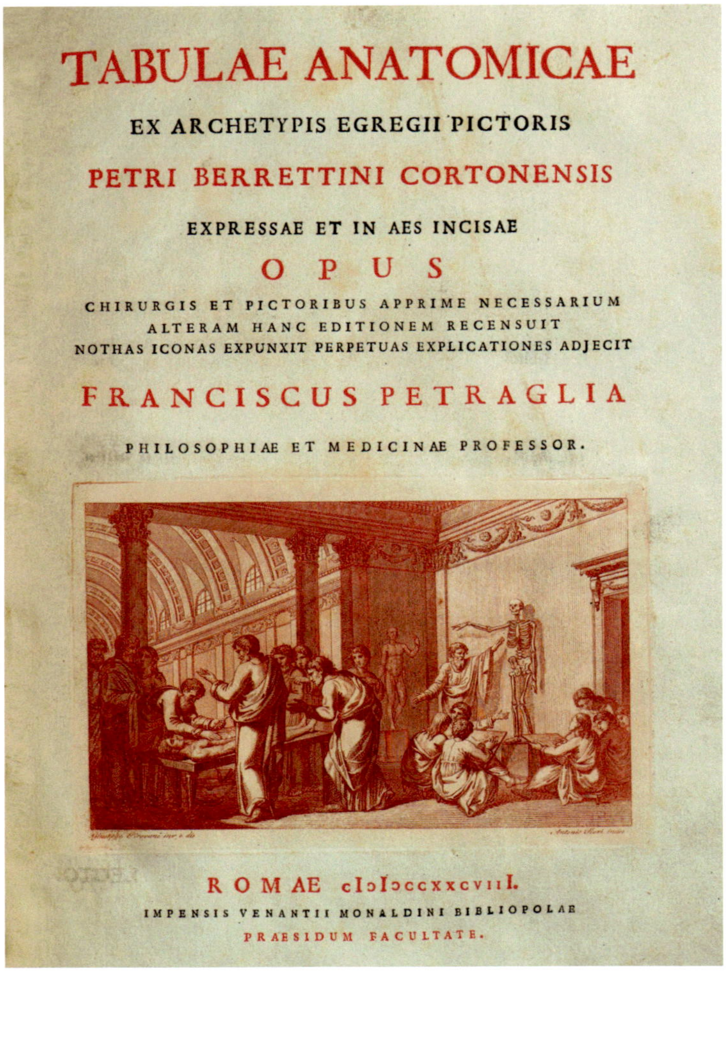
TABULAE ANATOMICAE
EX ARCHETYPIS EGREGII PICTORIS
PETRI BERRETTINI CORTONENSIS
EXPRESSAE ET IN AES INCISAE
O P U S
CHIRURGIS ET PICTORIBUS APPRIME NECESSARIUM
ALTERAM HANC EDITIONEM RECENSUIT
NOTHAS ICONAS EXPUNXIT PERPETUAS EXPLICATIONES ADJECIT
FRANCISCUS PETRAGLIA
PHILOSOPHIÆ ET MEDICINÆ PROFESSOR.
ROMAE cIↃIↃccxxcviiI.
IMPENSIS VENANTII MONALDINI BIBLIOPOLAE
PRAESIDUM FACULTATE.

FIG. 3.7
ANTONIO FIORI
(Italian, active 1788), after Giuseppe Pirovani (Italian, ca. 1755–ca. 1788).
Title-page vignette showing an ancient anatomy school, engraving.
From Francisco Petraglia, *Tabulae anatomicae ex archetypis egregii pictoris Petri Berrettini Cortonensis expressae et in aes incisae: Opus chirurgis et pictoribus apprime necessarium* (Rome: Venanzio Monaldini, 1788).
Vienna, Österreichische Nationalbibliothek, 69.C.5.

for example, that offered in the late nineteenth century under Thomas Eakins at the Pennsylvania Academy of the Fine Arts (cat. no. 48). Yet artists continue to be fascinated by the subject, and anatomy books for artists are still catering to their interest. One hallmark of modern and contemporary art is that the representation of anatomy has itself become a form of expression and a means of referencing the body (cat. nos. 46, 47). This is done with biomorphic images in dialogue with early anatomical illustration in Shelagh Keeley's artist's book (cat. no. 29). The multiple layers of James Allen's book excavation sift through the very history of anatomical illustration to create a work of art that carries the viewer through time (cat. no. 56).

Notes

1 Vesalius 2015, 227. As an alternative to sculptors, the "sculptoribus" that Vesalius mentions could be in reference to the woodblock cutters that had been under his employ.

2 Vesalius 1543a, 171, 175.

3 Vesalius 1543a, 171; and Vesalius 2014, 1:337.

4 First appearing in Berengario 1521, 520v.

5 Vasari 1568, pt. 3, 2:209.

6 Condivi 1553, between fols. 42, 43.

7 For the preparatory drawing and print with earlier bibliography, see Hapoienu in Aymonino and Lauder 2015, 94–99, nos. 4ab; Baroni and Sellink 2012, 227–28, no. 21; and Nanobashvili in Heilmann et al. 2015, 299–301, no. 54. For the mid-sixteenth-century print of Baccio Bandinelli's academy with skeleton parts strewn in the foreground, see Hapoienu in Aymonino and Lauder 2015, 85–87, no. 2.

8 See Lukehart 2015.

9 Condivi 1553, between fols. 42, 43.

10 Daly Davis 1982; Kornell 1989b; Kornell 1998; Haddjeri 2018; and Nanobashvili 2018, 21–80.

11 See Sellink 1992. For Giulio Bonasone's mid-sixteenth-century series of anatomical prints, which appeared without text and may have been produced for artists, see Massari 1983, 1:109–11, nos. 171–84.

12 Bolten 1985, 19, 94, 233–39; Heilmann et al. 2014; Heilmann et al. 2015; and Remond 2019, 297–98, figs. 8, 9.

13 See Kornell, "The Living Dead," this volume, p. 19n13.

14 For surveys of anatomy books for artists, see Choulant 1945; Duval and Cuyer 1898; Kornell 1996; and Röhrl 2000.

15 Genga's birth date, which has been variously reported, is 28 February 1636. This is indicated by an inscription below his portrait engraved by Jean-Charles Allet after a drawing by Guillaume Sarrabat and dated 1694: "Bernardinus Genga de Mundulpho natus Pried. Kal. Martij MDCXXXVI." The portrait is found in Genga's commentary on aphorisms of Hippocrates (Genga 1694), facing the dedication in the copy in the Biblioteca Nazionale Centrale di Roma (12.13.K.22). Genga was dead by 11 October 1695, when his heirs accepted payment of his salary from the hospital of Santo Spirito (Savio 1971, 167n79).

16 Kornell 1998, 174–77; and Rinaldi 2016, 184, figs. 25, 26.

17 Carducho 1633, 2r–v. For anatomical drawings attributed to Bresciano, see Ciardi and Tongiorgi Tomasi 1984, nos. 12–14.

18 Caturla 1968–69, 180, 191; and Baglione 1642, 42.

19 On Allori's dialogue, see Kornell, "Surface Anatomy," this volume, pp. 63–65. For artists' ownership of the 1543 and 1555 editions of the *Fabrica* in the sixteenth and seventeenth centuries, see Margócsy, Somos, and Joffe 2018, 33.

20 Bellori 1672, 412. Bellori also reports that Poussin had earlier studied anatomy at a hospital in Paris, indicating a previous experience with dissection.

21 Vancouver, University of British Columbia Library, WZ250 B58. The date of 28 September 1693 appears after Hawksmoor's signature on the front pastedown.

22 Kornell 2016b, 46; and Kopp 2017, 193–228.

23 La Teulière to Colbert, 6 January 1693, in Montaiglon 1887, 346, no. 379.

24 La Teulière to Colbert, 17 February 1693, in Montaiglon 1887, 363, no. 388.

25 Vesalius 1560. Christie's, London, 2020, lot 99. I am grateful to Dániel Margócsy for kindly examining this copy and supplying me with photographs of it. On the edition, see Baudrier 1901, 26–27; De Feyfer 1914, 468, nos. 38–39; and Cushing 1962, 131, VI.D.-2. There are eleven plates in an incomplete copy in Paris, Bibliothèque Nationale de France, RES 4-TA12-64. The title in this copy has been corrected to "onze" from the original "douze."

26 Vesalius 1559; Haller 1774–77, 1:184; De Feyfer 1914, 468, no. 37; and Cushing 1962, 131, VI.D.-1. There is a copy in Rome, Biblioteca Palatina, Stamp.Pal.III.141.

27 Coiter 1955, 163.

28 For ownership of the book by artists and physicians, see Travers 2019, 273–74.

29 Martínez 1740.

30 Genga and Lancisi 1723. On James Douglas, see Flis 2018, 62–67; and Kornell, "'As Large as Nature,'" this volume, p. 51. For the role of the man-midwife, the term for men practicing obstetrics in the eighteenth century, see Gamer 2018, 112–13; and cat. no. 42, this volume.

31 Tinney 1743; and Russell 1987, 192, no. 816.

32 Petraglia 1788.

33 See Hooper 1807, with plates after Bernhard Siegfried Albinus.

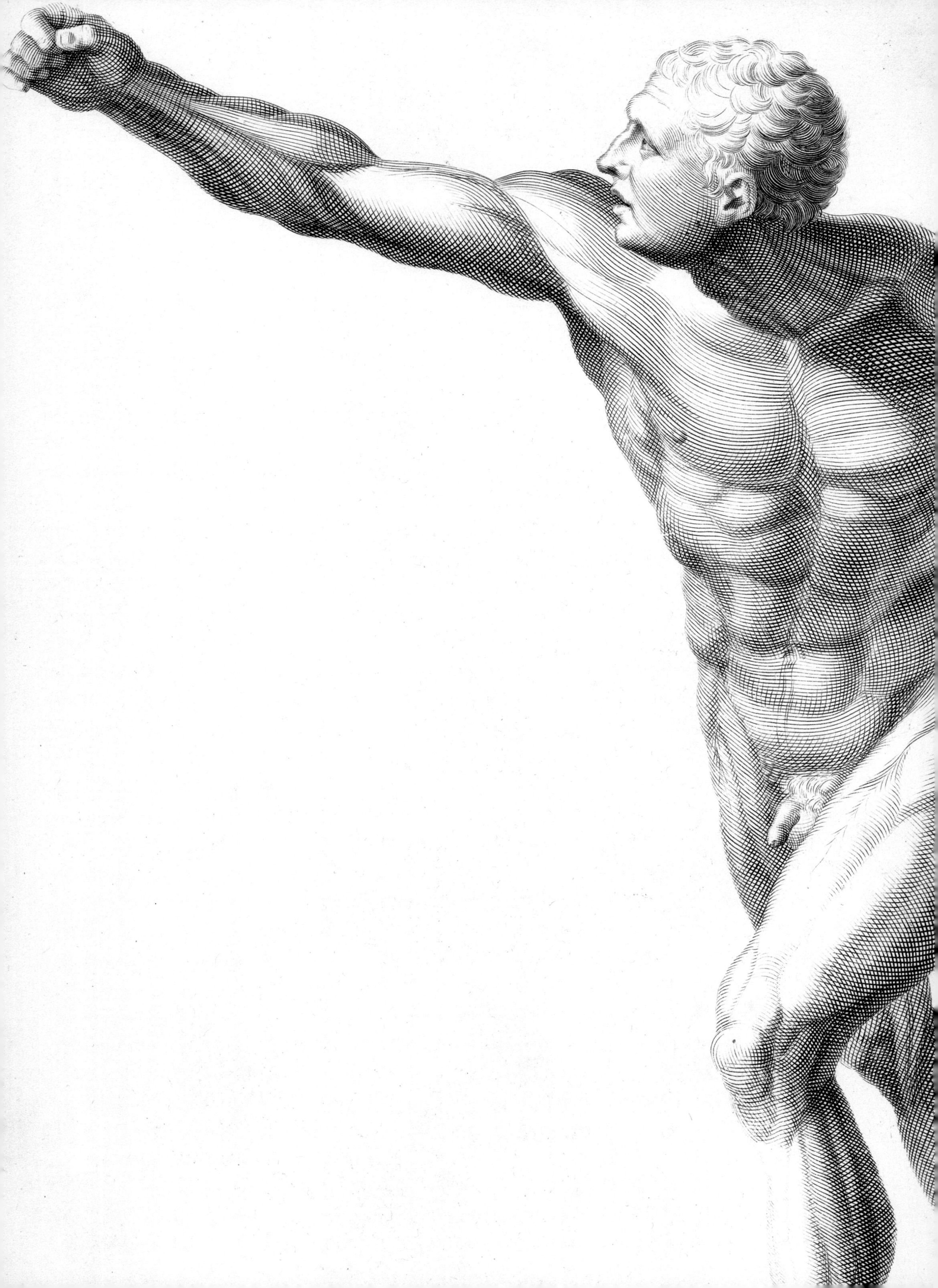

chapter four

ANATOMY AND THE ANTIQUE

Monique Kornell

[Anatomy] alone can teach us to conceive and penetrate the beauties of the Antique and of Nature, this aim of perfection which we all seek to approach.

—Charles Monnet, *Études d'anatomie à l'usage des peintres*

Admired and collected since the Renaissance as a tangible link to ancient Greece and Rome, antique sculpture was universally considered as having achieved the ideal representation of the human body in all its aspects, including beauty and proportion as well as anatomy. A notable theme in anatomical illustration from the sixteenth century onward is the anatomizing of ancient sculpture, suggesting that the surface stone could be peeled back to expose the muscles and that deeper cuts would reveal an inner anatomical structure.[1] This conceit implies that the secrets of the perfection of antique sculpture could be revealed by what was, after all, a fictive dissection on paper.

An early and influential example of this is found in the fifth book of Andreas Vesalius's *De humani corporis fabrica libri septem* (1543), in which abdominal anatomy is displayed encased in torsi, the limbs of which are broken off in a manner that evokes ancient sculpture (cat. no. 30). Even though internal organs, intestines, and generative systems are revealed, no bone or muscle is seen at the breaks, thereby maintaining the fiction of a sculptural fragment. Treating ancient sculpture as a living thing and examining its structural layers was analogous to what Vesalius and his colleagues were doing in studying and teaching from texts of ancient authors.[2] The *Fabrica,* as well as being a description of the human body, is a commentary on the famed second-century Greek physician Galen, whose observations were primarily based on the dissections of apes and other animals, rather than humans.

The torsi of the *Fabrica* signal to the reader Vesalius's knowledge and appreciation of the culture of the ancient world, as do the ruins and obelisks that dot the landscapes in the second book, in which Vesalius's flayed and gradually disintegrating muscle figures are set.[3] In reference to the ideal normative body to be used for dissection, Vesalius cites the fifth-century BC Greek sculptor Polykleitos, who wrote a treatise on symmetry and proportion.[4] Vesalius looked at antique sculpture with a critical eye. In a lecture in Bologna in 1540, he commented that the large muscles of the back, upper arm, and shoulder were more beautifully represented in ancient works of art than by contemporary painters.[5] Furthermore, in the *Fabrica,* after evocatively likening the appearance of the closed tricuspid valve of the heart to the trigrooved point of a Turkish spear, he goes on to compare the open valve with its membranes and fibers to the crowns of heads of ancient statues excavated in Rome.[6]

One may say of the Vesalian torsi that they have been reconstructed in order to be deconstructed, for there are no marks or surface losses and, apart from where the stone skin has been resected, they are unblemished. Aside from a figure in book five, which is similar to the Belvedere Torso in Rome,[7] antique sources for the other torsi have not been securely identified. One torso with a strong contrapposto pose (cat. no. 30) was adapted for three of the figures, with small changes introduced, such as the length of a thigh or upper arm—an indication that the torsi of book five were not exact records of specific marbles.[8] For one of the torsi, Vesalius reversed the process and turned flesh into stone. The female cadaver shown lying on the dissecting table at the center of a great crowd of spectators populating the temporary anatomy theater (see fig. 1.1) is the basis for a marble torso seen in the same perspective but in reverse direction in the twenty-fourth figure.[9]

In the Renaissance, fragments of antique sculpture, regardless of their incomplete state, were highly esteemed and collected (cat. no. 30, fig. I).[10] Vesalius's encasement of exposed abdominal anatomy in a pristine antique torso helped to neutralize and ameliorate repugnant connotations. For contemporary readers, an attractive and desirable antique marble was a far more pleasant association than a limbless, eviscerated body.[11] The practice of quartering convicted criminals was a public corporal punishment meted out for serious crimes. In the final stages, the body was decapitated, the limbs were cut off, and the entrails were exposed.[12] Vesalius records taking advantage of this type of execution in Padua in order to view the still-beating heart of a body.[13] Two figures at the beginning of the sixth book—torsi with heads, one still in breeches and with a noose tied around his neck—are of executed criminals. Aside from the title page, they are the only decidedly dead bodies in the book. There was also a functional purpose to the use of the torsi for the layout of the book, in that they provided a stylish space-saving method of display. Whereas an entire page is devoted to the illustrations that deal with abdominal anatomy in the books of Berengario and Estienne and La Rivière (cat. no. 8; see fig. 2.3), using sculptural fragments allowed Vesalius to interweave text and image on the page while still depicting the dissected area on a clearly discernible scale.[14]

The anatomist Juan Valverde de Amusco was one of Vesalius's most successful copyists, in terms of number of editions and adaptations. The painter and sculptor Gaspar Becerra (ca. 1520–ca. 1570)—like Valverde, a Spaniard working in Italy—adapted the Vesalian illustrations and also provided original ones for Valverde's anatomy book, first published in Spanish in 1556 as *Historia de la composicion del cuerpo humano.* Some of these embrace the *all'antica* style, as seen in the remarkable anatomized pregnant Venus pudica (cat. no. 31), the Apollo Belvedere in écorché, and Roman cuirasses peeled back to reveal dissected abdomens.[15] Overall, Becerra preferred depicting a living body to stone, and, in this respect, he is Pygmalion to many of the *Fabrica* torsi.

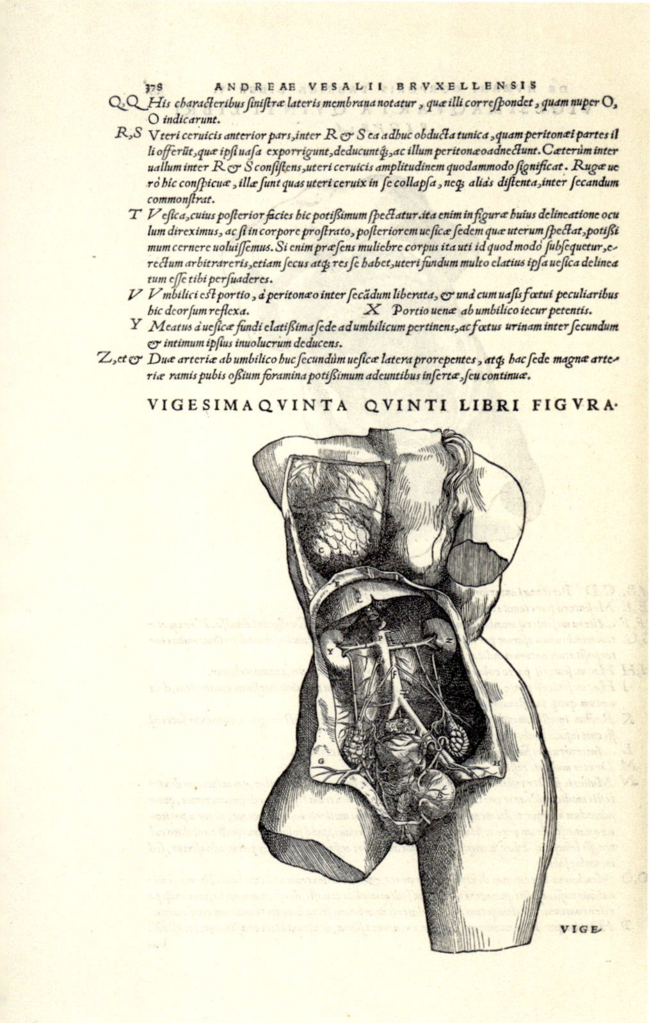

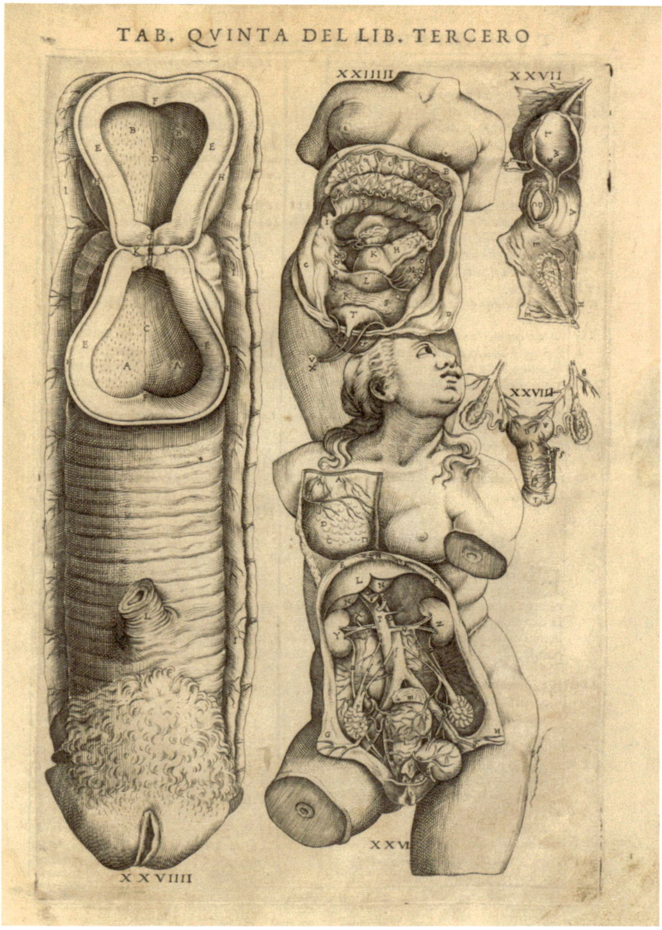

FIG. 4.1
AFTER JAN STEVEN VAN CALCAR
(North Netherlandish, ca. 1515–ca. 1546).
Female generative system and breast dissection, woodcut.
From Andreas Vesalius, *De humani corporis fabrica libri septem* (Basel: J. Oporinus, 1543), bk. 5, p. 378 [478], fig. 25.
Los Angeles, Getty Research Institute, 84-B27611.

FIG. 4.2
NICOLAS BEATRIZET
(French, 1507 or 1515–ca. 1565), after Gaspar Becerra (Spanish, ca. 1520–ca. 1570), after Jan Steven van Calcar (North Netherlandish, ca. 1515–ca. 1546).
Female abdominal dissections and generative organs, engraving.
From Juan Valverde de Amusco, *Historia de la composicion del cuerpo humano* (Rome: Antonio Salamanca and Antonio Lafrerij, 1556), bk. 3, pl. 5.
Bethesda, National Library of Medicine, 2294023R.

Several figures have had their heads and arms restored.[16] Far from being inert sculpture, the first three torsi taken from the *Fabrica*'s fifth book are so lively that they participate in their own dissection by holding back resected flaps from their bodies (see fig. 2.9). For the final torso of the series, a female figure with a dissected breast and an abdomen displaying the generative system, Valverde's illustrator was intent on presenting a body of flesh rather than one of marble (figs. 4.1, 4.2). The newly added head awkwardly turns to look toward the top right of the page, where an illustration of the uterus, the bladder, and an ovary from the *Fabrica* has been turned sideways to fit onto the plate (see fig. xxvii in fig. 4.2).[17] Additionally, the ends of the arm and the leg exposed to the viewer now have an indication of severed bone, which, if less detailed than in the

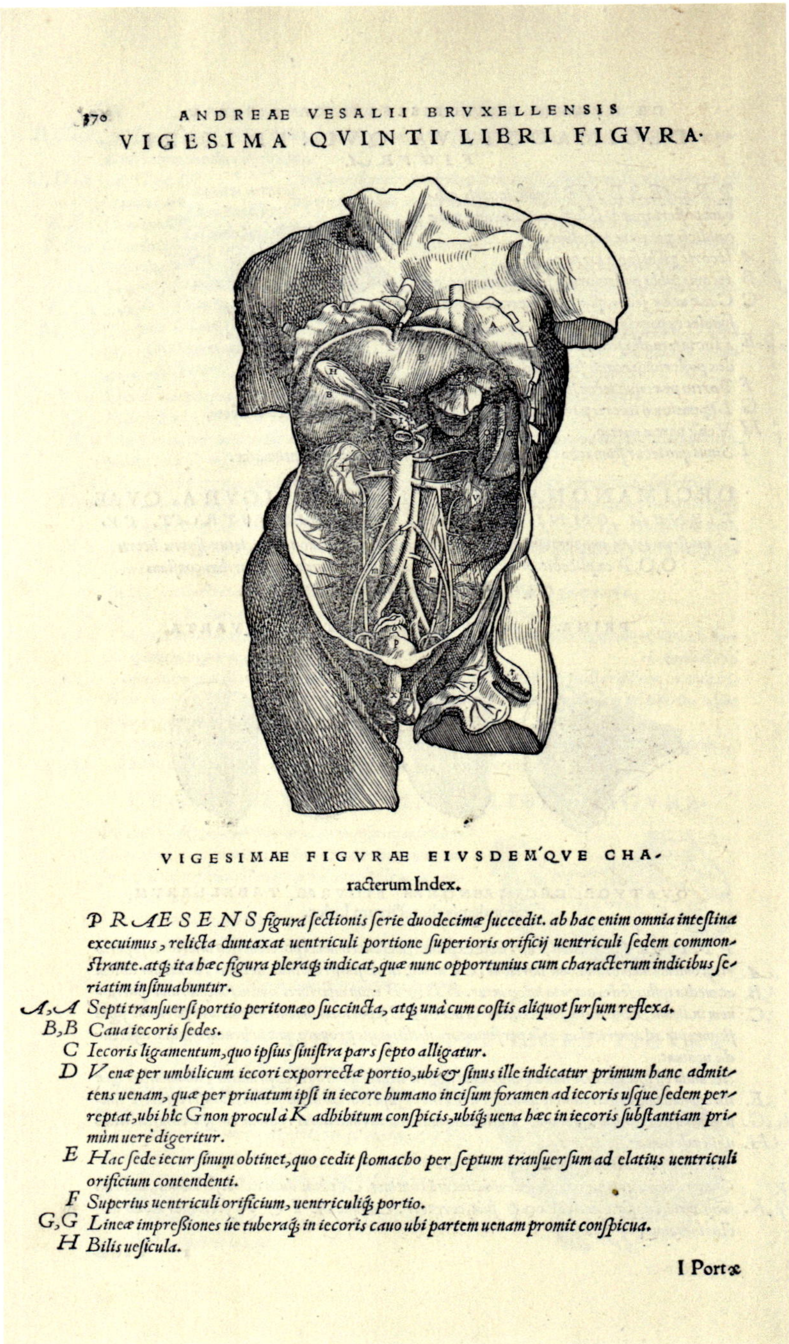

370 ANDREAE VESALII BRVXELLENSIS

VIGESIMA QVINTI LIBRI FIGVRA.

VIGESIMAE FIGVRAE EIVSDEM'QVE CHAracterum Index.

PRAESENS figura ſectionis ſerie duodecimæ ſuccedit. ab hac enim omnia inteſtina execuimus, relicta duntaxat uentriculi portione ſuperioris orificij uentriculi ſedem commonſtrante. atq; ita hæc figura pleraq; indicat, quæ nunc opportunius cum characterum indicibus ſeriatim inſinuabuntur.

A,A Septi tranſuerſi portio peritonæo ſuccincta, atq; unà cum coſtis aliquot ſurſum reflexa.
B,B Caua iecoris ſedes.
C Iecoris ligamentum, quo ipſius ſiniſtra pars ſepto alligatur.
D Venæ per umbilicum iecori exporrectæ portio, ubi & ſinus ille indicatur primum hanc admittens uenam, quæ per priuatum ipſi in iecore humano inciſum foramen ad iecoris uſque ſedem perreptat, ubi hîc G non procul à K adhibitum conſpicis, ubiq; uena hæc in iecoris ſubſtantiam primùm uerè digeritur.
E Hac ſede iecur ſinum obtinet, quo cedit ſtomacho per ſeptum tranſuerſum ad elatius uentriculi orificium contendenti.
F Superius uentriculi orificium, uentriculiq; portio.
G,G Lineæ impreſſiones ùe tuberaq; in iecoris cauo ubi partem uenam promit conſpicua.
H Bilis ueſicula.

I Portæ

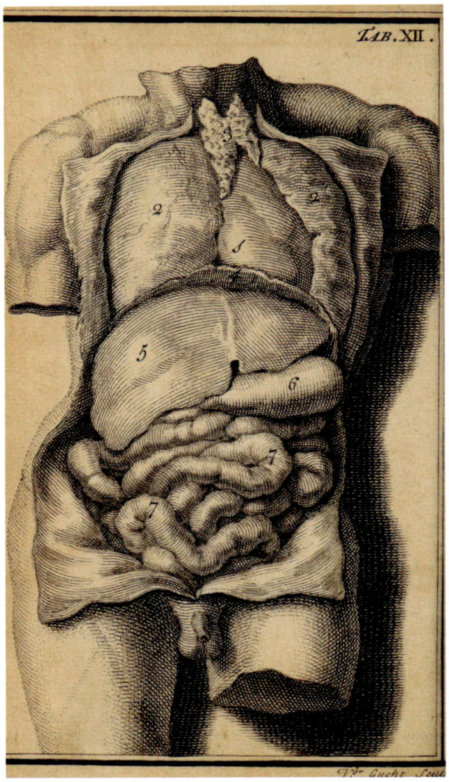

FIG. 4.3
AFTER JAN STEVEN VAN CALCAR
(North Netherlandish, ca. 1515–ca. 1546).
Abdominal dissection, woodcut.
From Andreas Vesalius, *De humani corporis fabrica libri septem* (Basel: J. Oporinus, 1543), bk. 5, p. 370 [470], fig. 20.
Los Angeles, Getty Research Institute, 84-B27611.

FIG. 4.4
ATTRIBUTED TO GERARD VANDERGUCHT
(English, 1696–1776).
Abdominal dissection, etching.
From William Cheselden, *The Anatomy of the Humane Body,* 2nd ed. (London: S. Collins, 1722), pl. 12.
London, Wellcome Collection, EPB/B/17543/1.

truncated legs of a female subject in William Hunter's the *Gravid Uterus* (1774) (cat. no. 42), is no less insistent on the depiction of an actual body of flesh and bone, rather than stone.[18]

A century and a half later, the English surgeon and anatomist William Cheselden was looking closely at the *Fabrica* illustrations.[19] Among the new illustrations added to the second edition (1722) of Cheselden's enormously successful *The Anatomy of the Humane Body* are two illustrations of anatomized stone torsi. The sculptural frames were copied directly from Vesalius, but the internal anatomy is quite different (figs. 4.3, 4.4).[20] Cheselden's twelfth figure depicts an earlier stage of dissection with intact viscera and a more extensive cut that exposes the lungs, yet care has been taken to preserve the identical breaks at the arms, neck, and leg. Cheselden's torsi drew the censure of the surgeon and anatomist John Bell, who forcefully promoted realism in anatomical illustration and railed against the "vitious practice of drawing from imagination" exemplified by these dissected torsi. Referring to the Belvedere Torso, he laments that "anatomists have, with one consent, agreed to borrow the celebrated Torso for putting their bowels into, to explain them there; a practice which has descended from the time of Vesalius down to Chesselden [*sic*], and from him to the systems of the present day."[21] Continuing the *all'antica* theme, Cheselden would later select the poses of the Apollo Belvedere and the Venus de' Medici for the male and female skeletons of his *Osteographia* of 1733 (cat. no. 33).

In the seventeenth century, the analysis of the proportions and anatomy of ancient statuary may be seen as a desire for the quantification of the antique. Charles Errard, first director of the Académie de France in Rome, investigated both.[22] In *Anatomia per uso et intelligenza del disegno* (1691)—produced in collaboration with the academy's long-serving anatomy professor, Bernardino Genga, and the anatomist and papal physician Giovanni Maria Lancisi—Errard provided designs for a series of antique sculptures "considered anatomically" (cat. no. 32). In doing so, two core subjects in the curriculum of the artist—the antique and anatomy—were combined. Unlike the majority of the Vesalian torsi, the focus was on famous works to be found in Rome, such as the Laocoön and the Borghese Gladiator. These are depicted in multiple views in a documentary approach that afforded a complete view of the exposed muscles. The anatomical rendering of these sculptures has the peculiarly forceful effect of highly familiar objects presented in a foreign yet still recognizable way, with the fancy that beneath their surface lay anatomical layers to explore for those who picked up a chisel.[23]

Antique sculpture was held to be the result of ancient sculptors selecting judiciously from nature, arriving at the ideal form of a human body from the observation of many. Skeletons or casts of flayed bodies that were set in the poses of antique statues, however, only captured one specific body. In 1776, William Hunter (1718–83), professor of anatomy for the Royal Academy of Arts in London, flayed the body of an executed criminal, which the sculptor Agostino Carlini (ca. 1718–90), who served as keeper of the academy, had arranged in the pose of the Dying Gaul and cast in plaster.[24] In the early nineteenth century, the French surgeon Jean-Galbert Salvage (1770–1813) dissected bodies of soldiers and made plaster casts of them in the pose of the Borghese Gladiator. These were the basis for his book *Anatomie du gladiateur combattant* (1812) (cat. no. 34). The same motivation of emulating the antique with a desire to capture its essence in nature is seen in the practice of drawing from life models set in poses of ancient statues.[25] For his 1829 book on surface anatomy, in which many examples of antique sculpture are cited, the French physician Pierre-Nicolas Gerdy (1797–1856) deliberately illustrated a human model set in the pose of the Discobolus instead of the sculpture itself, because, as he avers, the book's subject is not antique statues; rather, it is "man himself."[26]

In Michael Sweerts's *A Painter's Studio* (fig. 4.5), dated to the mid-seventeenth century and painted while the Flemish artist was in

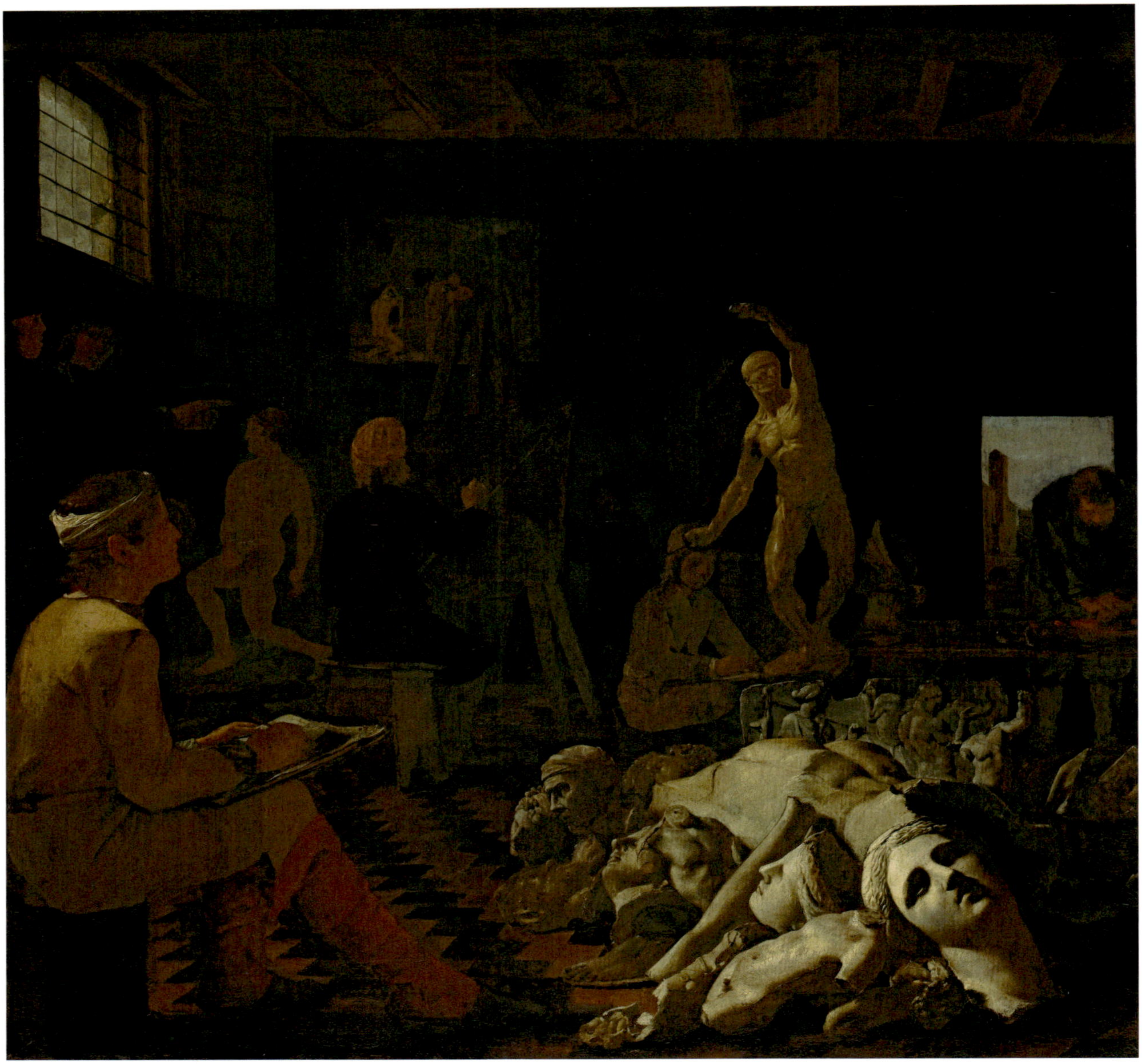

FIG. 4.5
MICHAEL SWEERTS
(Flemish, 1618–64).
A Painter's Studio,
ca. 1646–50, oil on canvas,
71 × 74 cm.
Amsterdam, Rijksmuseum,
SK-A-1957.

Rome, an écorché model stands amid antique casts and fragments.[27] Two youths make drawings of them, while, in the background, an artist paints from a life model. These activities are elements of a sequence of study recommended by the critic and artist Roger de Piles in the *Abregé d'anatomie* (1668) that starts with the student learning the bones and muscles, then drawing multiple views of an écorché model, and, as the final step, relating these studies to "the most beautiful Antiques and with Nature herself."[28] In Nicolas Dorigny's engraving after Carlo Maratti's late seventeenth-century design for an academy of drawing (fig. 4.6),[29] the place of anatomy in the hierarchy of artistic study—represented here by Leonardo da Vinci leading an anatomy lesson before an écorché model—is made clear by the inscriptions: anatomy as well as perspective and geometry are labeled "tanto che basti" (as much as is sufficient), whereas antique sculpture is labeled "non mai abastanza" (never enough).

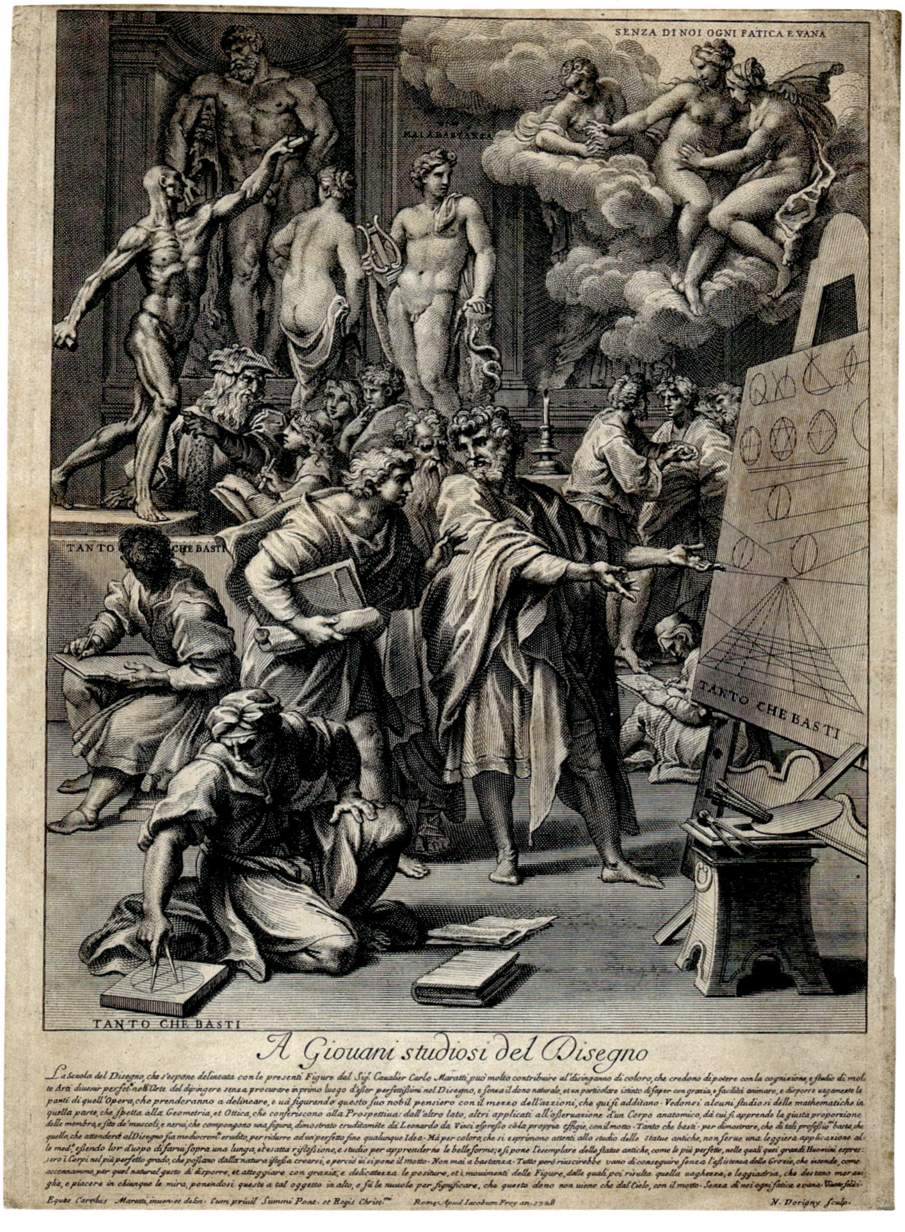

FIG. 4.6
NICOLAS DORIGNY (French, 1658–1746), after Carlo Maratti (Italian, 1625–1713).
The Academy of Drawing, 1728, etching and engraving, second state, platemark: 46.8 × 32.1 cm.
London, Wellcome Collection, no. 32740i.

The only way to grasp the beauty of the antique and the truth of nature, according to de Piles, is through a sure understanding of anatomy, an opinion also shared by the artist Charles Monnet (1732–after 1808), who was the author of a book of anatomy for artists (cat. no. 23). At the same time, the observation of antique sculpture, being the ideal representation of nature, was generally considered a necessary modifier of anatomical studies. Cornelis Ploos van Amstel—the printmaker, collector, and codirector of the Stadstekenacademie in Amsterdam—puts forward the selective expression of anatomy by antique sculptors in their figures as a corrective model, one that would keep the artist from falling into the trap of displaying too much of their anatomical knowledge, a failing he perceived in Michelangelo.[30] Furthermore, Salvage says that by consulting both antique sculpture and the life model, the artist will avoid a dryness of style that comes with a "parading" of anatomy,

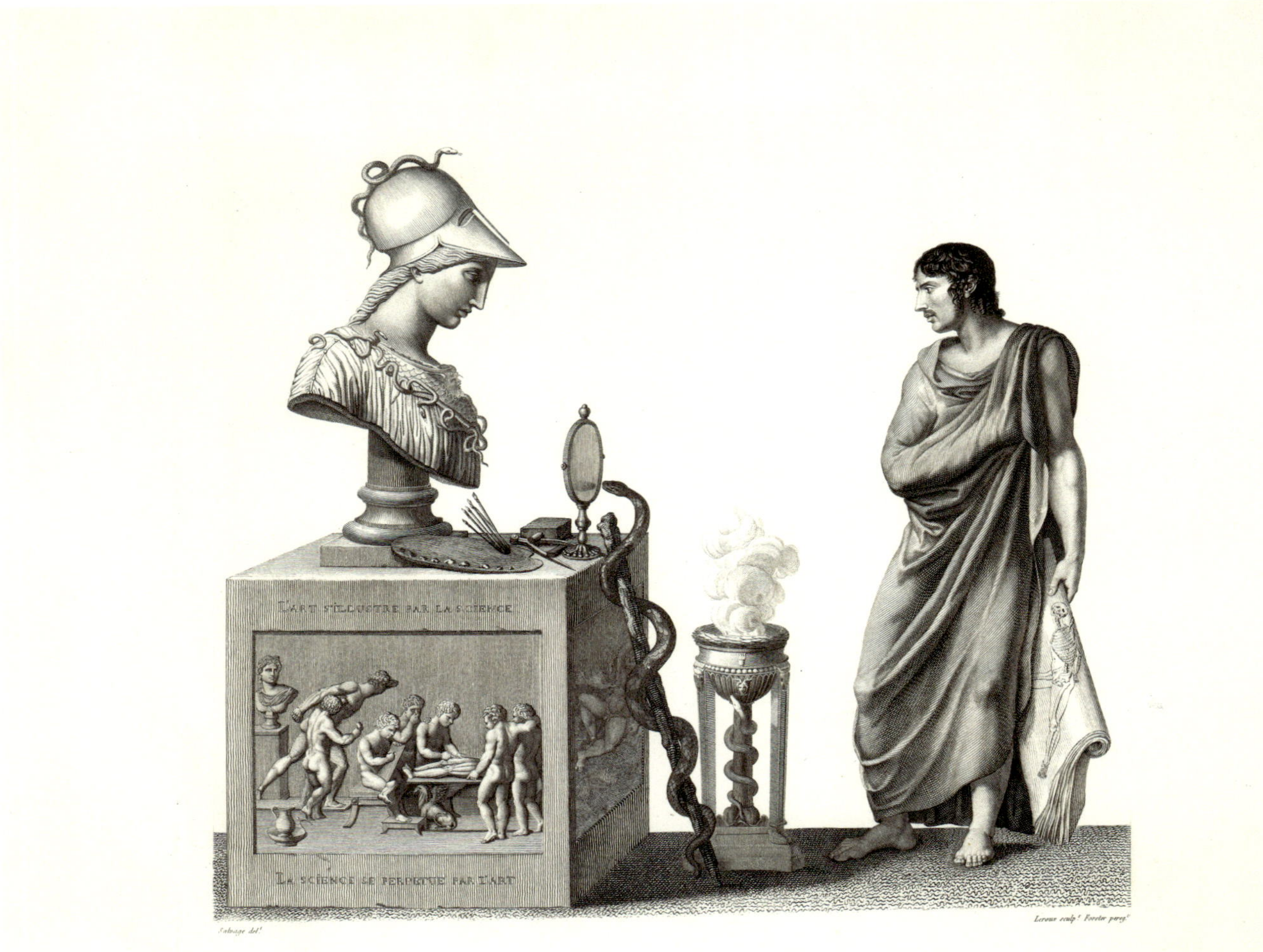

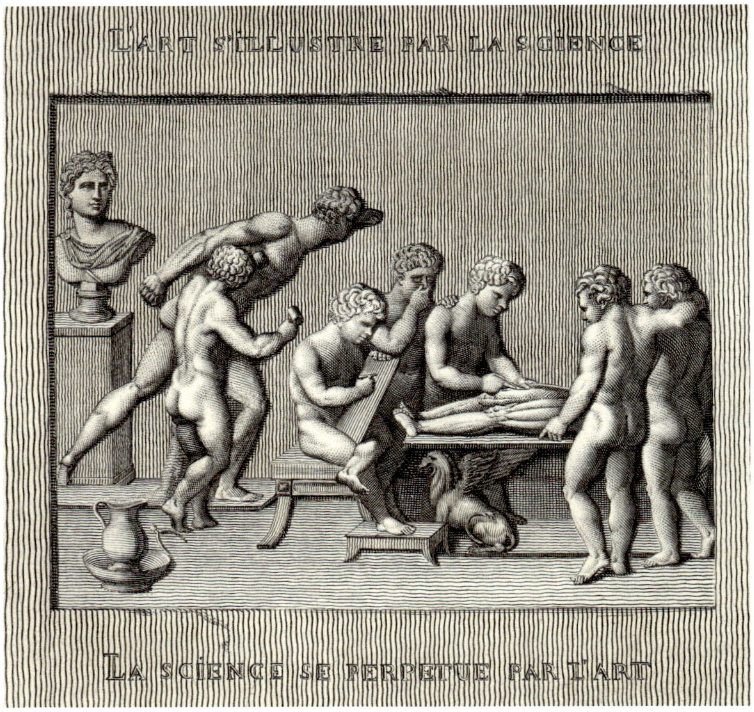

FIG. 4.7
JEAN MARIE LEROUX (1788–1813) and François Forster (1790–1872), after Jean-Galbert Salvage (French, 1770–1813).
The sculptor Agasias before the head of Minerva, etching and engraving.
From Jean-Galbert Salvage, *Anatomie du gladiateur combattant, applicable aux beaux arts; ou, Traité des os, des muscles, du mécanisme des mouvemens, des proportions et des caractères du corps humain* (Paris: Jean-Galbert Salvage and l'imprimerie de Mame, 1812), frontispiece.
Los Angeles, Getty Research Institute, 85-B12146.

FIG. 4.8
Putti engaged in dissection and the carving of the Borghese Gladiator, etching (detail of fig. 4.7).

a knowledge of which will instead be used "to animate his figures and give them the expression of truth."[31]

In his influential survey of ancient art, published in 1764, Johann Joachim Winckelmann (1717–68) identifies a lack of understanding of anatomy due to strict taboos concerning cadavers as a factor impeding the development of artists in Egypt.[32] However, a study of anatomy is not cited as contributing to Greek excellence in figural art; rather, Winckelmann attributes it to the beauty of the Greek people, the country's climate, and the gymnasiums, where beautiful and naked youths wrestled and practiced other sports—for these "were the schools where the artists examined the beauty of the body's build."[33] If Winckelmann did not make an overt connection between anatomical dissection and the mastery of the human figure in Greek art, the Roman surgeon Giuseppe del Medico (1772–1825) did. In his *Anatomia per uso dei pittori e scultori* (1811) (cat. no. 27), Medico refers to Winckelmann's assessment of Egyptian art, pointing out that only the Egyptian priests had access to corpses in the process of embalming, but he goes on to say that the Greeks did not have this restriction, citing the example of Erasistratus of Alexandria (ca. 304–ca. 250 BC), who is known to have practiced human dissection.[34] John Flaxman also made the connection between the Alexandrian school of anatomy and the greater attention to anatomy in Greek sculpture.[35] Alexandria, however, was the exception rather than the rule in terms of human dissection in the ancient world.[36] Galen, for example, based many of his anatomical observations on primates and dogs.

Salvage declares at the outset of his *Anatomie du gladiateur combattant* that the success of those modern artists who had made a profound study of anatomy and, moreover, that the evidence of the works of ancient artists were proof enough of the necessity of a knowledge of the subject.[37] Salvage's frontispiece of Agasias, the sculptor of the Borghese Gladiator, makes a visual declaration of the belief that antique sculpture was the result of study by dissection (fig. 4.7). Holding a transparent anatomy drawing of the figure that relates to one in the publication, Agasias stands before a bust of Minerva on a plinth, which is decorated on the front with a relief of putti dissecting a body (fig. 4.8). One holds his nose from the stench; another draws the cadaver; and a third sculpts the Borghese Gladiator beside a bust of the Apollo Belvedere, both works that feature in the book.

The perceived supreme mastery of the nude by ancient artists was later the subject of a heated debate on how artists should study anatomy. Sir Anthony Carlisle (1768–1840), professor of anatomy at the Royal Academy from 1808 to 1824, was of the opinion that in the ancient world, Greek knowledge of anatomy was limited and that Greek artists achieved their excellence in the human figure through the observation of bodies in action rather than dissection.[38] Accordingly, Carlisle's lectures featured living models, and, on one occasion, "he had six or eight naked Life-guardsmen going through their sword exercise, exhibiting the varied muscular action of the human body."[39] Arguing for dissection was the history painter Benjamin Robert Haydon (1786–1846), who had supported the Scottish surgeon and anatomist Sir Charles Bell (1774–1842) when he stood against Carlisle in his election for the Royal Academy post. Another opponent of Carlisle was the surgeon and amateur artist James Birch Sharpe (1789–1863), who felt it was imperative for a young artist to study anatomy by dissection and that true knowledge of the structure of the body could "be obtained by sight and touch alone."[40] In his *Elements of Anatomy; Designed for the Use of Students in the Fine Arts* (1818) (cat. no. 35), Sharpe frequently refers the reader to anatomical structures in specific antiquities and concludes the section on the muscles by commending the left hand of the Laocoön and the foot of the Farnese Hercules for the arrangement of the veins, observing that "they will teach the student the importance which the ancients attached to the minutest parts of anatomic expression."[41]

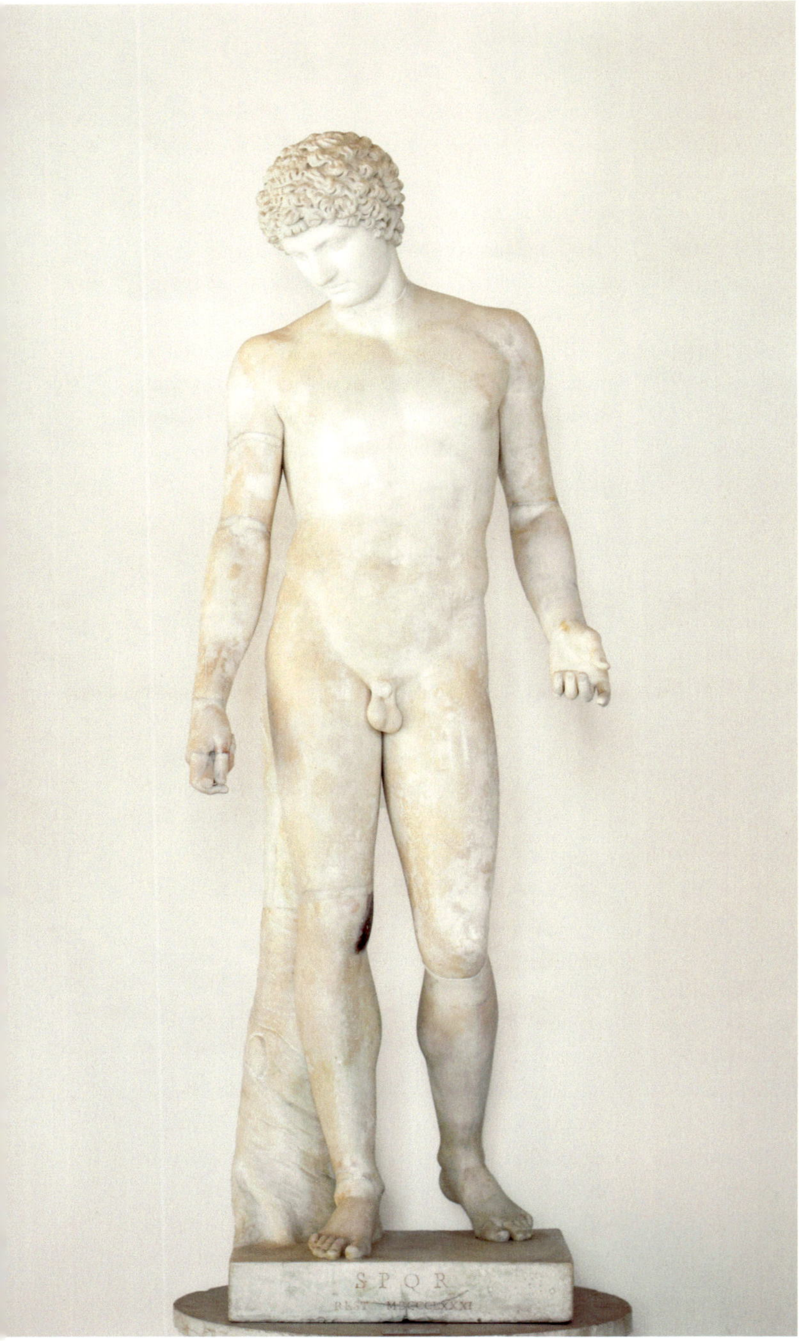

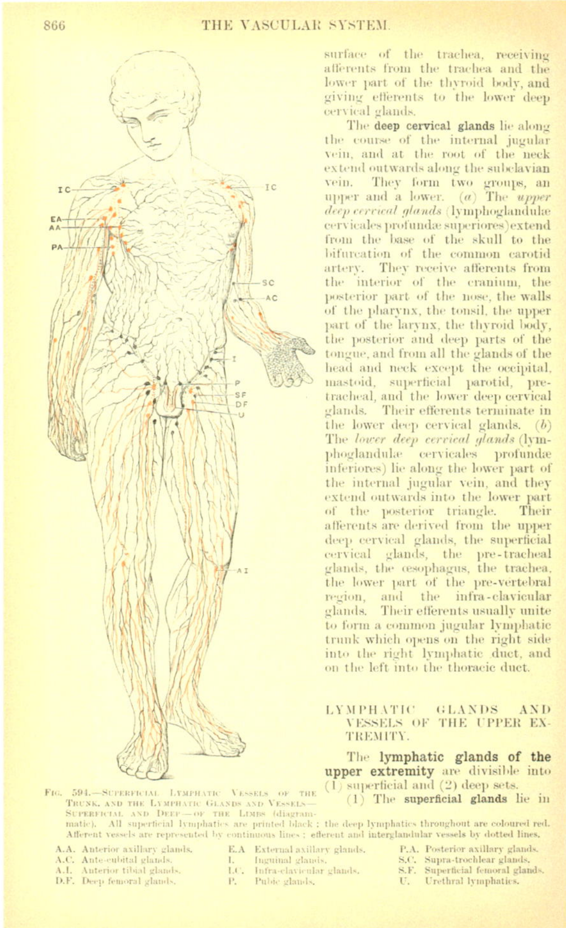

866 THE VASCULAR SYSTEM.

surface of the trachea, receiving afferents from the trachea and the lower part of the thyroid body, and giving efferents to the lower deep cervical glands.

The **deep cervical glands** lie along the course of the internal jugular vein, and at the root of the neck extend outwards along the subclavian vein. They form two groups, an upper and a lower. (*a*) The *upper deep cervical glands* (lymphoglandulæ cervicales profundæ superiores) extend from the base of the skull to the bifurcation of the common carotid artery. They receive afferents from the interior of the cranium, the posterior part of the nose, the walls of the pharynx, the tonsil, the upper part of the larynx, the thyroid body, the posterior and deep parts of the tongue, and from all the glands of the head and neck except the occipital, mastoid, superficial parotid, pre-tracheal, and the lower deep cervical glands. Their efferents terminate in the lower deep cervical glands. (*b*) The *lower deep cervical glands* (lymphoglandulæ cervicales profundæ inferiores) lie along the lower part of the internal jugular vein, and they extend outwards into the lower part of the posterior triangle. Their afferents are derived from the upper deep cervical glands, the superficial cervical glands, the pre-tracheal glands, the œsophagus, the trachea, the lower part of the pre-vertebral region, and the infra-clavicular glands. Their efferents usually unite to form a common jugular lymphatic trunk which opens on the right side into the right lymphatic duct, and on the left into the thoracic duct.

LYMPHATIC GLANDS AND VESSELS OF THE UPPER EXTREMITY.

The **lymphatic glands of the upper extremity** are divisible into (1) superficial and (2) deep sets.

(1) The **superficial glands** lie in

FIG. 594.—SUPERFICIAL LYMPHATIC VESSELS OF THE TRUNK, AND THE LYMPHATIC GLANDS AND VESSELS—SUPERFICIAL AND DEEP—OF THE LIMBS (diagrammatic). All superficial lymphatics are printed black; the deep lymphatics throughout are coloured red. Afferent vessels are represented by continuous lines; efferent and interglandular vessels by dotted lines.

A.A. Anterior axillary glands.
A.C. Ante-cubital glands.
A.I. Anterior tibial glands.
D.F. Deep femoral glands.
E.A External axillary glands.
I. Inguinal glands.
I.C. Infra-clavicular glands.
P. Pubic glands.
P.A. Posterior axillary glands.
S.C. Supra-trochlear glands.
S.F. Superficial femoral glands.
U. Urethral lymphatics.

FIG. 4.9
Capitoline Antinous, Roman copy of 4th-century BC Greek statue, marble, 180 cm.
Rome, Palazzo dei Conservatori, Musei Capitolini, MC0741.

FIG. 4.10
J. T. MURRAY
(Scottish, active 1902).
Superficial Lymphatic Vessels of the Trunk, and the Lymphatic Glands and Vessels—Superficial and Deep—of the Limbs, colored wood engraving.
From D. J. Cunningham, *Text-Book of Anatomy* (Edinburgh: Young J. Pentland, 1902), 866, fig. 594.
Washington, D.C., Smithsonian Libraries, QM23 .C97 1902.

Sharpe continued the convention of presenting antique statues anatomically, and two of his six plates are of flayed antiquities: the Ludovisi Gaul and the Borghese Gladiator. Examples are also found in the anatomy books for artists by Bertinatti (1837–39) and Seiler (1850).[42] A late example in a medical textbook appears in the first edition of the popular *Text-Book of Anatomy* (1902), by D. J. Cunningham (1850–1909), the Scottish anatomist who also oversaw dissections for a stereoscopic anatomy atlas (cat. no. 55). The statue of the Capitoline Antinous (fig. 4.9)[43] served as the inspiration for the figure displaying the lymphatic vessels and glands (fig. 4.10). In a period when whole figures were largely abandoned in anatomical book illustration, it speaks to the lingering and powerful appeal of antique sculpture and the appreciation of its depiction of the human form.

Epigraph: Monnet 1774, advertisement.

Notes

1 For considerations of this theme, see Harcourt 1987; Kornell 1996, 65–68; San Juan 2008; Schwartz 2008; Ciardi 2009, 30; Kusukawa 2012, 215–18; Guest 2014; and Aymonino 2015, 48–52.

2 For Vesalius's contribution to the Giunta edition of Galen's works (1541–42), see O'Malley 1964, 101, 104–8.

3 Additionally, the industrious putti of the decorated initial letters of the *Fabrica* and the cross-legged pose of the skeleton in profile (see fig. 2.8) have their sources in ancient sarcophagi reliefs.

4 On norm and variance in the *Fabrica*, see Harcourt 1987, 39–43; Pigeaud 1990; Siraisi 1994; Kusukawa 2012, 214, 217; and Massey 2017, 76–80.

5 Heseler 1959, 139.

6 Vesalius 1543a, 592; and Vesalius 2014, 2:1198. See Roth 1892, 104.

7 Janson 1946, 51; and Kusukawa 2012, 215, figs. 10.9, 10.10. The muscle figures in book 2 have also eluded exact parallels to antique works (Harcourt 1987, 44, 60n49).

8 Vesalius 1543a, bk. 5, figs. 10, 12, 20, with the latter being mirror images of each other.

9 For the suggestion that this is the same body, see Roth 1892, 179; O'Malley 1964, 143; and Park 2006, 207.

10 Bober and Rubenstein 2010; and Haskell and Penny 1981.

11 A similar employment of "visually pleasing containers" (Talvacchia 1999, 163) is seen in the adaptation of Giovanni Jacopo Caraglio's print series *Loves of the Gods* for the obstetric figures in Estienne and La Rivière's *De dissectione* (Paris, 1545); see cat. no. 1, this volume.

12 This was a fate suffered by Battista da Vercelli, a physician who had been implicated in a plot to kill Pope Leo X. His quartering is compared to fragments of antique sculpture by Annibale Caro in a humorous prose poem dated to circa 1536. See Simons and Kornell 2008, 1076–79, 1088–93.

13 An emendation in the 1555 edition makes clear it is a quartering that occurred (Vesalius 1543a, 584; and Vesalius 1555, 728: "ex his quos vivos in quatuor partes dissecant").

14 Similarly, sculptural torsi were employed to illustrate a text on bandages ascribed to Galen in Vidius 1544, 444–60. Although the torsi are notably absent in the Byzantine manuscript on which the woodcuts are partly based, the influence of the *Fabrica* published the year before may not be easily claimed, as it has been argued that preparatory drawings attributed to Francesco Salviati (1510–63) for Vidius's *Chirurgia* date to 1541, for which see Hirst 1969.

15 San Juan 2008.

16 As noted by Herrlinger 1970, 124; and Harcourt 1987, 52.

17 Vesalius 1543a, bk. 5, figs. 25, 26.

18 An unusually anatomically detailed cross section of the ankle appears in Crispijn van de Passe's drawing manual (Van de Passe 1643–44, pt. 1, 45). Albrecht von Haller, who eschewed the convention of the animated cadaver, used a torso to display arteries in his *Icones anatomicae*, with breaks at the arms and legs to show cross sections of the bones (Haller 1753, pl. 1). The signature of the artist and engraver Joes Paul Kaltenhofer (1716–77) makes clear that the illustration was drawn from life: "Kaltenhofer del. ad nat. et sc. Gott."

19 In the preface to his *Osteographia*, Cheselden faulted the proportions of the skeletons in the *Fabrica* (Cheselden 1733).

20 Vesalius 1543a, bk. 5, figs. 2, 20; and Cheselden 1722, tables 12, 13.

21 Bell 1794, vi.

22 Coquery 2013, 37–40; and Aymonino 2015, 46–49.

23 For sixteenth-century examples, see Kornell in Cazort, Kornell, and Roberts 1996, 140, no. 33; 146–48, no. 38; and Bober and Rubenstein 2010, 174.

24 Gamer 2019; and Black 2007b, 95–96.

25 Cuzin, Gaborit, and Pasquier 2000, 415–29, nos. 219–34; Aymonino 2015, 23, 50; and Gasparotto 2018, 249, 250, fig. 81.

26 Gerdy 1829, 327. On Gerdy, see Kornell, "Surface Anatomy," this volume, p. 69n21.

27 Adriano Aymonino and Anne Varick Lauder, in Aymonino and Lauder 2015, 134–39, no. 12, with earlier references.

28 Tortebat and de Piles 1668, address to reader, [iv]r. Cornelis Ploos van Amstel offers a variant of de Piles's course of study by suggesting that "the book of Carlo Errard" be consulted as well as the best plaster casts after the antique (Ploos van Amstel 1783, 103). He specifies the edition of 1691, which secures an identification with Genga and Lancisi's *Anatomia per uso et intelligenza del disegno*.

29 Adriano Aymonino, in Aymonino and Lauder 2015, 148–52, no. 15; Pierguidi 2017, 12–21.

30 Ploos van Amstel 1783, 108-9.

31 Salvage 1812, iv.

32 Winckelmann 2006, 130.

33 Winckelmann 2006, 197.

34 Del Medico 1811, 3.

35 Flaxman 1829, 116. For Flaxman and anatomical training at the Royal Academy, see Hainy 2019.

36 Staden 1992; and Nutton 2013, 130–41.

37 Salvage 1812, i. For the argument that Salvage was relying on views of the archaeologist and art theorist Toussaint-Bernard Émeric-David (1755–1839), see Shedd 1991, 98–100.

38 Carlisle 1807.

39 Bewick 1871, 1:141. For a discussion of Carlisle's lectures at the Royal Academy, see Darlington 1990, 1:344–54.

40 Sharpe 1818, 1.

41 Sharpe 1818, 49.

42 See Choulant 1945, 344–46, 410.

43 Haskell and Penny 1981, 143–44, no. 5.

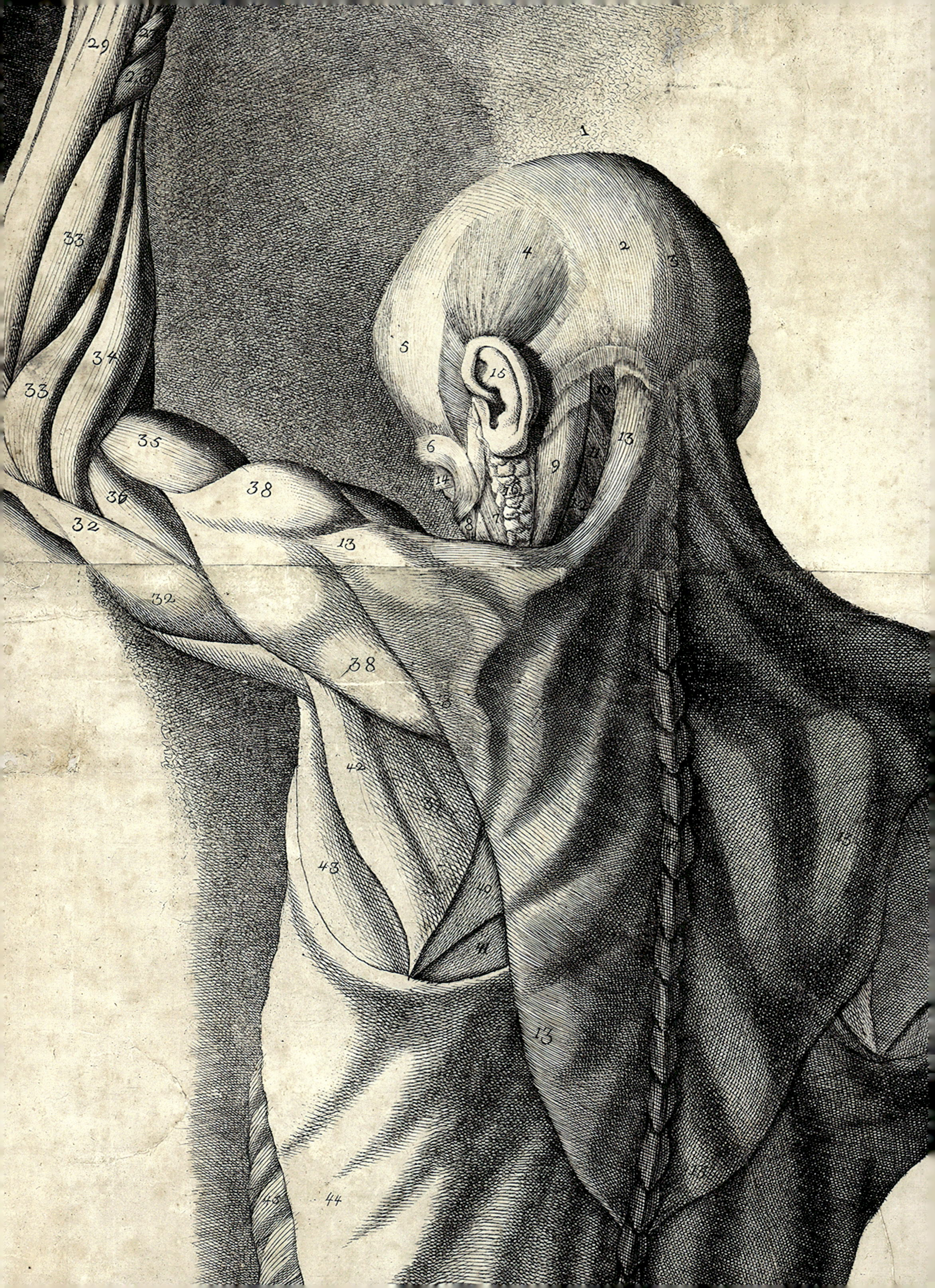
29
1
33
4
2
3
5
34
33
15
35
13
6
9
14
36
38
32
8
13
32
38
42
43
13
44

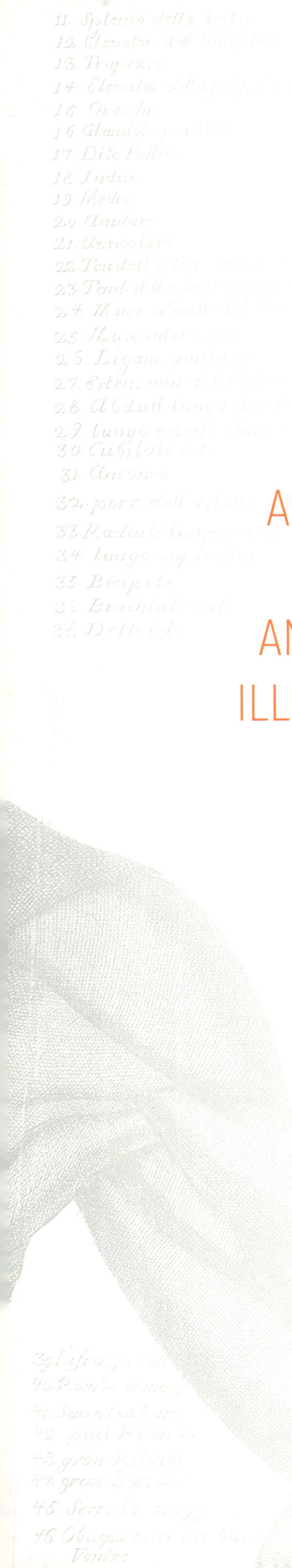

chapter five

"AS LARGE AS NATURE": LIFE-SIZE ANATOMICAL ILLUSTRATION

Monique Kornell

The depiction of the anatomy of the entire body on a one-to-one scale is an extravagant and physically astonishing way of capturing its structure.[1] The three life-size figures by the Bolognese printmaker Antonio Cattani (active 1777–ca. 1790) (cat. nos. 36–38) can be seen as part of a trend in the eighteenth century of life-size anatomical display in print and sculpture, one that had roots in the sixteenth and seventeenth centuries. Its apex may be considered to be the *Anatomia universa* (1823–31), by the anatomist Paolo Mascagni (1755–1815) (cat. no. 45), a book that was too large to use and prohibitively expensive to buy.[2] The eighteenth century also saw the establishment of collections of colored wax anatomies in Bologna, Dresden, Florence, and other cities.[3] Often life-size, these works convey a sense of the pliable softness and glistening textures of the dissected body and, like the miraculous bodies of saints, never decompose (see fig. 8.3). Mascagni served as an adviser to Felice Fontana (1730–1805) on the creation of waxes for Pietro Leopoldo I, grand duke of Tuscany, and, with the life-size engravings for his *Anatomia universa,* Mascagni succeeded in rivaling wax anatomies in scale, color, and effect.[4]

The *Anatomia universa* was admired both for the effort that went into producing it and for its visual impact. Like John James Audubon's (1785–1851) *The Birds of America* (1827–38), which also featured colored life-size illustration, Mascagni's book required dedicated furniture for viewing and was difficult for one person alone to handle. It is not surprising that reduced versions of both books were immediately planned.[5] Shortly after the final volume of the *Anatomia universa* appeared, Mascagni's main artistic collaborator, Antonio Serantoni (1780–1837), issued a

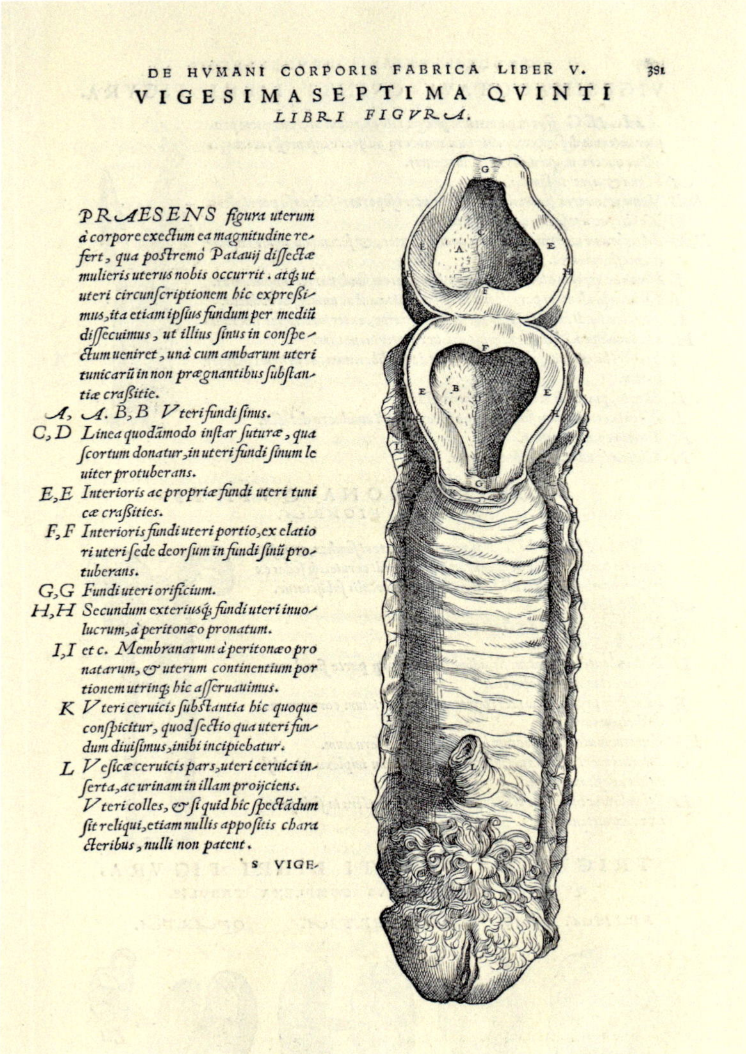
DE HVMANI CORPORIS FABRICA LIBER V. 381

VIGESIMASEPTIMA QVINTI LIBRI FIGVRA.

PRAESENS figura uterum à corpore exectum ea magnitudine refert, qua poſtremò Patauij diſſectæ mulieris uterus nobis occurrit. atq; ut uteri circunſcriptionem hic expreſsimus, ita etiam ipſius fundum per mediū diſſecuimus, ut illius ſinus in conſpectum ueniret, unà cum ambarum uteri tunicarū in non prægnantibus ſubſtantiæ craßitie.

A, A. B, B Vteri fundi ſinus.

C, D Linea quodāmodo inſtar ſuturæ, qua ſcortum donatur, in uteri fundi ſinum leuiter protuberans.

E, E Interioris ac propriæ fundi uteri tunicæ craßities.

F, F Interioris fundi uteri portio, ex elatiori uteri ſede deorſum in fundi ſinū protuberans.

G, G Fundi uteri orificium.

H, H Secundum exteriusq; fundi uteri inuolucrum, à peritonæo pronatum.

I, I et c. Membranarum à peritonæo pronatarum, & uterum continentium portionem utrinq; hic aſſeruauimus.

K Vteri ceruicis ſubſtantia hic quoque conſpicitur, quod ſectio qua uteri fundum diuiſimus, inibi incipiebatur.

L Veſicæ ceruicis pars, uteri ceruici inſerta, ac urinam in illam proijciens.

Vteri colles, & ſi quid hic ſpectādum ſit reliqui, etiam nullis appoſitis characteribus, nulli non patent.

S VIGE-

FIG. 5.1
AFTER JAN STEVEN VAN CALCAR
(North Netherlandish, ca. 1515–ca. 1546).
Sectioned uterus, vagina, urethra, and vulva, woodcut.
From Andreas Vesalius, *De humani corporis fabrica libri septem* (Basel: J. Oporinus, 1543), bk. 5, p. 381 [481], fig. 27.
Los Angeles, Getty Research Institute, 84-B27611.

folio-size edition that combined the text and plates in a smaller and more accessible format (cat. no. 45). Serantoni observed that the magnificence of the original work caused it to be beyond the reach of students' pockets and that the size of the plates made their storage and handling difficult, particularly if they were in continuous use, "as it happens for those who wish to possess them for instruction rather than for luxury."[6]

Andreas Vesalius had favored large-size illustration, and his *Tabulae anatomicae sex* (1538), *Fabrica* (1543), and *Epitome* (1543) were the largest illustrated anatomy books to have appeared until that time.[7] In a letter to Johannes Oporinus, the publisher of his *Fabrica,* Vesalius rails against the copyists of the *Tabulae anatomicae sex* and, in particular, an unnamed Strasbourg plagiarist who "has done the worst disservice to medical study because he has so disgracefully reduced illustrations which could never be made large enough for students."[8] Vesalius is recorded as having used woodcuts from the *Tabulae anatomicae sex* as visual aids in his crowded public lectures, and their size would have helped them to be visible at a distance (see fig. 1.1).[9] In the *Fabrica,* many of the illustrations of separate bones, such as the vertebrae, were by far the most accurate representations to date and were rendered at actual size. Vesalius himself specifically points out that the twenty-seventh figure of the fifth book is a same-size depiction of the uterus of a woman whom he had recently dissected in Padua and who had claimed a false pregnancy in order to avoid execution (fig. 5.1).[10] The reason for portraying it at actual size is made clear in the second edition of 1555, when Vesalius adds that the uterus is abnormally large and is from a tall woman who had given birth multiple times.[11]

By 1600, Hieronymus Fabricius ab' Acquapendente, a professor of anatomy and surgery in Padua, had prepared three hundred tables of human and comparative anatomy for a projected work on anatomy. He describes them in the dedication of his *De visione, voce, auditu* (1600) as surpassing the illustrations of

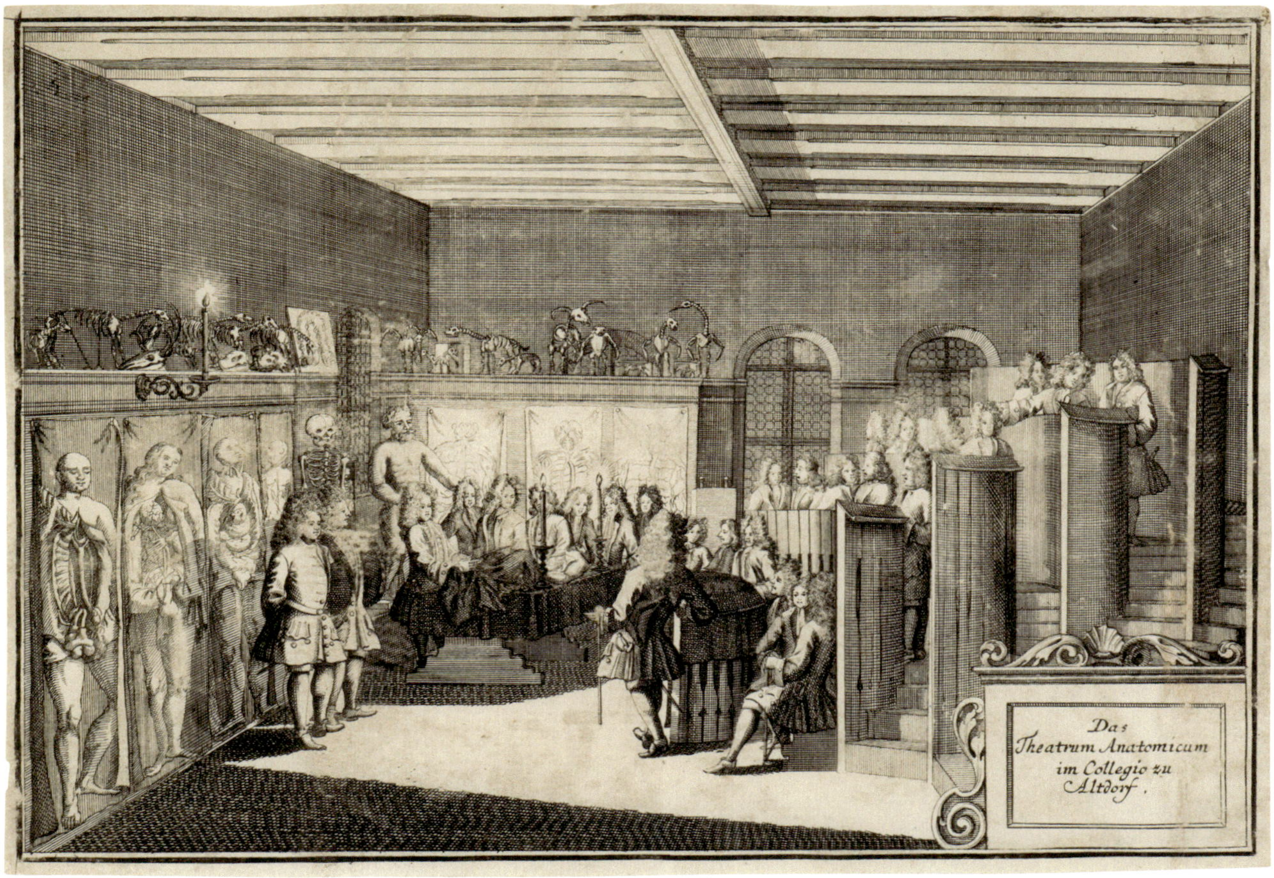

FIG. 5.2
JOHANN GEORG PUSCHNER (German, active ca. 1700–1750). *The Anatomy Theater in Altdorf,* 1718, etching, sheet: 19.4 × 27.5 cm. Braunschweig, Germany, Herzog Anton Ulrich Museum, JGPuschner AB 3.1.

Vesalius not only in quality and precision but also in being rendered life-size and in color.[12] Although his treatise was never published, the preparatory paintings and prints survive in the Biblioteca Marciana in Venice, having been willed by Fabricius to the Venetian state.[13] Life-size illustrations of sections of the body appear in Fabricius's *De venarum ostiolis* (1603), a book on the veins that influenced his pupil William Harvey (1578–1657).[14]

There were also life-size anatomical illustrations being done in northern Europe in connection with universities. For the Universität Helmstedt, Heinrich Julius, Duke of Brunswick-Lüneburg (1564–1613), commissioned his court painter, Christoph Gertner, to carry out colored life-size illustrations after dissections by the physician and moral philosopher Henning Arnisaeus.[15] Work on the illustrations, now lost, would have started sometime between 1605, when Arnisaeus and Gertner were both in Helmstedt, and the duke's death in 1613. And in Leiden in the early 1650s, the anatomist Johannes van Horne (1621–70) worked with the artist Marten Sagemolen (ca. 1620–69) on life-size and large-scale anatomical drawings, some colored, which were funded by the Universiteit Leiden.[16] An instructional use of early large-scale anatomical wall hangings is indicated by those that decorated the Altdorf anatomy theater, founded in 1650 (fig. 5.2).[17] They are early antecedents to the life-size wall charts of anatomical figures used in nineteenth- and twentieth-century anatomy theaters and art academies.[18]

News that the Dutch physician, anatomist, and playwright Govard Bidloo had prepared life-size engraved plates of anatomy reached the Royal Society in London in 1682. The Dutch physician Anton Nuck (1650–92)

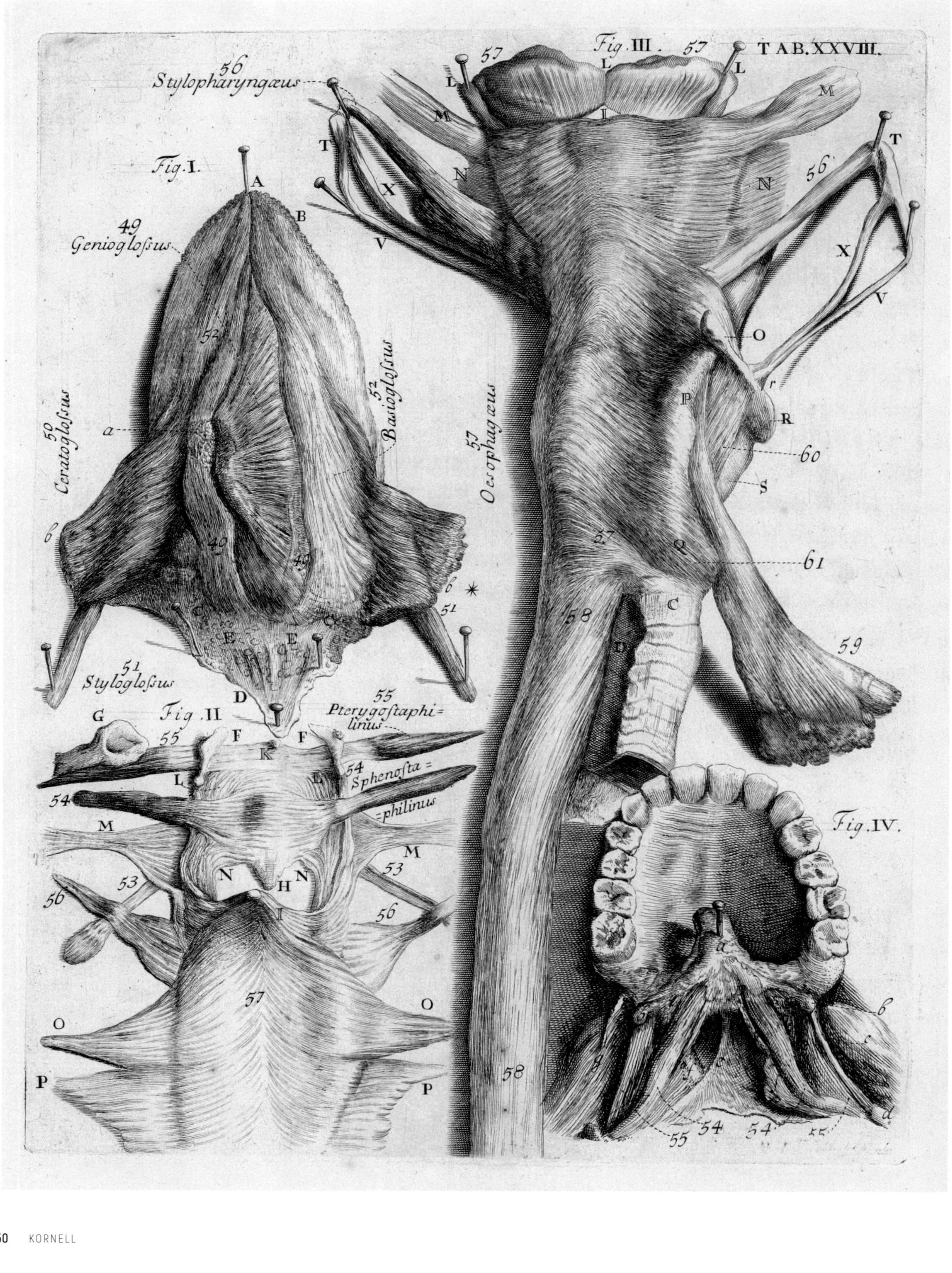

Fig. III.
TAB. XXVIII.
56
Stylopharyngæus
Fig. I.
49
Genioglossus
50
Ceratoglossus
52
Basioglossus
57
Oesophagæus
51
Styloglossus
Fig. II
55
Pterygostaphi=
linus
54
Sphenosta=
=philinus
60
61
59
58
Fig. IV.

FIG. 5.3
MICHAEL VANDERGUCHT (Flemish, 1660–1725, active in England), after William Cowper (English, 1666/67–1710).
Muscles of the tongue, parts of the mouth, and the esophagus, etching and engraving.
From William Cowper, *Myotomia reformata; or, An Anatomical Treatise on the Muscles of the Human Body, Illustrated with Figures after the Life*, 2nd ed. (London: Printed for Robert Knaplock, and William and John Innys, 1724), pl. 28.
Paris, Bibliothèque Interuniversitaire de Santé, 1892.

reported that "Mons. Bidloo, a skillful chirurgeon of Amsterdam, had newly shewed him about 100 anatomical figures of the parts of a man as big as the life, ingraven on copper, with a description of the parts, but not of their use."[19] These were published in 1685 in Bidloo's *Anatomia humani corporis* (cat. no. 40). In his address to the reader, Bidloo's irascible and spirited nature comes through in his description of the plates, which he says were drawn "as far as possible in proper size and from the life" by Gérard de Lairesse (1640–1711) and not copied from others, adding "I hate imitators, servile flock."[20] Like those of Fabricius, Bidloo's plates abandon the then dominant style of illustrating animate dissected figures in their entirety and depict instead large segments of the body, many life-size and shown quite dead. Bidloo's attention to scale is further demonstrated in those images that were drawn under magnification.[21]

Both the scale of Bidloo's illustrations and their presentation of the body as carefully observed dissected preparations had an influence on English anatomy atlases. In 1698, the illustrations were published with a new English text by the surgeon and anatomist William Cowper.[22] For the second edition of Cowper's own *Myotomia reformata* (1724) (cat. no. 13), an asterisk is used to denote those images "as big as the Life,"[23] as seen in his illustration of the muscles of the tongue (see fig. 1 in fig. 5.3), where following Bidloo's example, the pins and props of the preparations are included. The first edition of the *Myotomia reformata* was printed in octavo, but before his death, Cowper prepared copperplates for a folio edition, allowing him more space to accommodate larger-scale images.[24]

The portrayal of the bones "large as the life" is a feature that William Cheselden, Cowper's former pupil, mentioned in the notice of his forthcoming work on the bones, in the preface to the 1726 edition of *The Anatomy of the Humane Body* and in the announcement of a subscription for the book the following year.[25] The opening sentence of Cheselden's *Osteographia; or, the Anatomy of the Bones*, which eventually appeared in 1733 (fig. 5.4) (cat. no. 33), makes it clear that the author considered its illustrations, either life-size or reduced, more effective than words in describing the bones: "Every bone in the human body being here delineated as large as the life, and again reduced to lesser scales, in order to shew them united to one another; I thought it useless to make long descriptions, one view of such prints shewing more than the fullest and best description can possibly do."[26]

Cheselden should be recognized not only for overseeing the production of particularly fine illustrations in which notable care was paid in rendering the texture of the bones[27] but also for successfully getting his book to press, since several eighteenth-century projects for books on osteology failed to appear. The Scottish physician, man-midwife, and anatomist James Douglas advertised the imminent publication of his *Osteographia* in 1725 with life-size illustrations, but he left it unfinished at his death in 1742, perhaps discouraged by Cheselden's publication of 1733.[28] This may be the same reason John Fotherby's manuscript on the bones (1729–30), which was accompanied by several life-size illustrations, was never published.[29] The brevity of Cheselden's description and the high cost of the book were two of the many complaints made by the surgeon John Douglas (d. 1743), brother of James, in his pamphlet *Animadversions on a Late Pompous Book, Intituled, Osteographia; or, The Anatomy of the Bones by William Cheselden Esq.* (1735).[30] Douglas's method for operating for bladder stones had been superseded by Cheselden's, earning him Douglas's enmity. At the end of his extended and at times petty critique is a proposal for his own book on the bones, which ultimately never appeared. It promises thirty copperplates that would excel in representation through the excellence of the engraving, rather than size, asserting that they would be "so neatly etched and touched up with the Graver, that they will convey as good an Idea of the Figures and Surfaces of the Bones, as if they were as large as the Life."[31] The *Tabulae osteologicae* of the physician Christoph Jacob Trew

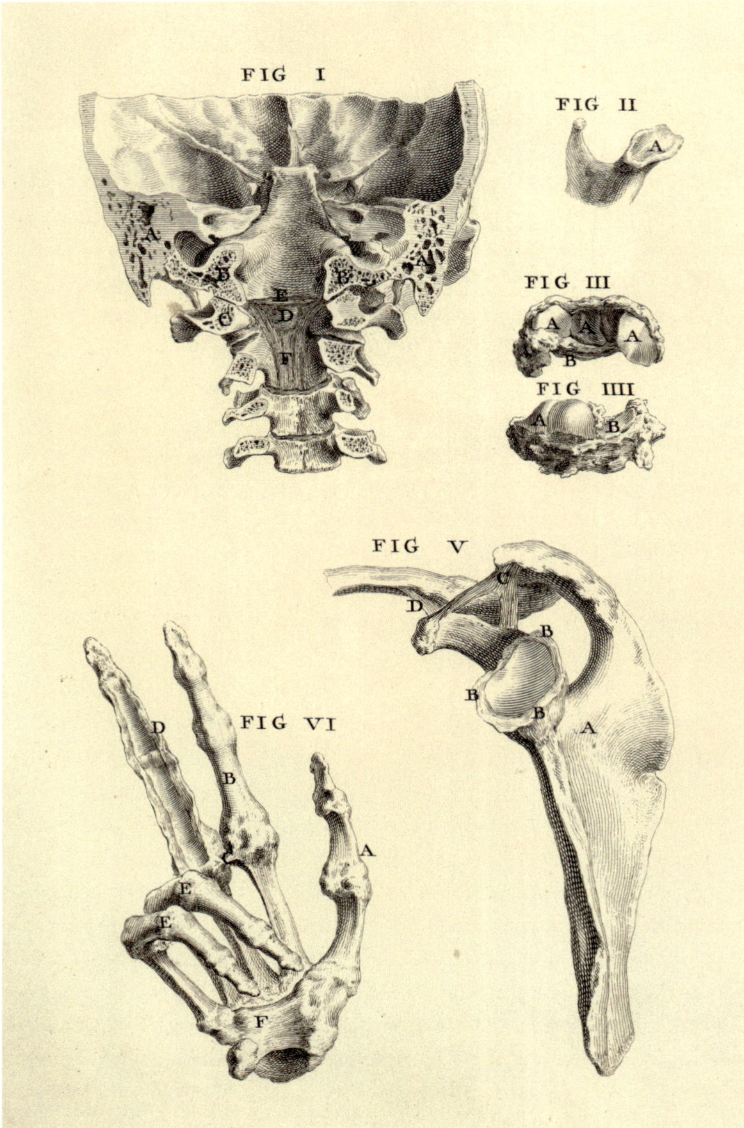

FIG. 5.4
GERARD VANDERGUCHT (English, 1696–1776). Cross section of the skull and cervical vertebrae, with ligaments and cartilages of the jaw, wrist, scapula, and hand, etching.
From William Cheselden, *Osteographia; or, the Anatomy of the Bones* (London: n.p., 1733), pl. 38. Bethesda, National Library of Medicine, 2651377R.

(1695–1769), with engravings of mostly life-size bones, appeared in Nuremberg in 1767, but it had a troubled publication history. Part of a larger project on anatomy that began in 1733, the engraver Georg Lichtensteger (1700–1781) and the draftsman Nikolaus Friedrich Eisenberger (1707–71), who had worked with Trew on the botanical prints after Georg Dionysius Ehret (1708–70), tired of delay and published the book themselves in an unauthorized edition two years before Trew's death.[32]

In the preface to the *Gravid Uterus* (cat. no. 42), a project begun in the 1750s though not published until 1774, the Scottish anatomist and physician William Hunter proclaimed Bidloo as his model in the illustration of a specific example rather than an idealized amalgam. In this respect, for Hunter, "it carries the mark of truth, and becomes almost as infallible as the object itself."[33] Hunter, who is remarkably forthcoming here about the production of illustration and the relative merits of style, asserts that if the aim of anatomical illustrations was to show, in his words, "the true nature" of what is being represented, they "should certainly be large; otherwise the smaller component parts cannot be distinctly represented; and if the natural size of the object be tolerably fit for an engraving, that must be of all others the very best, as it has the advantage of shewing such an important circumstance."[34] The mezzotint illustrations to

Charles Nicholas Jenty's obstetric atlas published in London in 1758 are described on the title page as being "As large as NATURE."[35] Like Hunter's *Gravid Uterus* and William Smellie's *A Sett of Anatomical Tables* (1754), life-size scale was achieved by showing only the relevant part of the pregnant body, as in the life-size plates to Folkert Snip's atlas of obstetric anatomy, which had been prepared by 1767 but were not published until 1793.[36]

While accuracy and legibility were factors in choosing to produce life-size illustration, the grandeur that reflected on the author was also an important motivating factor in taking on the expenditure involved in producing large illustrated books.[37] Even though he readily admitted that such size allowed a better representation of the body, John Bell, writing in 1794, dryly remarks that the purpose of a "stately anatomical figure of full six feet high" is to "establish a reputation for its author; which, if not high, will not fail to be at least of a lasting kind; neither apt to be forgotten, nor liable, like other discoveries, to go astray."[38] The cost of plates, paper, and artists' salaries resulted in prices that made these books into luxury objects that were unaffordable to impecunious medical students. Hunter acknowledges that with smaller, fewer, and less carefully engraved figures, "a great part of the expense might have been spared, and the work thereby rendered of more general use."[39] James Jurin (1684–1750), the secretary of the Royal Society, describes Cheselden's *Osteographia* (1733) (cat. no. 33) as "beautifully produced, at an exorbitant price."[40] The expense of these books and their unwieldiness put them at the opposite spectrum in terms of affordability and ease of use from small, portable anatomy books. If Cheselden lost great sums in producing his *Osteographia* (1733), he had a best seller in *The Anatomy of the Humane Body,* an octavo that first appeared in 1713, which, unlike the *Osteographia,* went through several editions.[41]

Life-size illustration of the complete body could be more easily perused on a wall than in the multiple openings of a book that it required, something Cattani grasped in advising that his separate prints be glued together to form a single figure in "natural measure."[42] A subscription announcement of 1779 (appendix C) presents plans for four such anatomical figures, although only three are known to have been issued (cat. nos. 36–38). Contiguous sections of a body had previously been offered by the illustrations in Fabricius's *De venarum ostiolis* (1603).[43] A more immediate example for Cattani would have been the life-size colored mezzotints of Jacques Fabien Gautier Dagoty, who claimed his *Anatomie générale des viscères en situation* [1752] presented for the first time ever entire life-size colored figures.[44] In the Huntington Library, there is an example of plates from this publication joined to show a complete figure of female anatomy (cat. no. 41). A modern counterpart to Gautier Dagoty's and Cattani's joined prints is Robert Rauschenberg's (1925–2008) *Booster* (1967), based on multiple X-rays of the artist's body that were misjoined to form a fractured anatomical self-portrait (cat. no. 46).

Tavares Strachan (b. 1979) also goes below the surface of the skin for his life-size anatomical portrait of 2018 in neon and glass of Robert Henry Lawrence Jr. (1935–67), the first African American astronaut (cat. no. 47). Unlike *Booster* or the female bodies and fetuses in Hunter's *Gravid Uterus, Robert* is not an exact likeness of the heretofore forgotten man it immortalizes. In choosing to represent the hidden interior of the body, Strachan makes visible the unique history of Lawrence while demonstrating an inner structure that equalizes all men. Anatomy, with *Booster* and *Robert,* becomes a medium of expression and a signifier of the body itself, rather than an object of study.

Cattani's prints were executed after the sculptures of Ercole Lelli (1702–66), who, following a rather undistinguished early career as a painter, made his name as an anatomical sculptor with the life-size écorché caryatids in the Bologna anatomy theater (see fig. K for cat. nos. 36–38). In a similar fashion, the French sculptor Jean-Antoine Houdon (1741–1828) launched his career in 1767 with a

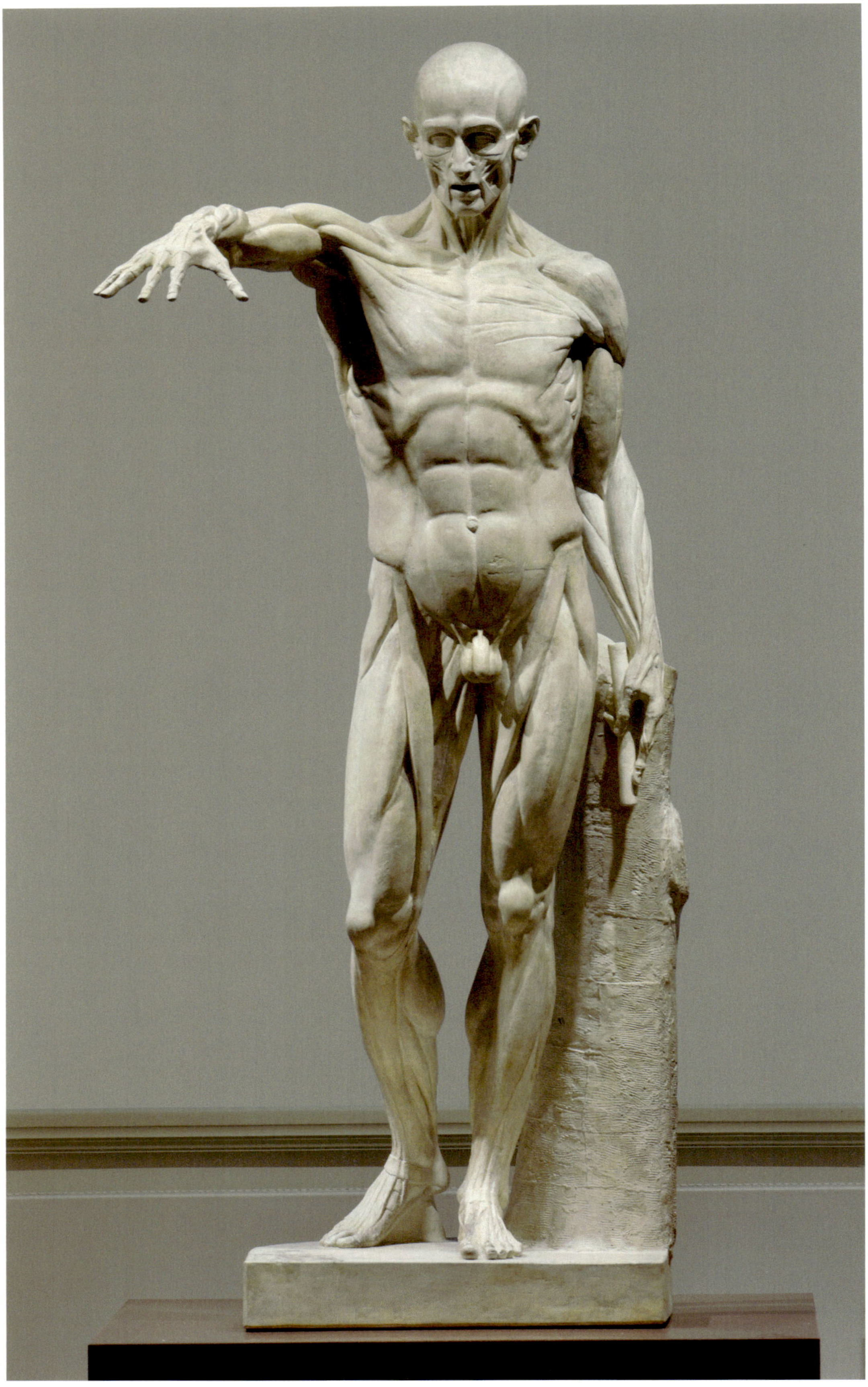

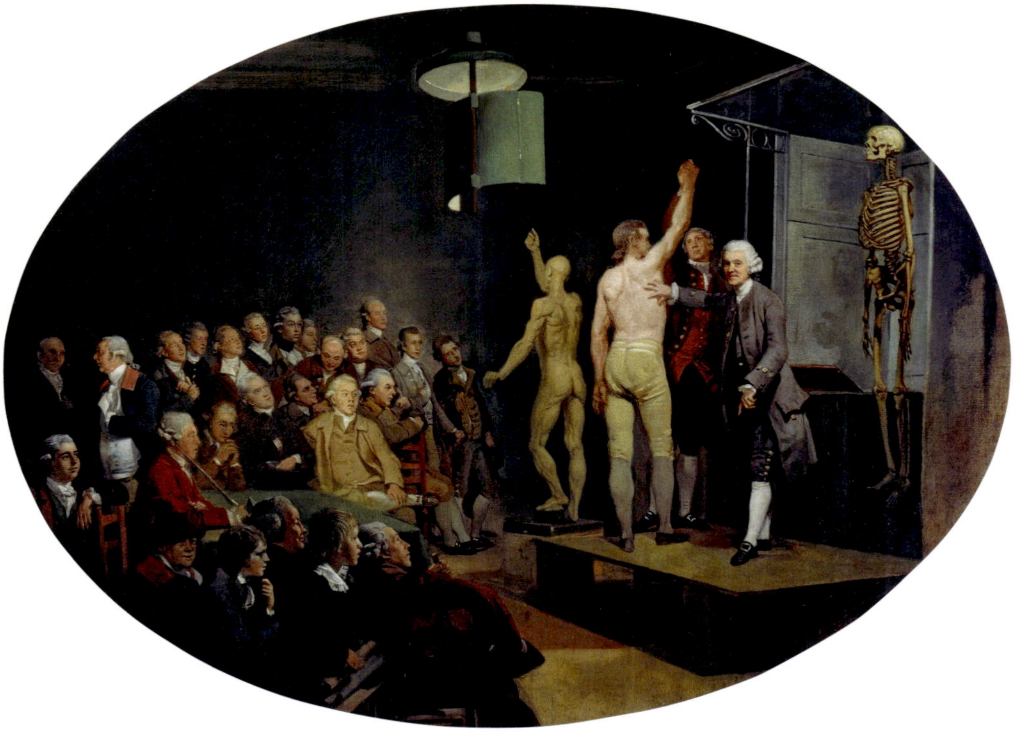

FIG. 5.5
JEAN-ANTOINE HOUDON
(French, 1741–1828).
Écorché, with outstretched arm, 1767, plaster, 184 cm (with base).
Gotha, Germany, Schlossmuseum Gotha, inv. P 25.

FIG. 5.6
JOHAN ZOFFANY
(German, 1733–1810).
Dr. William Hunter Lecturing at the Royal Academy of Arts, ca. 1772, oil on canvas, 77.4 × 103.5 cm.
London, Royal College of Physicians, X142.

life-size écorché created in Rome; casts in plaster and bronze of the first of two versions were soon to be found in academies across Europe (fig. 5.5).[45] The écorché bust (cat. no. 43) of a complete figure once associated with the sculptor Edme Bouchardon is another example of a life-size écorché model for the use of artists.[46] Like Marco d'Agrate's life-size écorché *Saint Bartholomew* (1562) in the Milan cathedral, Lelli and Houdon's écorchés were sculpted, but they should be considered alongside the tradition practiced by artists of making plaster casts after flayed bodies. What seems to be an early account of this is found in Baldasar Heseler's notes of Vesalius's lectures on dissection in Bologna in 1540: "Last year I saw our artists in Bologna thanks to the grace of the governor taking a plaster cast of a hanged man in Hospitale alla morte and sketching the superficial muscles, because the doctors at that time did not dissect, for the Rector, an Italian, was ill-disposed and mean."[47] Philips Angel, in his *Praise of Painting* (1642), mentions as a substitute for dissection life casts of flayed bodies by the Dutch artists Cornelis van Haarlem (1562–1638), Pieter Fransz de Grebber (ca. 1600–1652/54) and "Master Hendrick," probably Hendrick Goltzius (1558–1617).[48] In the seventeenth century, casts of dissected bodies were used in the teaching of anatomy at the Académie de Peinture et de Sculpture in Paris.[49] The statutes of 11 February 1666 for the Académie de France in Rome called for a mold to be taken of the dissection that was to be done each winter for the benefit of the painters and sculptors.[50] A cast from one of these dissections is illustrated in Bernardino Genga and Giovanni Maria Lancisi's anatomy book of 1691, dedicated to the Académie de France.[51]

Around 1750, while a lecturer at St Martin's Lane Academy in London, Hunter had a plaster cast made of a flayed body. This écorché, which no longer survives, is seen in Johan Zoffany's painting of Hunter teaching anatomy at the Royal Academy, where it stands next to a living model in a similar pose on which Hunter is indicating the form of the upper back (fig. 5.6). In annotations to a biography of Hunter, his brother, the surgeon John Hunter

(1728–93), relates that a body of an executed criminal was obtained from Tyburn, which William Hunter allowed to stiffen in a pose and then flayed so that by the following morning it was ready for a mold to be taken from it.[52] The partly closed, raised right hand of the écorché indicates that it had been arranged around something, perhaps the rope used to suspend the cadaver. The left arm was originally supported by a staff. This is not seen in Zoffany's depiction of the cast, but it appears in a print after Anthony Walker (1726–65) dated to 1783 in *The Artist's Pocket Companion*, by his nephew, John Walker (active 1776–1802) (cat. no. 25), who reports that the drawing for it was made "of the figure dissected by the late Dr. Hunter, for the Academy, (at that time in St Martin's Lane) before the mold was taken, in which the figure now at the Royal Academy was cast."[53] As indicated by the title, John Walker's book is of a size that was meant to fit into the pocket as a vade mecum of anatomical instruction. This telescoping of Hunter's life-size plaster cast to a pocket-size version in print was motivated by the same desire that led Serantoni to shrink the copperplate engravings for Mascagni: to deliver anatomical instruction to the student in a portable and affordable size.

Notes

1 For a discussion of this theme in anatomical illustration, see Cazort in Cazort, Kornell, and Roberts 1996, 225–33, nos. 113–21; Kemp and Wallace 2000, 50–52; and Kemp 2010, 193, 202–3.

2 Mascagni 1823–31.

3 For wax anatomies, see Gensler, "Visualizing the Body in Three Dimensions," this volume, pp. 85–86.

4 See illustration in Roberts and Tomlinson 1992, 395.

5 Audubon describes the octavo edition as answering the demand for a publication "similar to my large work, but of such dimensions, and at such price, as would enable *every* student or lover of nature to place it in his Library." Audubon 1840–44, 1:1 (emphasis Audubon).

6 Mascagni 1833, preface, n.p.

7 For their sizes, see Cushing 1962, 14, 80, 112. Comparing copies in the Getty Research Institute, Vesalius's *Fabrica* (1543) is, at 44 centimeters, approximately double the height of Berengario's portable textbook, *Isagogae breves* (1523) (cat. nos. 8, 30). For their sizes, see Cushing 1962, 14, 80, 112; and Goree 2014, 5–6.

8 Vesalius 1543a, 5v; and Vesalius 2014, 1:13. Cushing identifies the plagiarist as Walther Hermann Ryff (Cushing 1962, 2–28). For Vesalius's complaints about Thomas Geminus reducing his images, see Vesalius 2015, 232; and Geminus 1959, 21.

9 Heseler 1959, 237, 253, 283.

10 Vesalius 1543a, 381 [481].

11 Vesalius 1555, 585. The contradictory passage in the first edition that describes this figure as showing "the size of the uterus that is frequently found in nonpregnant women" is deleted in the second edition (Vesalius 1543a, 533; and Vesalius 2014, 2:1072). Figure 24 in book 5 is also of the same subject (Vesalius 1543a, 539). Regarding the Paduan woman, see Park 2006, 207–19.

12 Fabricius ab' Acquapendente 1600, [ii] v.

13 See Rippa Bonati and Pardo-Tomás 2004; and, in particular, Kemp 2004.

14 Fabricius ab' Acquapendente 1933; Putscher 1972, 28; Roberts and Tomlinson 1992, 251–53; and De Caro 2018, 211–13.

15 Conring 1687, 179. See also Vincent and Perrot 2016, 9n24.

16 On the recently rediscovered drawings in the Bibliothèque Interuniversitaire de Santé in Paris, see Vincent and Perrot 2016. I am grateful to Erin Travers for sharing her unpublished research on the drawings.

17 Wolf-Heidegger and Cetto 1976, 351–52, no. 310. For a print of the theater dated to ca. 1650–60, see idem, 350, no. 308.

18 See François Sallé's *The Anatomy Class at the École des Beaux-Arts* (1888), in Callen 2018, 31, fig. 1.1.

19 For Nuck's letter to Dr. Frederick Slare, dated 18 May 1682, see Birch 1757, 151.

20 Bidloo 1685, address to reader, [*4]r. On copyists, Bidloo here paraphrases Horace, *Epistles*, 1, 19.

21 Fend 2019, 316–25.

22 Cowper 1698. Russell records four editions of Cowper's version, two in English and two in Latin (Russell 1987, 54–55, nos. 211–14). For editions of Bidloo, see Margócsy 2014, 161–64.

23 Cowper 1724, preface, n.p. The same wording is used by Nuck with regard to Bidloo's illustrations (see note 19 above) and on the title pages of both volumes of Hans Sloane's *A Voyage to the Islands Madera, Barbados, Nieves, S. Christophers, and Jamaica* (London, 1707–25) to describe the engraved illustrations of the natural history of these islands. I'm grateful to Sachiko Kusukawa for alerting me to Sloane's use of life-size illustrations.

24 Kornell 2019, 493.

25 Nichols 1812, 365; and Russell 1954, 32.

26 Cheselden 1733, address to reader, n.p.

27 Bertoloni Meli 2015, 178–83.

28 He was at work on this from 1715, and drawings, plates, and proofs survive in the University of Glasgow Library. See Roberts and Tomlinson 1992, 431–33; Flis 2018, 62–67; and Flis in Campbell and Flis 2018, 218, no. 34.

29 John Fotherby, *Anatomy* (1729–30), Bethesda, National Library of Medicine, 101498641. Tables 35–36, 38, and 40 have the annotation "Magnitudo omnium naturalis." Tables 28 and 29 are after skeletons by William Cowper's first edition of the *Myotomia reformata* (Cowper 1694, figs. 17, 18), reprinted in James Drake's *Anthropologia nova,* where the skeleton in posterior view is identified as female (Drake 1707, vol. 2, pls. 21, 22). Fotherby, who according to an inscription on folio 1 was from Great Grimsby, was likely a descendant of Charles Fotherby (1549–1619), dean of Canterbury, whose bone-encrusted sarcophagus is depicted on table 28 with a view of Canterbury in the background.

30 Douglas 1735. Wragge-Morley 2016, 282–86.

31 This proposal is dated "London, Feb. 26. 1735," Douglas 1735; and Russell 1987, 70–71, no. 285.

32 Schnalke 1995. Alexander Monro (secundus) published *A Description of All the Bursae Mucosae of the Human Body* in Edinburgh in 1788, and like Trew's *Tabulae osteologicae,* it has folding plates to accommodate the life-size illustrations. Albinus avoided this by illustrating the skeleton in sections in his *Tabulae ossium humanorum* (1753) (Choulant 1945, 281).

33 Hunter 1774, preface, n.p. On Hunter and illustration, see Kemp 1993, 113–16.

34 Hunter 1774, preface, n.p.

35 Jenty 1758.

36 For these obstetric atlases, see Longo and Reynolds 2016; and Gamer 2018.

37 Kemp 1993, 100–105, 118; Carlino 2004, 76; Berkowitz 2015; and Margócsy, Somos, and Joffe 2018, 6–7.

38 Bell 1794, ii, see also xviii.

39 Hunter 1774, preface, n.p.

40 James Jurin, letter to Josias Weitbrecht (1702–47), 24 June 1737 (Jurin 1996, 417, no. 238). The *Osteographia* was originally offered to subscribers at four guineas and to regular purchasers at six guineas.

41 Cheselden 1713; and Russell 1954.

42 "Manifesto," 22 September 1779, appendix C.

43 Fabricius ab' Acquapendente 1933, 62, pls. 5–7.

44 Gautier Dagoty, J. F. [1752], "Première table explicative," n.p. See also a set of twelve oil paintings on canvas, dated 1764, of life-size figures by Gautier Dagoty, in the Wellcome Collection, London. For a multiple-plate engraving of a life-size recumbent memento mori skeleton by Pierre Landry (1630–1701), see Griffiths 2016, 415, and for large composite prints in general, 412–15. For the example of composite prints in the sixteenth century, see Landau and Parshall 1994, 43–45, 82–89, 231–37, 300; Silver and Wyckoff 2008; Schmidt 2011, 24–32; and Kornell 1989a for Battista Franco's unrealized project of composite anatomical prints.

45 Poulet 2003, 63–71, nos. 1, 2.

46 For a historical survey of écorché models, see Amerson 1975.

47 Heseler 1959, 139. On this hospital, see Heseler 1959, 305n26.

48 Hoyle and Miedema 1996, 247–48, 256.

49 Comar 2008a, 19; and Lichtenstein and Michel 2006, 489–92. For a cast made circa 1594 after a dissection carried out by the artist Federico Zuccaro (1540/41–1609) for the Accademia di San Luca in Rome, see Roccosecca 2009, 133.

50 Montaiglon 1887, 10, no. IX.

51 Genga and Lancisi 1691, pls. 21–23.

52 See Kemp 1975, 16; Brock 1983, 9; Postle 2004, 57–58; and Black 2007b, 93–94. See also Kornell, "Surface Anatomy," this volume, p. 67.

53 For this unnoticed account, see Walker 1787, 7. The staff is also included in a reduced wax model of the cast by Michael Henry Spang (d. 1762) dated to 1761 in the Hunterian, University of Glasgow (Campbell and Flis 2018, 206, no. 24), and in a drawing by J. M. W. Turner (1775–1851) of circa 1791 in the Tate Museum, London, D00124 (see Martin Postle in Bignamini and Postle 1991, 94, no. 89; and Wilton 2012).

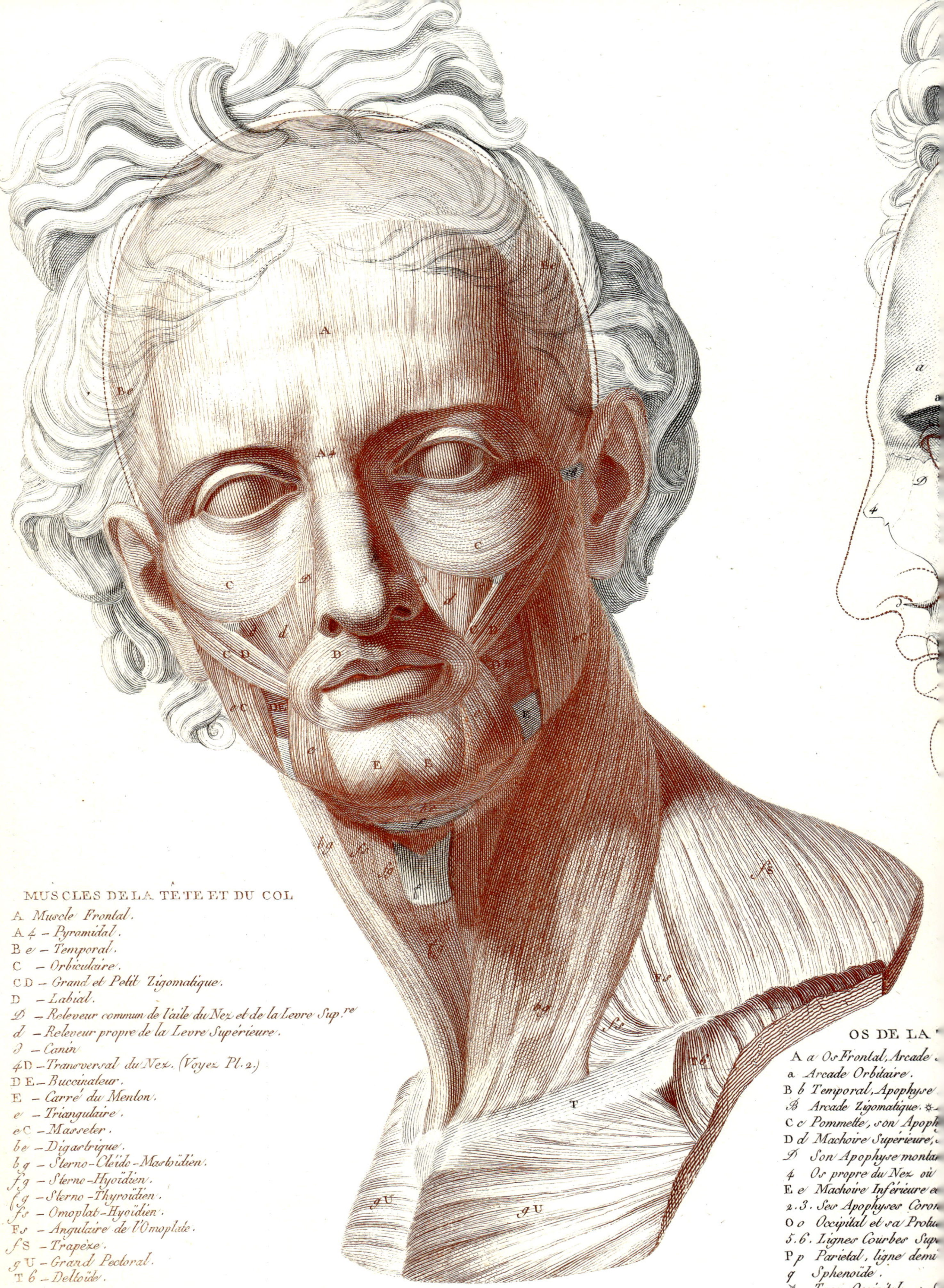

MUSCLES DE LA TÊTE ET DU COL

A Muscle Frontal.
A 4 – Pyramidal.
B e – Temporal.
C – Orbiculaire.
CD – Grand et Petit Zigomatique.
D – Labial.
𝒟 – Releveur commun de l'aile du Nez et de la Levre Sup.re
d – Releveur propre de la Levre Supérieure.
∂ – Canin.
4D – Transversal du Nez. (Voyez Pl. 2.)
DE – Buccinateur.
E – Carré du Menton.
e – Triangulaire.
eC – Masseter.
be – Digastrique.
bg – Sterno-Cleïdo-Mastoïdien.
fg – Sterno-Hyoïdien.
f g – Sterno-Thyroïdien.
fs – Omoplat-Hyoïdien.
Fs – Angulaire de l'Omoplate.
fS – Trapèze.
gU – Grand Pectoral.
T 6 – Deltoïde.

OS DE LA

A a Os Frontal, Arcade
a Arcade Orbitaire.
B b Temporal, Apophyse
𝔅 Arcade Zigomatique. ✡
C c Pommette, son Apoph
D d Machoire Supérieure,
𝒟 Son Apophyse monta
4 Os propre du Nez où
E e Machoire Inférieure e
2.3. Ses Apophyses Coron
O o Occipital et sa Protu
5.6. Lignes Courbes Sup
P p Parietal, ligne demi
q Sphenoïde.
✡ Trou Occipital. + S
7 Fosse Orbitaire.

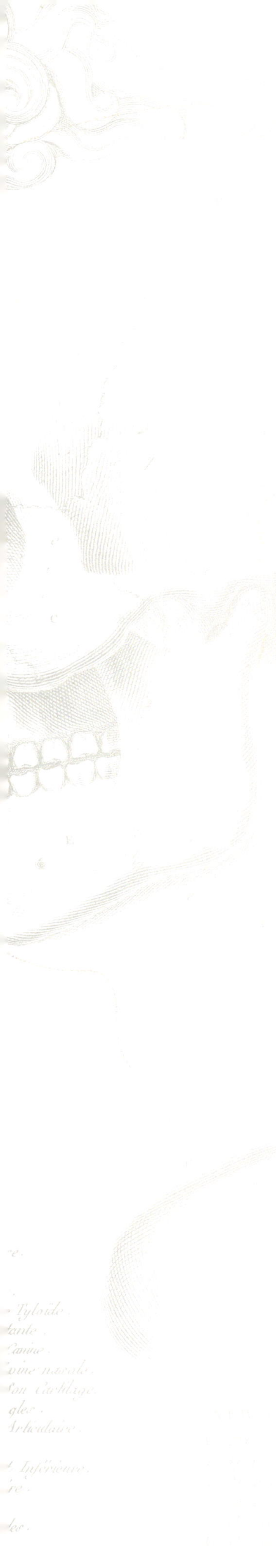

chapter six

SURFACE ANATOMY: FROM THE INSIDE OUT

Monique Kornell

According to Giorgio Vasari's biography of Raphael in the second edition of *Le vite* (*The Lives*) of 1568, the artist perceived a weakness in his own style after seeing Michelangelo Buonarroti's treatment of the nude and his use of foreshortening in the cartoon for the *Battle of Cascina* (1504–6). Raphael (1483–1520) sought to rectify this through an intense period of anatomical study, specifically through the comparison of muscles of a dissected body with those of the living nude.[1] Vasari's account is an early example of the simultaneous study of the dead and the living body in order to achieve a command of the representation of the nude, and it is one of many instances in *The Lives* where the study of anatomy leads to improvement in artistic skill. Moreover, it is noteworthy that the early training of an artist of such great stature as Raphael is cast as insufficient because he had only drawn the body from life, thereby neglecting what Vasari presents as a seminal facet of artistic study: the anatomy of the dissected corpse.

Knowledge of the interior of the body could be expressed in how the muscles, bone structure, and veins are depicted on the body's exterior at rest, and in the changes on its surface when in motion or exertion. Leonardo da Vinci (1452–1519), in his notes on anatomy for artists, stresses the importance of knowing and representing the muscles according to their different actions, and he decries those artists who depict all of them: "In order to appear great draughtsmen, [they] make their nudes wooden and without grace, so that they seem a sack full of nuts rather than the surface of a human being, or indeed, a bundle of radishes rather than muscular nudes."[2] The highly muscled bodies favored by sixteenth- and seventeenth-century

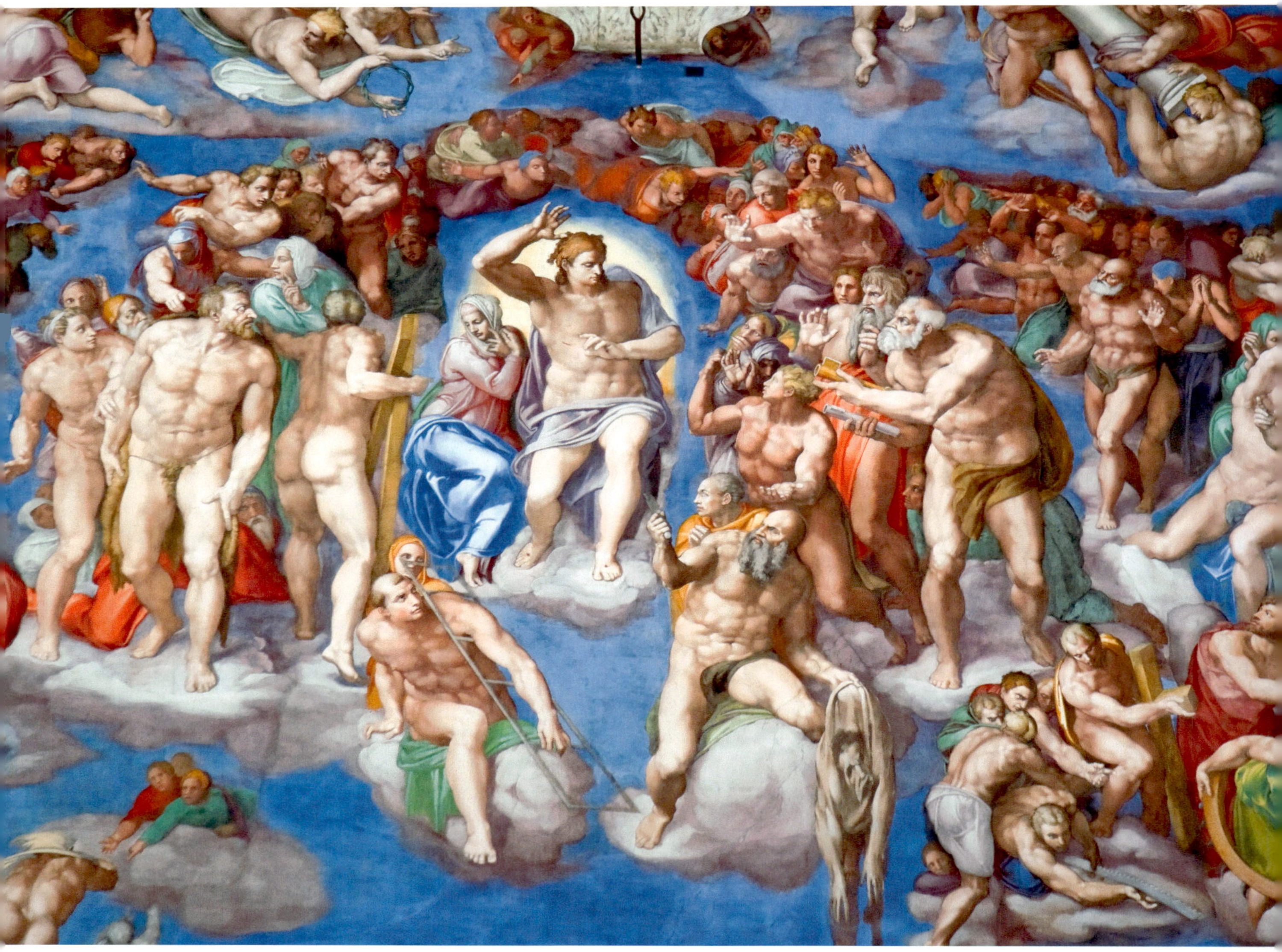

FIG. 6.1
MICHELANGELO BUONARROTI (Italian, 1475–1564).
The Last Judgment (detail), 1536–41, fresco.
Vatican City, Sistine Chapel.

artists were in part the product of a desire to display a knowledge of anatomy and to emulate the example of the exalted Michelangelo, whose muscular nudes were the figure school for generations of artists (fig. 6.1).

An artist who exemplified the vogue for a display of musculature was the Dutch painter and virtuosic engraver Hendrick Goltzius, who specialized in figures with remarkably developed physiques, as seen in his print series of Roman heroes from 1586 (fig. 6.2). The Roman consul and military commander Marcus Valerius Corvus wears a cuirass of anatomical detail far exceeding that seen in ancient examples. Only the ties of the leg coverings at his buttocks, the wrinkles of his hose, the decorative upper border of the cuirass, and the bunched drapery at the arm signify that he is clothed and not nude. A letter written by the Dutch anatomist Pieter Pauw (1564–1617) asking that the artist be notified of an upcoming dissection is an indication that friendship with Pauw facilitated Goltzius's study of anatomy by dissection. This would have allowed Goltzius access to flayed cadavers from which to make plaster casts that he is credited with.[3]

The Spanish sculptor and goldsmith Juan de Arfe y Villafañe promoted the study of

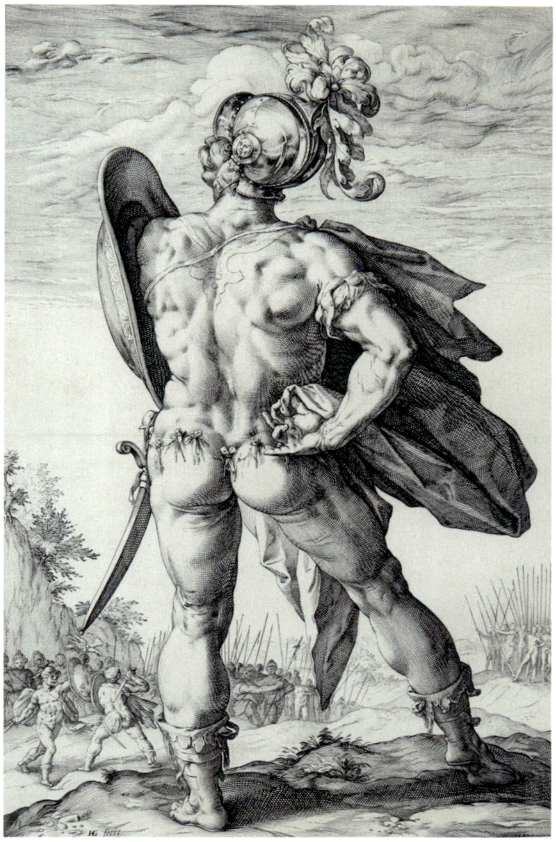

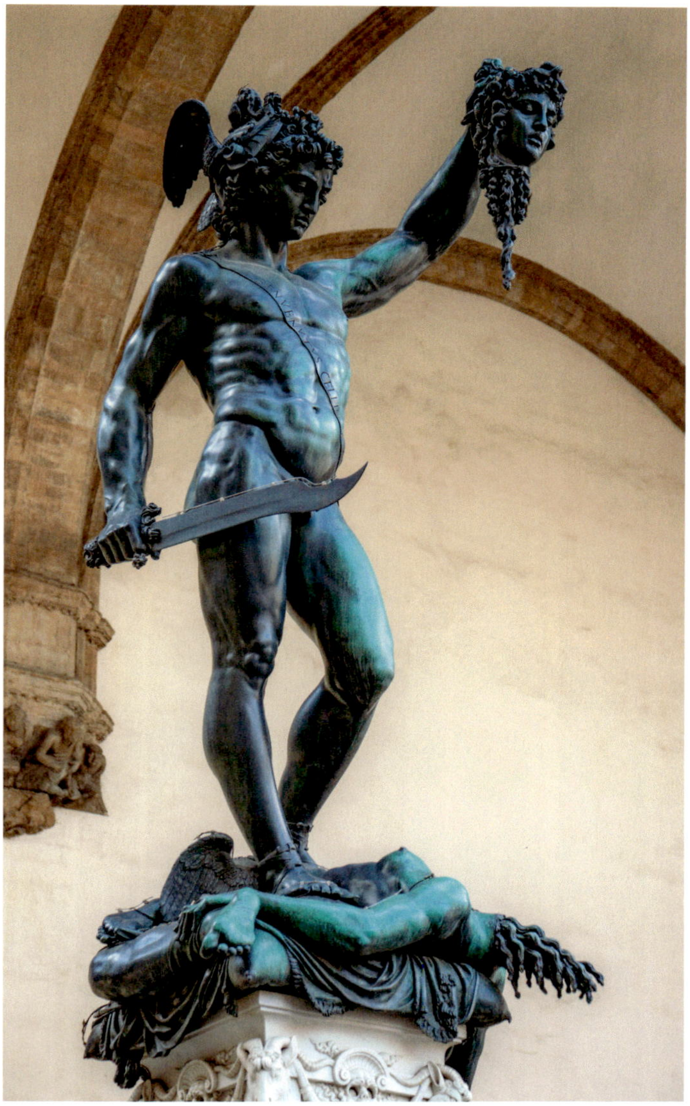

FIG. 6.2
HENDRICK GOLTZIUS
(Dutch, 1558–1617).
Marcus Valerius Corvus,
1586, engraving, plate:
36.8 × 23.2 cm.
Los Angeles, Hammer
Museum, UCLA Grunwald
Center for the Graphic
Arts, 1962.19.63.

FIG. 6.3
BENVENUTO CELLINI
(Italian, 1500–1571).
Perseus with the Head of Medusa, 1545–54,
bronze, height: 320 cm
(excluding base).
Florence, Loggia dei Lanzi.

anatomy but did not consider learning by dissection a necessity for artists and favored observation of the life model. In his *Varia commensuración para la esculptura, y architectura* (1585–87), Arfe y Villafañe remarks that some muscles can easily be discerned in the living body. He deemed the underlying muscles of the face as mainly the concern of physicians and surgeons and not of artists, and so he describes instead a relief pattern of *bultos redondos*[4] (raised bumps) on the surface of the skin (cat. no. 17). Benvenuto Cellini, another sculptor and goldsmith, was also a keen observer of surface anatomy. In his treatise *Sopra i principii e 'l modo d'imparare l'arte del disegno* (On the principles and the way of learning the art of drawing) of the mid-1560s, Cellini promotes the study of the skeleton as the ideal starting point for the education of the artist, something seen put into practice in Cornelis Cort's engraving after Stradanus, dated 1578 (see fig. 3.2). Cellini particularly admired the way the scapula (shoulder bone) is discernible under the skin and its "diverse and most beautiful actions" when the arm is in use, and he calls for the young artist to commit the appearance of the ribs to memory because they create "very beautiful reliefs and hollows" below the flesh and skin.[5] Such effects in this area were beautifully realized in Cellini's great bronze *Perseus with the Head of Medusa* (fig. 6.3). The same intricate pattern of the muscles over the ribs was later also carefully rendered in a charcoal study by the late

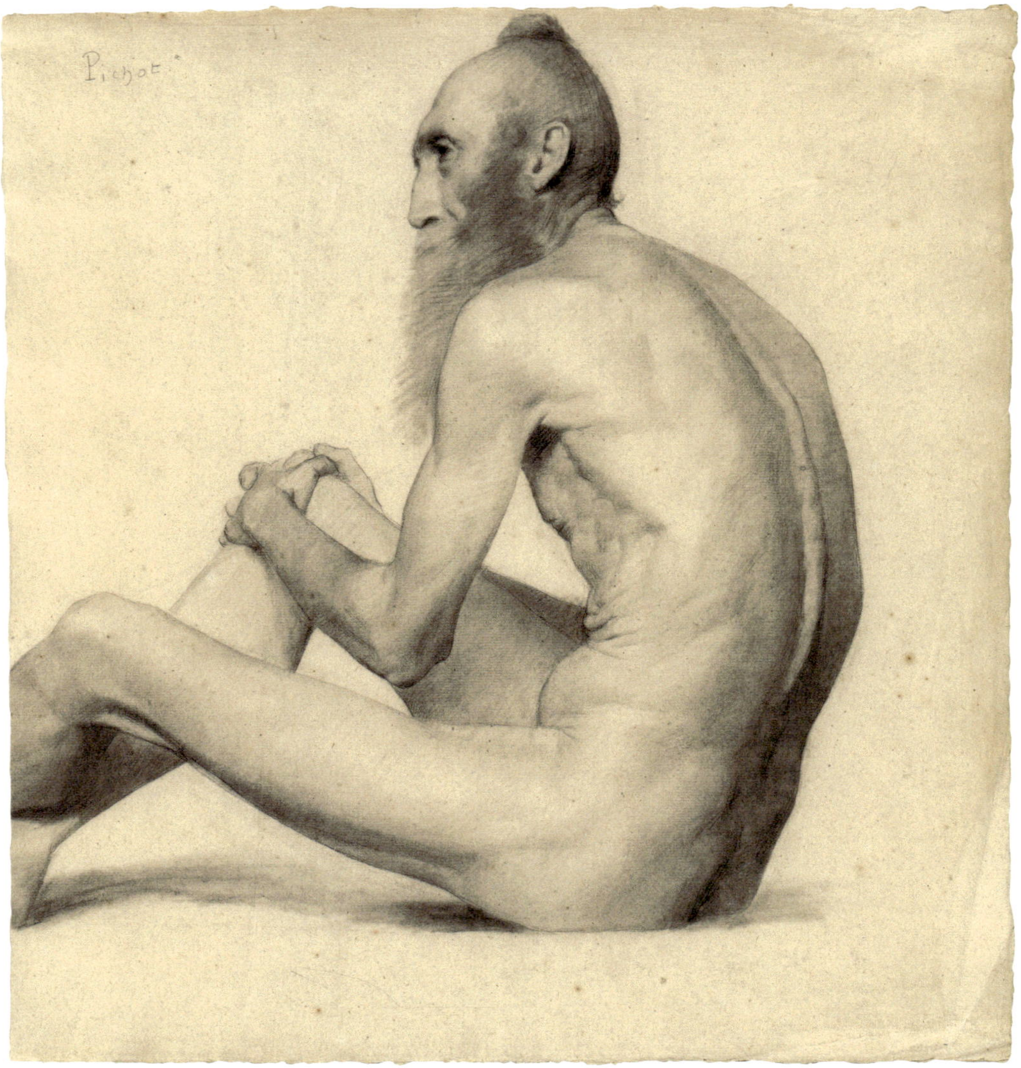

FIG. 6.4
ÉMILE-JULES PICHOT
(French, 1857–1936).
Nude study of an old man, charcoal and powdered vine charcoal with stumping and lifting on laid paper, ca. 1878–79, 46.5 × 43 cm.
Los Angeles, J. Paul Getty Museum, 2016.81.

nineteenth-century French artist Émile-Jules Pichot (fig. 6.4).[6] After studying anatomy, the artist and critic Roger de Piles recommends, in his *Abregé d'anatomie* of 1668, that the student consult both the most beautiful examples of the antique and a life model set in active poses, preferably one that is "extremely muscular and with little fat"[7]—a type of figure photographed by Thomas Eakins (1844–1916) as a study for *Wrestlers* (cat. no. 48, fig. P), and by Robert Mapplethorpe (1946–89) (cat. no. 50).

Vasari took the first-century Roman writer Pliny's *Historia naturalis* as a model in considering the accurate portrayal of anatomy as a technical innovation, one that contributed to the overall advancement of art. Pliny notes that in sculpture, Pythagoras of Reggio was the first "to show the sinews and veins" and that the painter Cimon of Cleonae, who showed the joins of the limbs, also "displayed the veins."[8] For Vasari, the portrayal of the veins was a distinguishing feature of *The Martyrdom of Saint Sebastian* completed in 1475 by Piero and Antonio del Pollaiuolo (fig. 6.5). In the altarpiece, Vasari singles out an archer bending over to load his bow as a figure in which Antonio sought "to imitate nature as much as he could" by effectively depicting the exertion of a body, as indicated by "the swelling of the veins and muscles and how the archer holds his

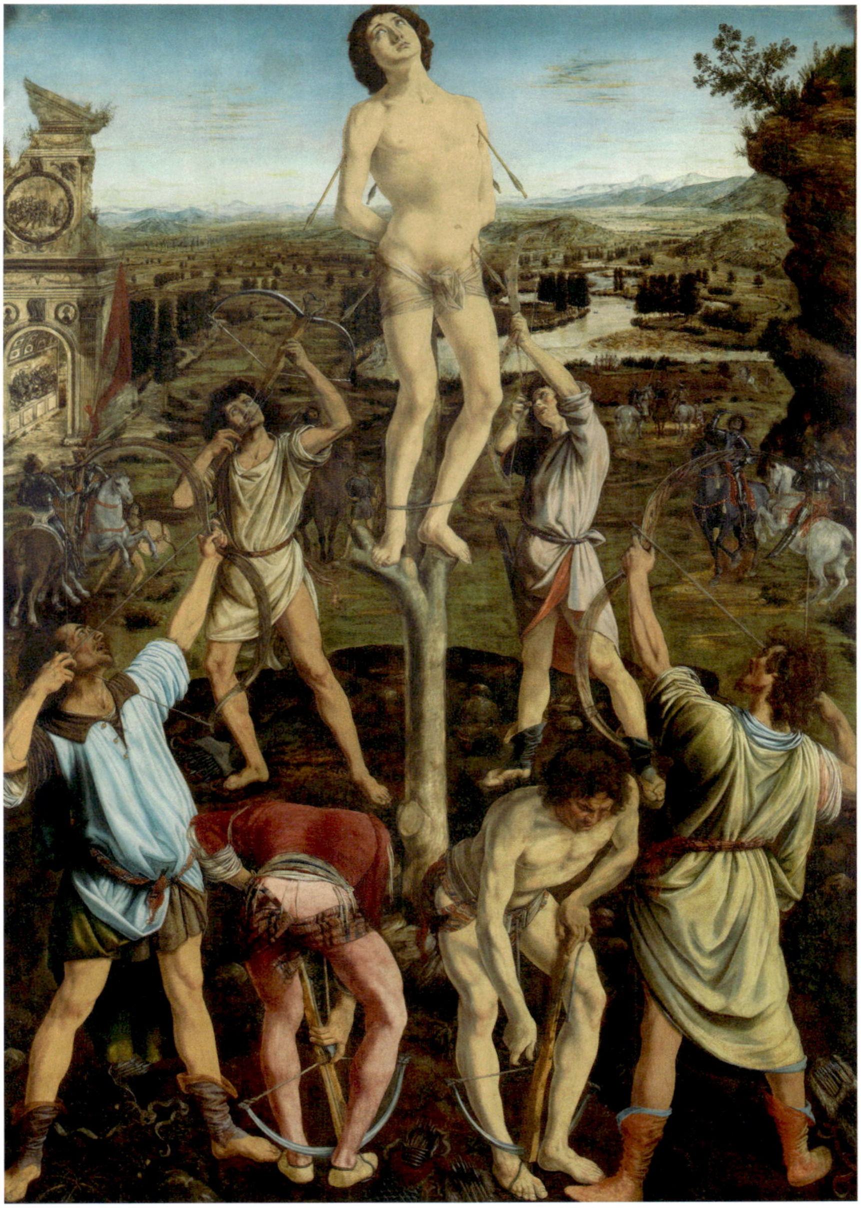

FIG. 6.5
ANTONIO DEL POLLAIUOLO
(Italian, ca. 1432–98) and Piero del Pollaiuolo (Italian, ca. 1441–ca. 1496).
The Martyrdom of Saint Sebastian, 1475, oil on wood, 291.5 × 202.6 cm. London, National Gallery, NG292.

breath in order to gain more strength."[9]

The Florentine painter Alessandro Allori, who kept a room in the cloisters of San Lorenzo for the purposes of dissection and who owned a copy of the first edition of Andreas Vesalius's *Fabrica* (1543),[10] provides specific advice for the depiction of the veins in his unfinished illustrated manuscript treatise of the late sixteenth century on the rules of design, *Delle regole del disegno*.[11] Written in the form of dialogues with interlocutors Allori, Florentine noblemen, and Allori's master, Agnolo Bronzino (1503–72), who had also studied anatomy through dissection, it begins with a then-standard method of artistic instruction: providing examples of the elements of a face, starting with the eye. Allori, however, quickly veers from these simple beginnings to the complex by turning to anatomy, which he considers a true grounding for the understanding of art. The manuscript is illustrated with drawings of the head in three

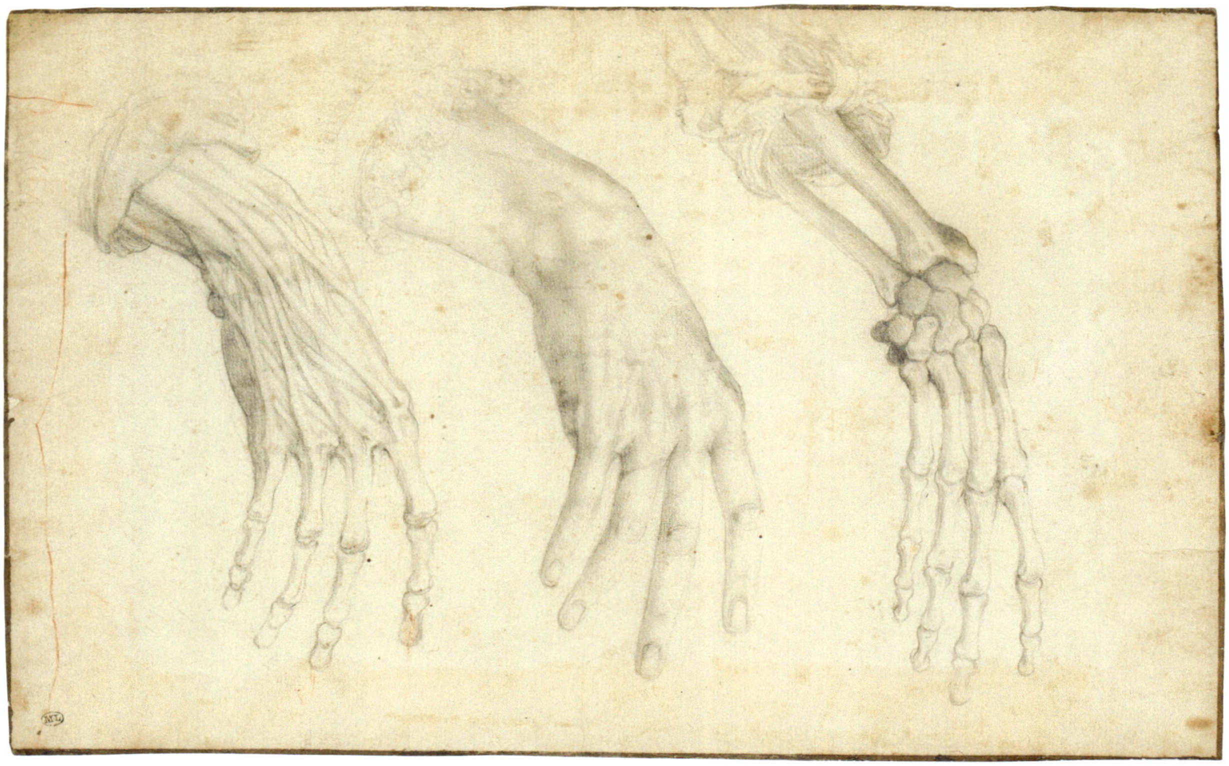

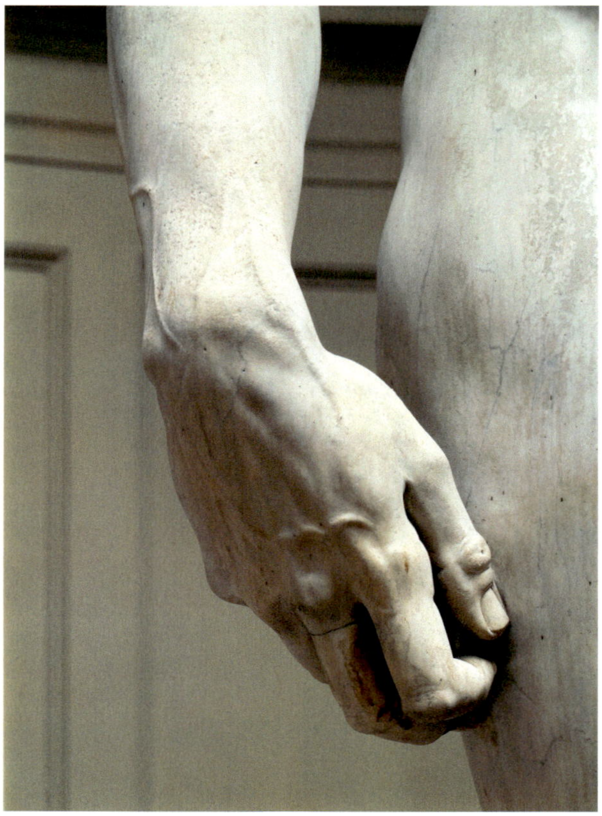

FIG. 6.6
ALESSANDRO ALLORI
(Italian, 1535–1607).
Three anatomical studies of a hand, ca. 1565–70, black chalk with traces of red chalk, 17.5 × 27.8 cm.
Paris, Musée du Louvre, 12.

FIG. 6.7
MICHELANGELO BUONARROTI
(Italian, 1475–1564).
David (detail of right hand), 1501–4, marble, 517 cm.
Florence, Galleria dell'Accademia, 1076.

anatomical levels, similar to a sheet in the Musée du Louvre comparing the bones, the muscles, and the surface of the hand (fig. 6.6). In *Delle regole del disegno*, Allori notes the visibility of the bones in the hand and goes on to remark that the veins and arteries are more easily seen in the hand than elsewhere in the body, particularly so when it is suspended and the veins appear swollen. He cautions that not all veins should be shown, "but rather certain principal ones such as one sees done by Michelangelo Buonarotti and Baccio Bandinelli and other very famous sculptors and painters."[12] This is in fact particularly well displayed in the statues of Michelangelo's *David* (fig. 6.7) and Bandinelli's (1488–1560) *Hercules and Cacus* in the Piazza della Signoria, Florence.

The hand of another sculpture by Michelangelo is the subject of a black chalk drawing attributed to Bronzino (fig. 6.8). It has been identified as the right hand of the statue of Duke Giuliano de' Medici (completed by 1534), part of the tomb decorations in the New Sacristy in Florence.[13] Given the viewpoint, this drawing is either after one of the contemporary casts made of the sculptures in the New Sacristy or by an artist who sought a higher vantage point, an activity seen in Federico Zuccaro's drawings of the 1570s.[14] In contrast to the original, and contrary to the advice of Allori, several veins have been made prominent and appear as a regularized pattern across the surface of the skin. Among the drawings on the verso are heads freely drawn in ink that appear to be after the sculptural group on the other side of the New Sacristy, the *Madonna and Child with Saints Cosmas and Damian*. If this identification is correct, the sheet most likely dates to after 1559, the year when the group was installed.[15] The careful, if stylized, rendition of the anatomy of the hand is perhaps the study made by a younger artist rather than by Bronzino, who was then an established artist in his midfifties.

Vesalius advised painters and sculptors to note how the appearance of the body changes when muscles are in use. He felt that it was not enough for them to know the location of the superficial muscles; they also needed to understand their function, and how they contract and bulge in action, and lengthen and subside at rest.[16] The way to understand this, according to the French sculptor Michel Anguier (1612–86), was to compare "dead and cooled muscles," which lack the action of life, with those of a living model.[17] In his *Manual of Artistic Anatomy* (1852), the Scottish anatomist and physician Robert Knox (1791–1862) states in

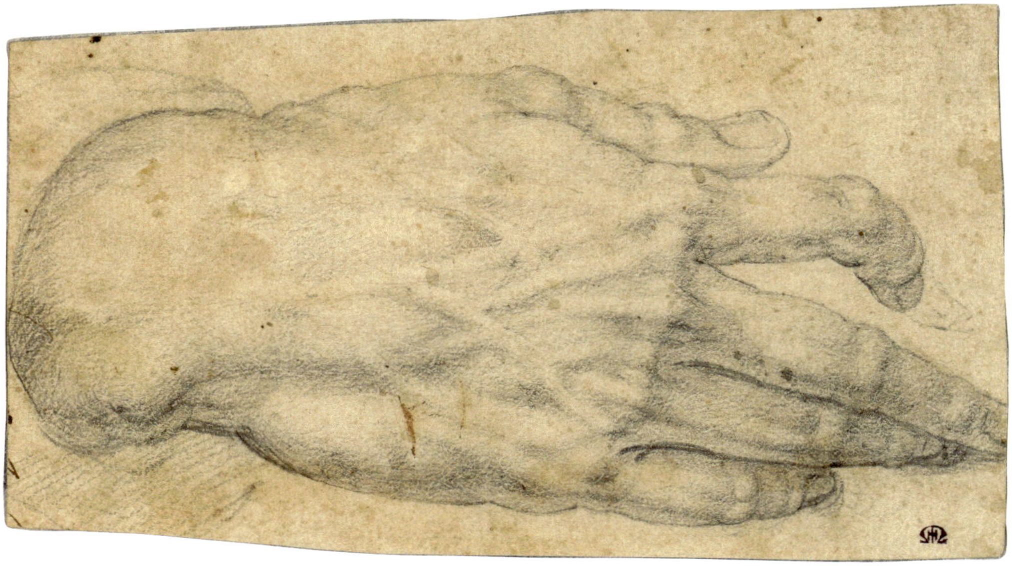

FIG. 6.8
ATTRIBUTED TO AGNOLO BRONZINO (Italian, 1503–72).
Study of a man's right hand, mid-16th century, black chalk (recto), pen and brown ink (verso), 7.6 × 15.2 cm.
Los Angeles, J. Paul Getty Museum, 92.GB.40.

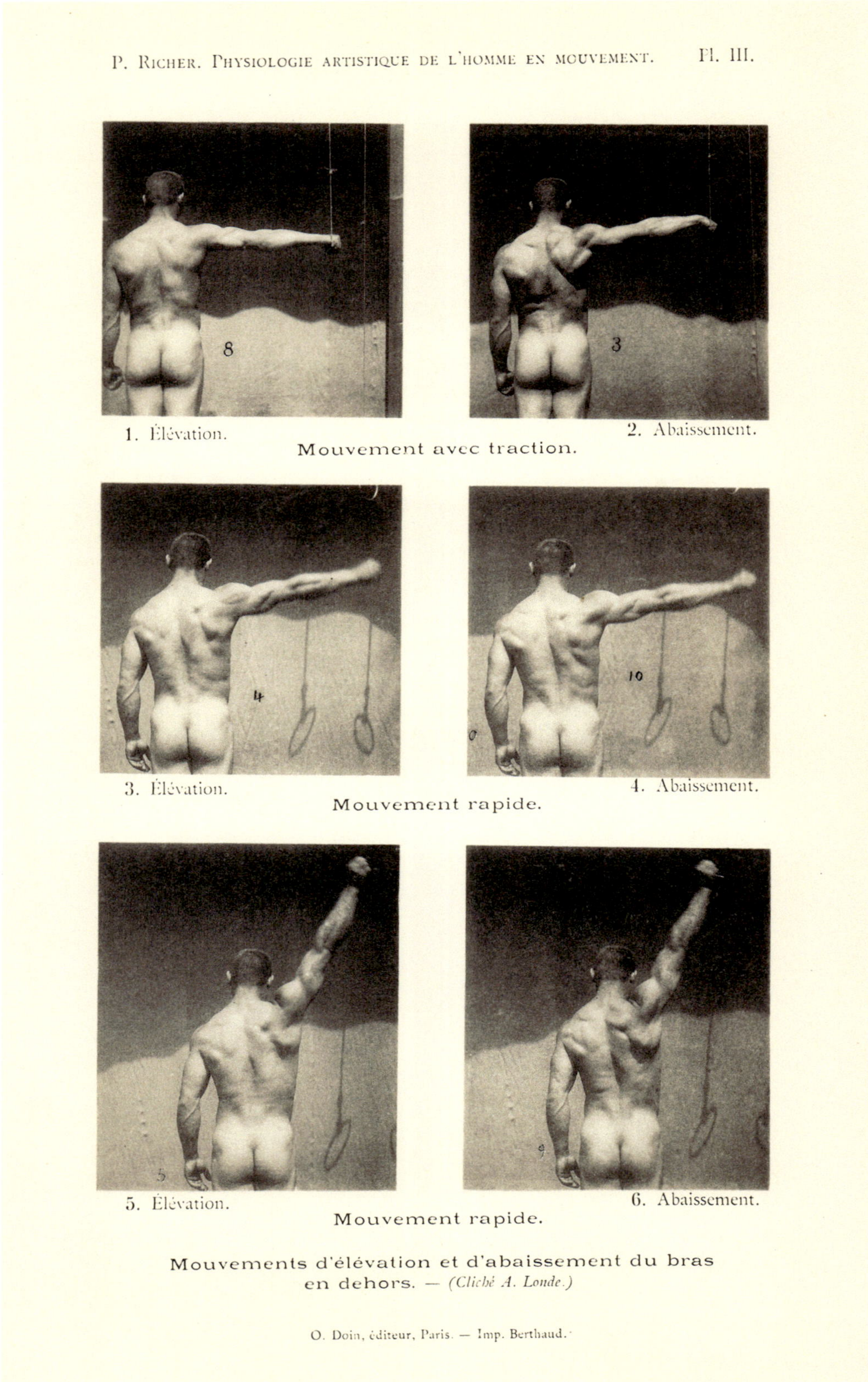

FIG. 6.9
ALBERT LONDE
(French, 1858–1917).
Outward Movements of the Arm, Elevated and Lowered,
photographs.
From Paul Richer,
Physiologie artistique de l'homme en mouvement
(Paris: Octave Doin, 1895),
pl. 3.
Los Angeles, Getty Research Institute,
91-B34994.

more polemical tones the case for the observation of the body in action. Knox labeled the method of drawing after dissection and anatomical illustrations "a theory and mode of study of the dead for the living" that is harmful to art, identifying it with the English artist Benjamin Robert Haydon and the Scottish anatomist and surgeon Sir Charles Bell (1774–1842), at whose anatomy theater Haydon's students drew from dissections.[18] Knox declared that the artist should never draw a dissected muscle, even though he conceded that observing and carrying out a dissection would provide knowledge of the body. Rather, the artist should always have as a corrective the example of the living body or an example of antique sculpture, since ancient sculptors were considered to have perfectly captured the living body:

> Place near you, at the time you inspect these dead muscles, *the living figure,* or, if that cannot be, place near the dead some first-rate plaster cast of an antique figure, and sketch the muscles you have just studied from the figure, but not from the dissected muscles. In this way will the student learn to draw living muscles, and not dead ones; in this way will the student learn to read the exterior of the living figure and of the antique statue.[19]

In the late 1870s, students at the anatomy-focused Philadelphia School of Art studied the muscles of a cadaver and a life model concurrently, as just one part of a course of anatomical instruction taught by Eakins and the anatomist William W. Keen (1837–1932)—a course for which, according to a contemporary commentator, "exhaustive" was too faint a description (cat. no. 48).[20] The same muscles demonstrated on a cadaver were correspondingly activated in a living model through the use of weights and suspended rings and other props.

Paul Richer (1849–1933)—the French neurologist, anatomist, physiologist, and sculptor—made a similar exploration of the changing appearance of the body when in action and in response to weight in his book on physiology applied to art (1895) (fig. 6.9), a follow-up to his book on anatomy for artists of 1890 and a successor to the books on surface anatomy for artists by Gerdy (1829) and Fau (1845).[21] Richer collaborated with Albert Londe (1858–1917), a photographer who had worked with Jean-Martin Charcot (1825–93) and Richer at the Salpêtrière hospital in Paris to document neurological cases, on the illustrations, most of which Richer drew after Londe's photographs. Richer also included a few plates of photographs at the end of the book to provide the reader "irrefutable proof of some new assertions contained in the text."[22]

In light of these examples, the action taking place in Johan Zoffany's painting of William Hunter lecturing at the Royal Academy of Arts becomes clearer (see fig. 5.6). A living model stands on a raised platform next to a plaster cast of a flayed body, one that the Scottish anatomist and physician had prepared for the St Martin's Lane Academy around 1750.[23] The model's right arm is raised, like the cast; his right hand is held by an assistant. Although the assistant's action has been described as helping the model retain his pose,[24] ropes and poles were commonly used as aids to avoid fatigue in life models in this period (fig. 6.10).[25] It is more likely that the assistant's hand provided a fixed point of resistance against which the model could push or pull so as to elicit a specific muscle contraction in the back. Hunter places his hand on the model's back to point out the resulting change in the appearance of the muscles under the skin and in the position of the scapula.[26]

Early anatomy books for artists alert the reader that drawing from a posed model could also lead to errors that a knowledge of the muscles could ameliorate. In the *Abregé d'anatomie* (1668), de Piles drew attention to the fact that when held in a pose for a long time, muscles fatigue and do not retain the same shape:

> It would be good if the muscles did not tire, and if the model could hold himself in the same attitude with the same vigor for a long time as when he first struck a pose; but hardly

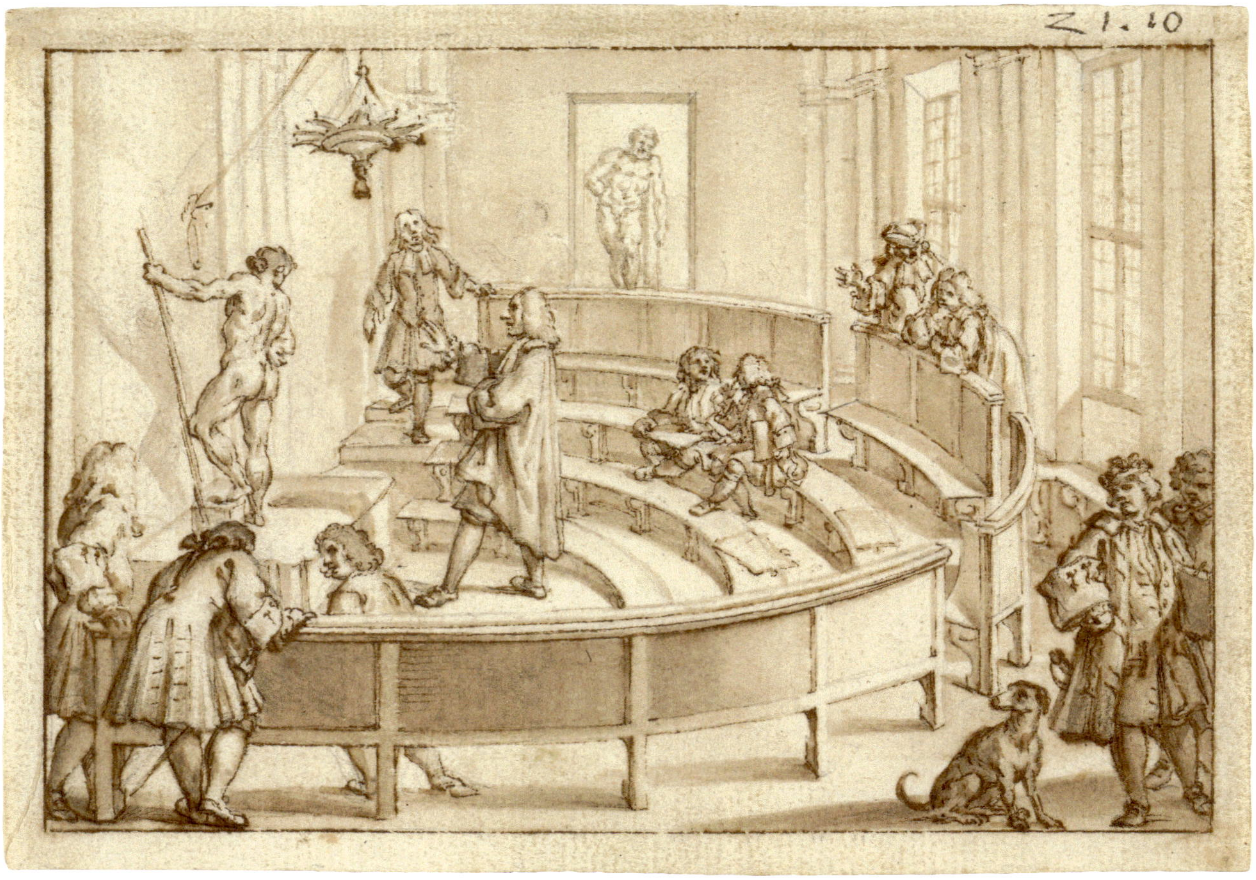

FIG. 6.10
GIAMPIETRO ZANOTTI (Italian, 1674–1765). *The Accademia Clementina Bologna, with a Nude Being Positioned by the Drawing Master, the Farnese Hercules Beyond*, 1739, drawing, pen and brown ink over black chalk; the outlines indented for transfer, 10.3 × 14.6 cm. Los Angeles, J. Paul Getty Museum, 96.GB.314.

> have you sketched your figure, when the model is compelled to seek the aid of a stick or a rope to continue his action, which then only shows loose muscles where they ought to be strongly pronounced. And this is what makes weighty mistakes for he who draws without knowledge of what he sees; so that most often he will make a flat contour where it must be swollen, and another swollen where it must be flat.[27]

The Dutch artist Samuel van Hoogstraten (1627–78) gives the same caution in his treatise on art of 1678 (cat. no. 5), observing that those with an understanding of muscles in motion will "have the judgment to reject wrong actions."[28] Anatomists with an interest in art were also aware of the problem. William Cowper (1666/67–1710), in his description of the appearance of the muscles on the surface of the body in *The Anatomy of Humane Bodies* (1698), remarks that "the Muscles fall, and Parts loose their necessary Appearance in Action, tho' the Posture is the same."[29] Hunter saw the solution in intense life study rather than dissection. In lecture notes for the Royal Academy, Hunter suggests that artists make "a general sketch of the figure" and then observe a figure in the same actions "a hundred times and a hundred times over: and the artist by thus confining his attention to one small part after another, will catch the expression of spirit and energy in each."[30] Hunter recognized, just as Raphael and artists to follow did, that observation of the body in motion and in life could not be separated from the study of its internal structure that could only be examined in death.

Notes

1 Vasari 1568, pt. 3, 1:84–85. For anatomical drawings attributed to Raphael, see Kornell in Cazort, Kornell, and Roberts 1996, 110–13, nos. 6, 7; and Laurenza 2012, 17–19. For Gian Paolo Lomazzo's advice to take the "example of dead and living bodies" when studying anatomy, see his *Idea del tempio della pittura* (Milan: Paolo Gottardo Ponto, 1590), 37.

2 Leonardo 1965, 136.

3 Hoyle and Miedema 1996, 247–48, 256.

4 Arfe y Villafañe 1585–87, bk. 2, fols. 25v–26r.

5 Cellini 1979, 1938–39. For a discussion and transcription of Cellini's manuscript in the Biblioteca Nazionale Centrale di Firenze, Autografi Palatini, Cellini, 19, see Kornell 1993, 94–107, and appendix B. On the discourse, see also Reilly 2004 and Nanobashvili 2018, 42–44.

6 For the Pichot drawing and one by Georges Seurat (1859–91) of the same slim model, see Hendrix 2015, 95, 104–5, nos. 59, 60.

7 Tortebat and de Piles 1698, address to reader ("Au lecteur"), [iv] v.

8 Pliny 1968, bk. 34, ch. 19, no. 59; bk. 35, ch. 34, no. 57.

9 Vasari 1568, pt. 2, 1:467.

10 Baldinucci 1974–75, 7:41; and Margócsy, Somos, and Joffe 2018, 193, no. I/108.

11 Allori's *Delle regole del disegno* has been most recently and comprehensively studied by Nino Nanobashvili, who dates the drafts for it to 1565–90 (Nanobashvili 2018, 21–80). See also Reilly 2004, 31–34.

12 Biblioteca Nazionale Centrale di Firenze, Fondo Palatino E.B.16.4, fol. 28r: "Dovendosi hor a trattar di far la mano, che apparisca con la pelle mi occorrono molti havvertimenti, i quali giudico necessarij e questi sono molto più importanti nella mano, che forse in alcun altra parte del corpo nostro, sendo che in detta parte le vene a l'arterie molto piu si veggano che altrove, e massime se tenghiamo la mano spenzolata, che si veggano ingrossar le vene che paiano gonfiate, a però bisogna molto havvertire di non metterle in opera così tutti, ma ben certe principali come si vede haver fatto M. Michelagnolo Buonarruoti, il cavalier Bandinelli e altri famosissimi e scultori e pittori."

13 Attributed to Bronzino by Goldner in Goldner 1990; and Goldner in Bambach, Cox-Rearick, and Goldner 2010, 188–89, no. 46. For reservations concerning the attribution, see Béguin 1996, 144–45; and Nicholas Turner and Carol Plazzotta, who connect the hand to the statue of Giuliano de' Medici, in Turner, Hendrix, and Plazzotta 1997, 18–20, no. 8.

14 Paris, Musée du Louvre, inv. nos. 4554r, 4555r. For drawings by sixteenth-century artists of Giuliano's hand, see Rosenberg 2000, NZ 322, NZ 325, NZ 329, NZ 330, NZ 331, NZ 333, NZ 334; and Lazzaro 2016, 15n64. A cast of the hand appears in reverse on a high shelf in Alberti's *Academia de' pitori* (see fig. 3.3). For more on casts of body parts from the chapel, see Lazzaro 2016, 15–16.

15 Rosenberg 2000, 135.

16 Vesalius 1543a, 171.

17 Conférence, Académie de France, 3 September 1672. See Lichtenstein and Michel 2006, 490.

18 Knox 1852, 10–11. On Haydon's school, see Cummings 1963.

19 Knox 1852, 74–75 (emphasis in original).

20 Brownell 1879, 747.

21 On Richer, see Comar and Barnard in Comar 2008b, 446–47, nos. 380–81; Ruiz-Gomez 2017; and Callen 2018. On Gerdy 1829 and Fau 1845, see Choulant 1945, 347, 349; Kornell 1996, 64, 66–67; Comar in Comar 2008b, 260–61, no. 117; and Fend 2017, 216–29.

22 Richer 1895, 3.

23 For a discussion of the casting of the figure, see Kornell, "'As Large as Nature,'" this volume, pp. 55–56.

24 Webster 2011, 397.

25 For the use of a rope by a life model in the Royal Academy of Arts depicted in another painting by Zoffany, see Webster 2011, 252, fig. 204. For French eighteenth-century examples of models posing with ropes and poles, see Desmas et al. 2016, p. 145, figs. 60, 61; p. 147, no. 59; and p. 153, no. 64.

26 For a similar demonstration, see Lockhart 1974, figs. 54–56.

27 Tortebat and de Piles 1668, address to reader, [iij]r. This issue was also briefly touched upon earlier by Jacob van der Gracht (Van der Gracht 1634, [A2]).

28 Van Hoogstraten 1678, 56; and Van Hoogstraten 2021, 109. For a similar comment taking into account this effect, see Walker 1787, 6.

29 Cowper 1698, explanation to pl. 3.

30 Kemp 1975, 44.

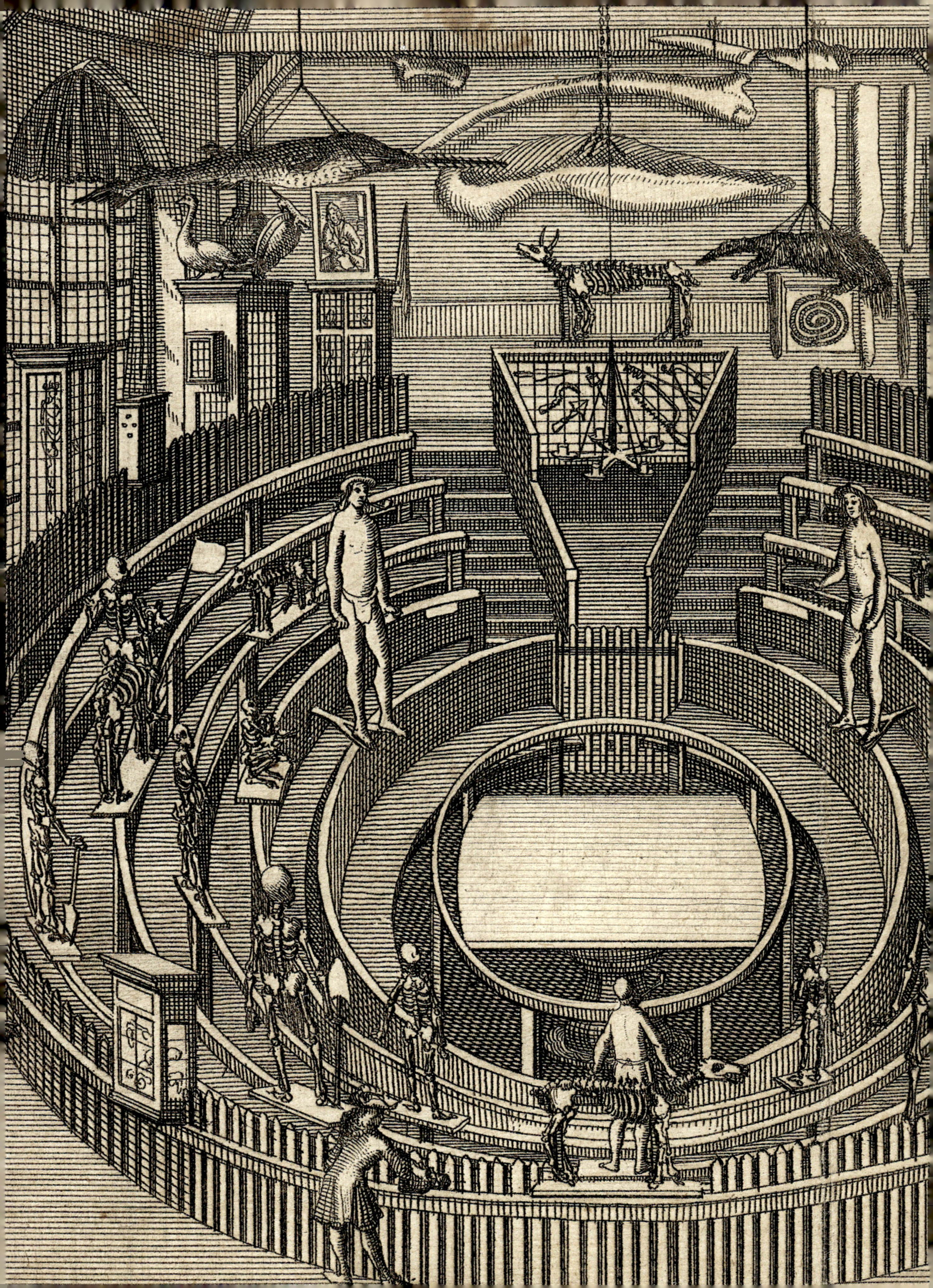

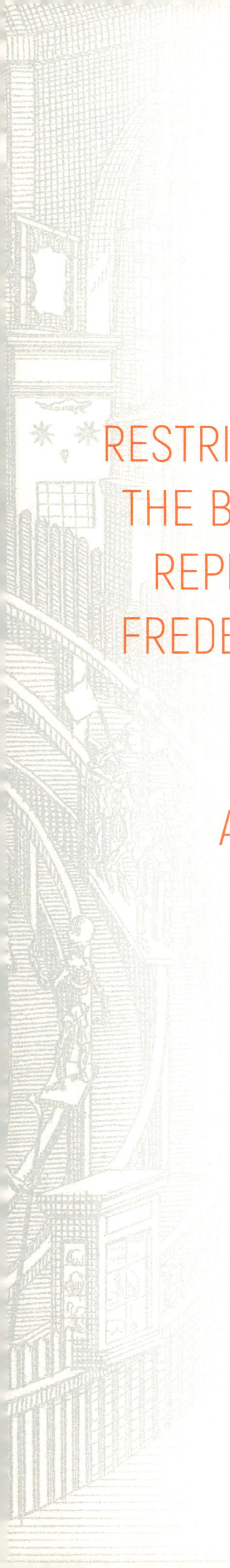

chapter seven

RESTRICTED ACCESS: THE BODY, SEX, AND REPRODUCTION IN FREDERIK RUYSCH'S ANATOMICAL COLLECTION AND CATALOGS

Erin Travers

Access to bodies, both dead and alive, was regulated in the early modern Netherlands through social, moral, and civic codes. In spaces of corporeal inspection and display, such as anatomical theaters and collections, specified parameters were put in place according to practical considerations and official decree. For example, dissections could only be held during the winter months, at which time it was cold enough to stave off decay. Statutes for the states of Holland and West-Friesland directed the remains of executed criminals to the Universiteit Leiden anatomy theater for dissection and dictated that those selected should not have lived within the city of their dissection, be without close relations, and be of average age and appearance.[1] However, the acquisition of a suitable body was often met with challenges. Repeated requests for the enforcement of this decree and complaints from several professors of anatomy indicate that annual dissections were often prevented or delayed due to a shortage of subjects.[2] Moreover, in the seventeenth century, the anatomy theater increasingly accommodated a nonspecialist audience, resulting in demonstrations that focused on a more generalized, philosophical, and moralizing discussion of the body (cat. no. 7).[3] In response, medical students and professors sought alternative methods for studying anatomy that allowed for greater regularity and satisfied their particularized interests, including pathological and nonnormative examples.

One solution that gained considerable traction and attention in the seventeenth century was the development of techniques for preserving the body that would enable repeat and comparative study not reliant upon traditional dissections.[4] Dry preparation methods had been in use since antiquity, but they were inflexible and did not replicate the color or shape of living tissue to a satisfactory degree. In the mid-seventeenth century, anatomists began to experiment with wet preparation techniques that entailed treating the specimen's tissues with a solution of high-proof alcohol and spices, injecting them with colored wax to replicate a natural flush, and, finally, containing the specimen and its preparation solution in glass vials.[5] The ability to perform this feat at Leiden in the 1660s was the product of several simultaneous developments among the students of Johannes van Horne (1621–70) and Franciscus Sylvius de le Boë (1614–72). Reinier de Graaf (1641–73) invented the clyster-pipe syringe, which enabled direct injection into the vessels, while Jan Swammerdam (1637–80) was the first to use this instrument to inject a prepared organ with warm wax, which became fixed upon hardening.[6]

These students also experimented with solutions of balsam, turpentine, and spirits, and a combination of these techniques eventually matured into the famous preparation method of the Dutch anatomist Frederik Ruysch (1638–1731) (cat. no. 6).[7] Today, several of Ruysch's preparations still survive and are housed in the anatomical collections at the universities of Leiden, Utrecht, and Groningen in the Netherlands and the Kunstkamera in Saint Petersburg, Russia (fig. 7.1). While these works may strike modern audiences as macabre, in the seventeenth and eighteenth centuries, these objects were praised for their lifelike appearance. Upon visiting Ruysch's collection, Peter the Great (1672–1725), the czar of Russia, is even reputed to have bent and kissed the cheek of a preserved child, who appeared to be merely sleeping.[8] To achieve this effect, Ruysch relied on colored pigments, fabrics, and the arrangement of his specimens, using several techniques developed by artists to enhance the veracity of his preparations. This approach was modified for the images included in his publications, which make use of pictorial devices to both elicit the environment of the cabinet and inform viewers' interpretations of these objects.

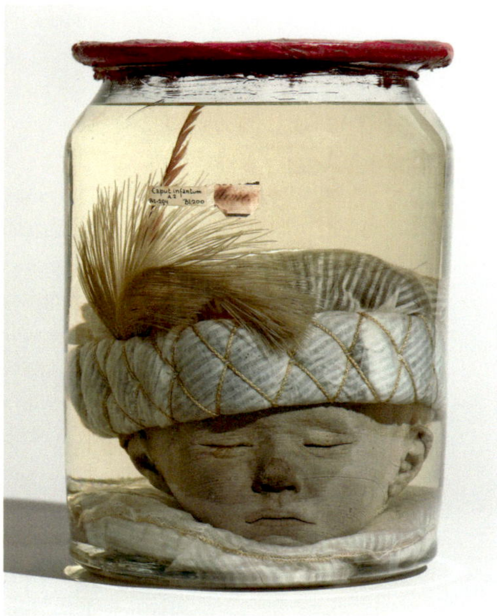

FIG. 7.1
Medical specimen prepared by Frederik Ruysch, ca. 1720, child's head, textile cap, and feathers. Utrecht, Museum Bleulandium, University Medical Centre.

Following his medical studies at Universiteit Leiden, Ruysch obtained a number of prominent positions in Amsterdam that facilitated his independent research, including praelector for the surgeon's guild (1667), municipal obstetrician (1672), and forensic physician (1672) (fig. 7.2).[9] These posts supplied Ruysch with materials for his preparations, which led to the creation of an anatomical collection that became the source of his most enduring fame. By 1673, Ruysch had constructed a purpose-built anatomical cabinet in Amsterdam, and by 1717 his collection had grown to more than 1125 preparations, which were displayed in fourteen cabinets arranged over five rooms in his home on the Bloemgracht.[10] To disseminate information about the cabinet and promote its contents, Ruysch published a series of ten catalogs, the *Thesauri anatomici* (1701–16), complete with more than forty images of his specimens.[11] The genre of the illustrated collection catalog was invented in the mid-sixteenth century and was particularly popular among natural historians and apothecaries.[12] Before beginning his training in Leiden, Ruysch owned an apothecary shop in The Hague, and his collection also encompassed a significant number of natural history subjects. Through this type of publication, medical professionals could communicate

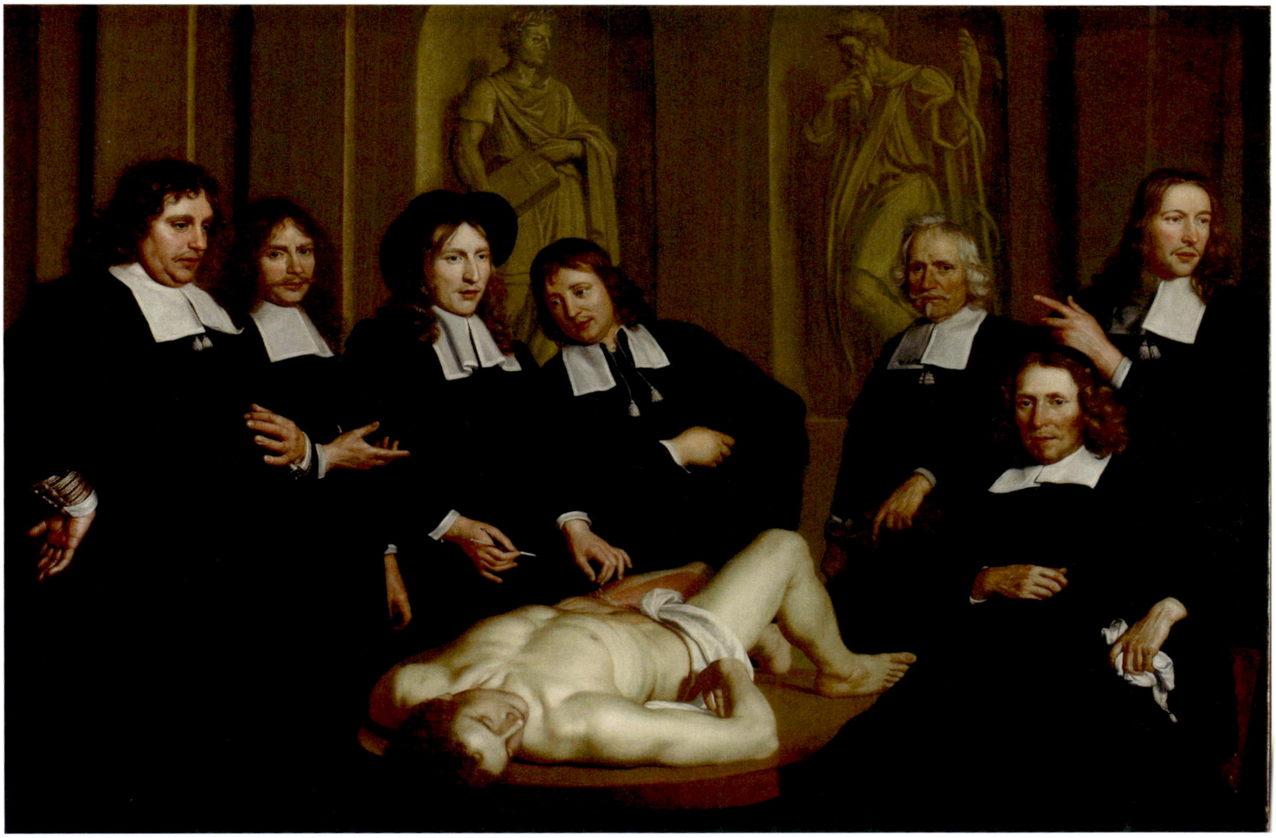

FIG. 7.2
ADRIAEN BACKER
(Dutch, 1635/36–1684).
Anatomy Lesson of Dr. Frederik Ruysch, 1670,
oil on canvas.
Amsterdam, Amsterdam Museum, SA 2000.

the contents of their collections with others and demonstrate the scope of their holdings in comparison to famed collections, a strategy that Ruysch used to his advantage.

Working through each cabinet as a contained entity, the cataloging of Ruysch's collection was interrupted in 1717 by its sale to Peter the Great for a sum of thirty thousand guilders. This event, together with Ruysch's reorganization of his cabinets while writing his *Thesauri*, resulted in the cataloging of only part of the collection in its entirety, though the *Thesauri* still provide an impressive amount of information for modern scholars. Breaking each cabinet into manageable parts, each *Thesaurus* progressed from the lowest shelf to the uppermost shelf, and the preparations were numbered to create a sense of order. As a result, Ruysch offered a tidy reference to his collection that could be surveyed quickly to gain familiarity with the objects in the anatomist's possession. Alternatively, for those unable to come to his home in Amsterdam, the catalogs supplied a virtual means of entrance.[13]

Today, the catalogs offer insight into the collection's contents, methods of display, and organization, including the ways in which these elements changed over the course of Ruysch's career.[14] However, we should be cautious in too closely conflating the collection and its catalogs, as Ruysch clearly regarded his printed works as functioning in service to his preparations. Moreover, in contrast to prepared specimens, works on paper were replicable and portable, which expanded Ruysch's audience. Whereas Ruysch could determine who entered his home and could mediate his visitors' interpretation of his specimens through selection, discussion, and display, the circulation of the catalog invariably reached a broader audience beyond the anatomist's direct control. To compensate, Ruysch and his artists made use of textual and pictorial strategies that guided readers' interpretations of the catalogs' contents.

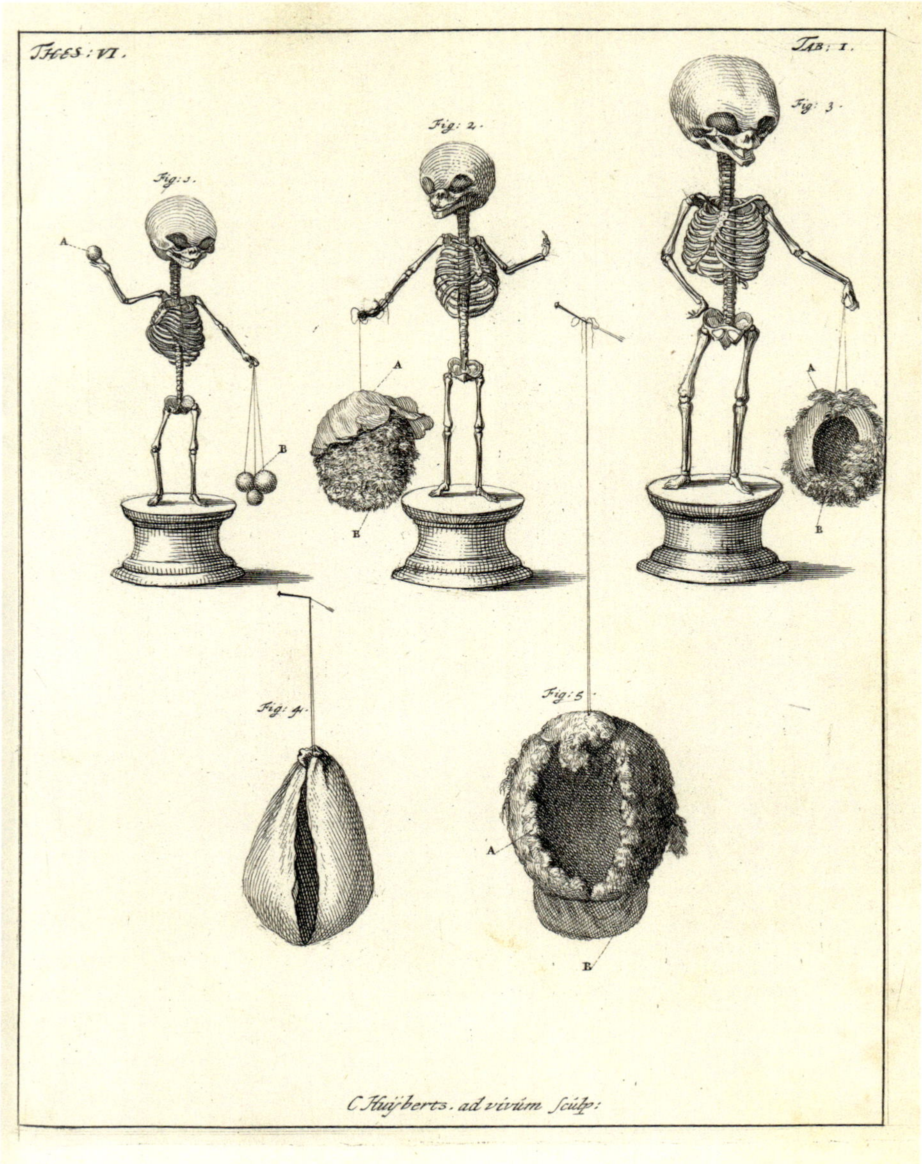

FIG. 7.3
CORNELIS HUYBERTS
(Dutch, 1669/70–ca. 1712).
Three fetal skeletons at progressive stages of growth; each holds an ovum, demonstrating the unfertilized to fertilized state; below are two specimens of the placenta from early pregnancy, etching and engraving, 1705.
From Frederik Ruysch, *Opera omnia anatomico-medico-chirurgica* (Amsterdam: Johannes Wolters, 1721), vol. 2, *Thesaurus VI*, pl. 1.
Los Angeles, Getty Research Institute, 94-B12267.

The names of five draftsmen and printmakers are included among Ruysch's plates. The majority of his images are the work of Cornelis Huyberts, whom Ruysch employed for his first eight catalogs published between 1701 and 1709. A student of Gérard de Lairesse (cat. no. 40), Huyberts made a name for himself as a draftsman and printmaker, working as an etcher and engraver who specialized in figural subjects, particularly allegorical and historical scenes.[15] Similar areas of specialization characterize the oeuvres of two other artists associated with the catalogs, Abraham de Blois (1655–1717) and Joseph Mulder (1658–1718/38), but these three artists do not seem to have had any other engagement with anatomical subjects beyond their commissions from Ruysch.[16] In contrast, Ruysch's later artists, Jacob Folkema (1692–1767) and Jan Wandelaar, had more experience working with medical professionals, and Wandelaar later became famous for his anatomical prints, produced for the physician Bernhard Siegfried Albinus (cat. no. 15).

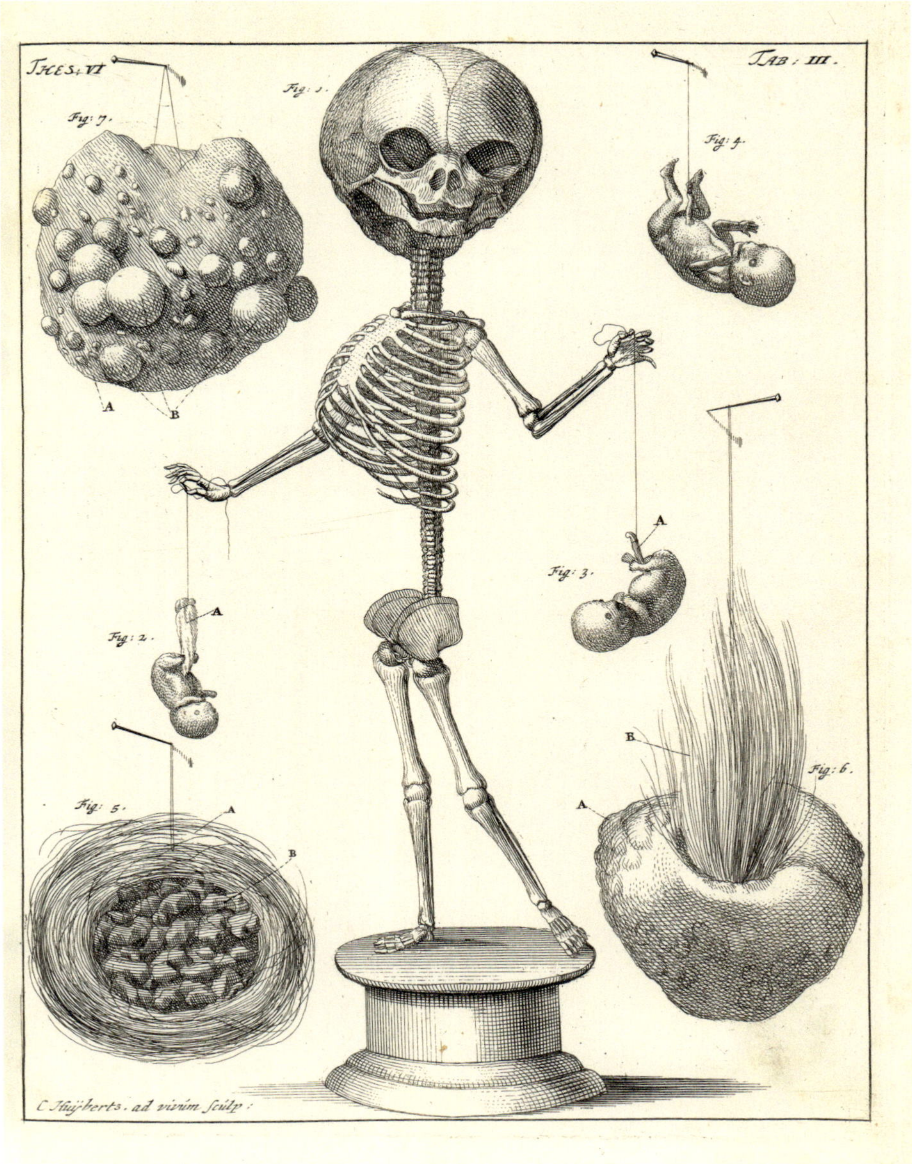

FIG. 7.4
CORNELIS HUYBERTS (Dutch, 1669/70–ca. 1712). The skeleton of a four-month-old child surrounded by specimens of embryos, fetuses, and afterbirth; figures 5 and 6 feature neck tumors, 1705, etching and engraving. From Frederik Ruysch, *Opera omnia anatomico-medico-chirurgica* (Amsterdam: Johannes Wolters, 1721), vol. 2, *Thesaurus VI*, pl. 3. Los Angeles, Getty Research Institute, 94-B12267.

Drawing on their expertise as book illustrators, Ruysch's artists permitted his readers to peek inside his cabinets remotely and view his prepared specimens through their printed plates. In the environment of the collection, objects could be arranged or shown together to encourage comparison and create a visual argument. Ruysch's lists of his cabinets' contents often indicate the location of specimens of similar subjects on the same shelf, which suggests that visitors could investigate multiple samples at once, a practice that is replicated in the printed images. At times, the same engravings even feature preparations that Ruysch's catalogs list as sitting side by side on the shelf. The anatomist's images document the relationship between preparations and make apparent their function as educative tools.

For example, Ruysch's sixth *Thesaurus* focuses on the subject of generation and includes five plates that address this process (figs. 7.3, 7.4). The first three prints include ova in states ranging from unfertilized to embryonic. They are suspended from nails that seem to pierce the page or are held aloft by fetal skeletons. These figures demonstrate different stages of growth as one moves progressively through the illustrations, culminating in a fetus at three months of development. Depictions of the placenta share the plates and also advance in size and development over the consecutive prints.[17] In the fourth and fifth plates, the placenta is shown alongside the uterus, fallopian tubes, and ovaries, including a specimen taken from a woman who died just after intercourse, resulting in the preservation of sperm within the uterus (fig. 7.5). This series of images allows the viewer to follow the development of a human embryo from conception to fetus and trace the growth of its skeleton in the months after birth. A similar type of activity likely took place during the lessons Ruysch conducted for medical students within his collection.[18]

In comparing these examples, we can see that Ruysch's artists present the anatomist's preparations in a range of formats and guises that at once invite the viewer into the space of the figurative collection and mark Ruysch's

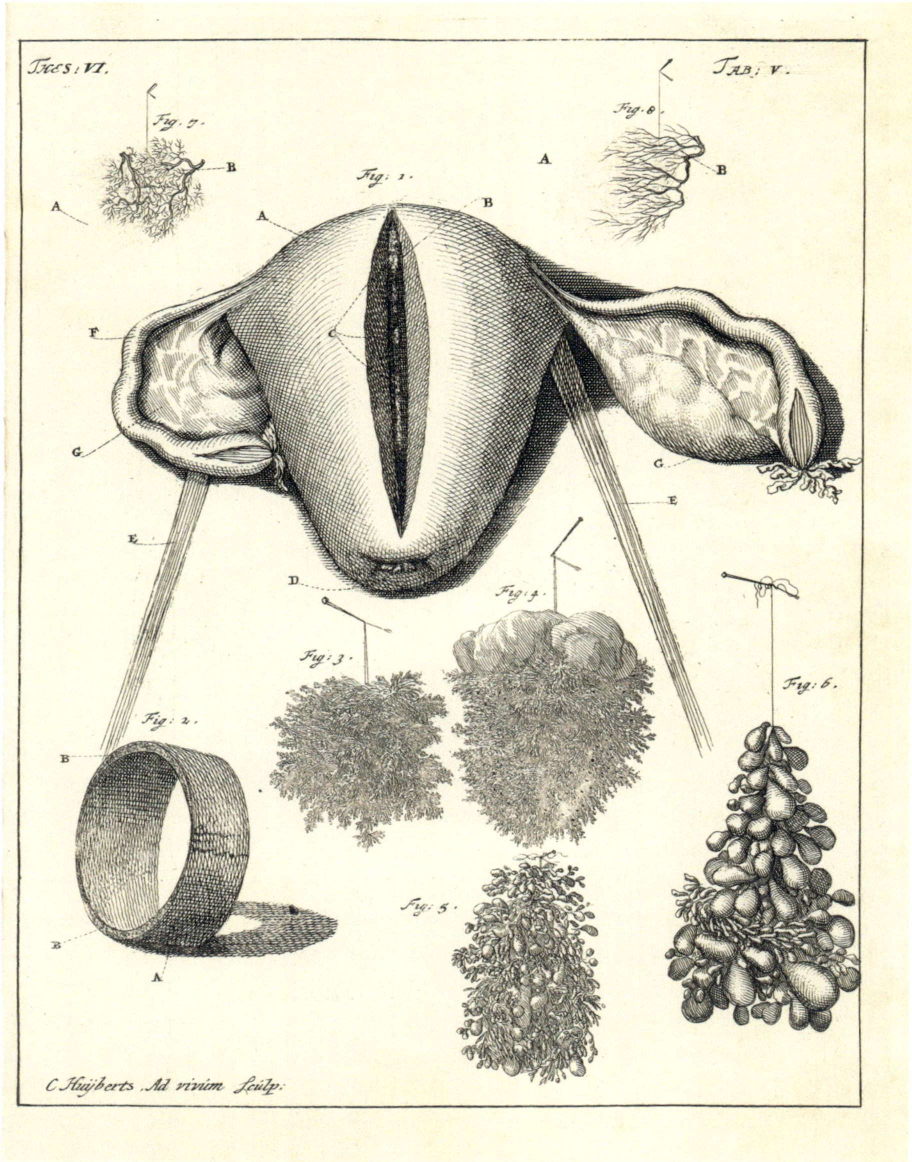

FIG. 7.5
CORNELIS HUYBERTS
(Dutch, 1669/70–ca. 1712).
The uterus of a woman who died shortly after intercourse, preserving semen in her body; this is surrounded by specimens of the intestine, placenta, and afterbirth, 1705, etching and engraving.
From Frederik Ruysch, *Opera omnia anatomico-medico-chirurgica* (Amsterdam: Johannes Wolters, 1721), vol. 2, *Thesaurus VI*, pl. 5.
Los Angeles, Getty Research Institute, 94-B12267.

catalogs as distinct from the cabinet. The fetal skeletons are shown on small pedestals, which cast a shadow as though they are sitting on the shelf within the cabinet. If these specimens were alone on the page, the viewer could more easily imagine a receding space and locate them in some kind of fictive three-dimensional environment. However, the presentation of additional specimens that hang from nails complicates our interpretation of Ruysch's images.[19] In contrast to the pedestals, which are situated in space, the nails appear to puncture the page and locate the dangling specimens on a plane that is closer to the viewer than the skeletons.[20] This device is found throughout Ruysch's *Thesauri* and likely references the hanging of specimens within vials or inside the doors of Ruysch's cabinets, as both methods of display are noted in the anatomist's catalogs.[21] These types of features evoke one's experience within the cabinet, while others were designed to curtail the viewer's access.

Reproduction was a subject of mounting interest in a century that saw the theorization of ova and the discovery of spermatozoa, but, given period codes of decorum, Ruysch was careful to mediate his viewers' engagement with this subject.[22] For example, the author uses language to restrict his readers' access to the reproductive organs in the original editions of his catalogs, which were printed with two columns—one in Latin, the other in Dutch. In some cases, entries that address this subject were printed exclusively in Latin and marked with an asterisk, which have been interpreted as signs that the female reader should skip these entries. The use of Latin throughout Ruysch's publications, however, was far more restrictive and excluded any reader, male or female, who could not read the language, leaving the author's identification and discussion of these objects a mystery to the less educated.[23] Even among learned colleagues, the subject of reproduction was treated with caution. In a letter written in Latin to the Royal Society, Antonie van Leeuwenhoek (1632–1723) disclosed the results of inspecting semen under a microscope but noted that if the Society found this study "either disgusting, or likely to seem offensive to the learned, I earnestly beg that [it] be regarded private and either published or suppressed as your Lordship's judgement dictates."[24]

A comparable concern with the illustration of the reproductive organs is found in the preface to Reinier de Graaf's *Alle de Wercken so in de Ontleed-kunde als andere deelen der medicyne* (1686). A friend and colleague of Ruysch, and a leading researcher in the study of generation, De Graaf writes,

> But here someone of a weaker mind or filthiness shall easily say, that if you had nothing other than such enticing and lust-provoking content, which at this time all too often possesses and dominates the hearts of youth, you could have kept this with you at home, and in this way not encourage the easily erring youth. To that I answered that firstly, if young and wanton people were inclined that way, they would be able to find other books of such content to satisfy their desire, whether in French, or whether, [one] says with regret, in our mother tongue, so that they will have no need of this [book].[25]

De Graaf acknowledges that some may take issue with the subject of his publication and its illustrations, but he dismisses these concerns as baseless given that more licentious material can be found without much difficulty, rendering his images relatively tame. Nevertheless, his compulsion to make such a statement points to period anxieties. Ruysch takes a different approach, and his use of Latin to discuss the reproductive organs ensures that the audience for his catalogs was prescribed by education.[26]

Generation was a particular area of interest for Ruysch, and the number of specimens in his collection and publications that feature the male and female reproductive organs makes evident his specialization.[27] It is notable that of the approximately one hundred specimens listed that address generation, only thirty are marked and explained exclusively in Latin.[28] In her article on the display of the anatomized female body in eighteenth-century Amsterdam,

Rina Knoeff identifies distinctions made within Ruysch's cabinet and finds that specimens of the pudenda, for example, were treated more carefully than the womb, which was connected more strongly to childbirth than conception.[29] Diseased and prepubescent specimens were also more visible in the collection and are rarely marked or restricted in Ruysch's catalogs, signaling that they did not carry the moral implications of healthy adult reproductive organs.[30] This could explain Ruysch's choice to include in his third catalog an illustration of a male infant's penis and testicles, in which the urinary tract is in an unnatural position (see fig. 1 in fig. 7.6). Several of the preparations Ruysch selected for illustration are marked by physiological irregularity, and in this case the deviation also seems to excuse Ruysch's representation of this subject.

Within the *Thesauri,* there are only two examples that depict "restricted" specimens, both found in the second plate of the eighth cabinet—specifically, a prolapsed uterus (see fig. 3A in fig. 7.7) and the intersex genitalia of a

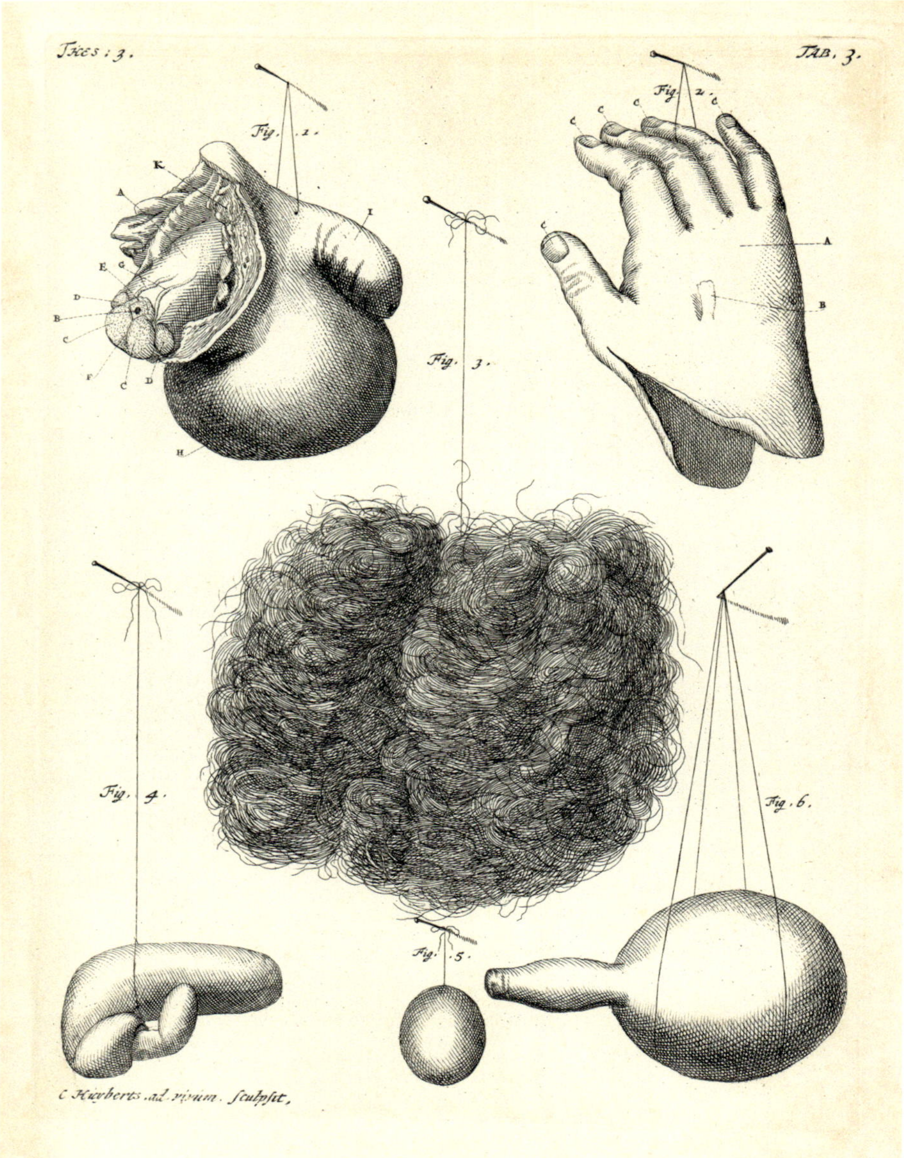

FIG. 7.6
CORNELIS HUYBERTS
(Dutch, 1669/70–ca. 1712).
The penis and testicles of a young boy, the skin from the hand of a young boy, a bundle of hair, and three chicken eggs, 1703, etching and engraving.
From Frederik Ruysch, *Opera omnia anatomico-medico-chirurgica* (Amsterdam: Johannes Wolters, 1721), vol. 2, *Thesaurus III*, pl. 3.
Los Angeles, Getty Research Institute, 94-B12267.

sheep (see fig. 5 in fig. 7.7). Notably, the written description of the uterus is provided in both Latin and Dutch, whereas the register for the printed illustration of this specimen is designated with an asterisk and appears only in Latin. In contrast, the sheep genitalia are discussed exclusively in Latin and consistently marked with an asterisk. As the only two marked entries in Ruysch's *Thesauri* to be illustrated, these particular specimens invite speculation concerning the reasons for their representation. In his *Anatomische en chirurgicale observatien* (1691), Ruysch discusses prolapsed uteruses several times with the assistance of images and, in his eighth *Thesaurus,* even directs his reader to this earlier work. Therefore, it does not seem that this subject was off-limits. The only element that distinguishes this image from other unrestricted representations of the uterus in Ruysch's catalogs is the inclusion of the external genitalia (see fig. 3A in fig. 7.7). However, the relatively schematic rendering of this subject, which is not easily interpreted without the accompanying labels,

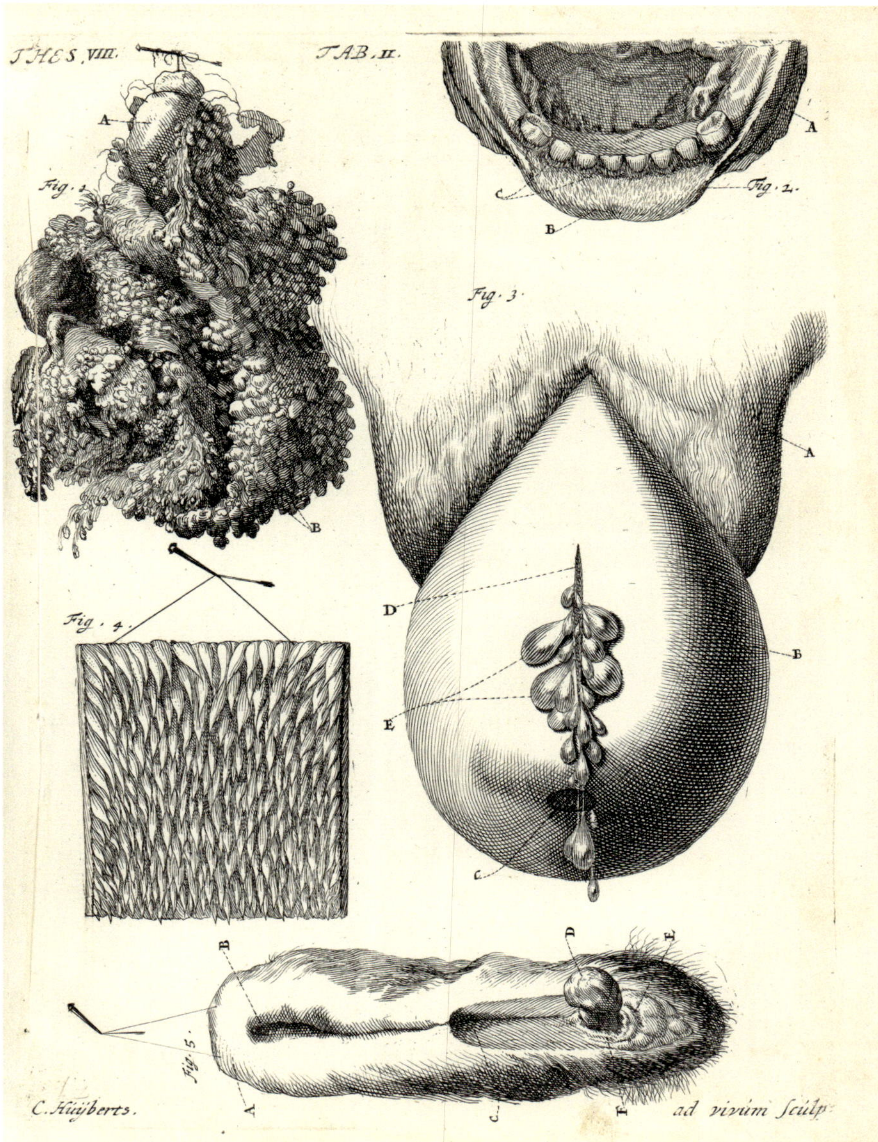

FIG. 7.7
CORNELIS HUYBERTS (Dutch, 1669/70–ca. 1712). Tissue from a uterus, a lower jaw, a prolapsed uterus, the esophagus of a turtle, and the intersexed genitalia of a sheep, 1709, etching and engraving. From Frederik Ruysch, *Opera omnia anatomico-medico-chirurgica* (Amsterdam: Johannes Wolters, 1721), vol. 3, *Thesaurus VIII,* pl. 2. Los Angeles, Getty Research Institute, 94-B12267.

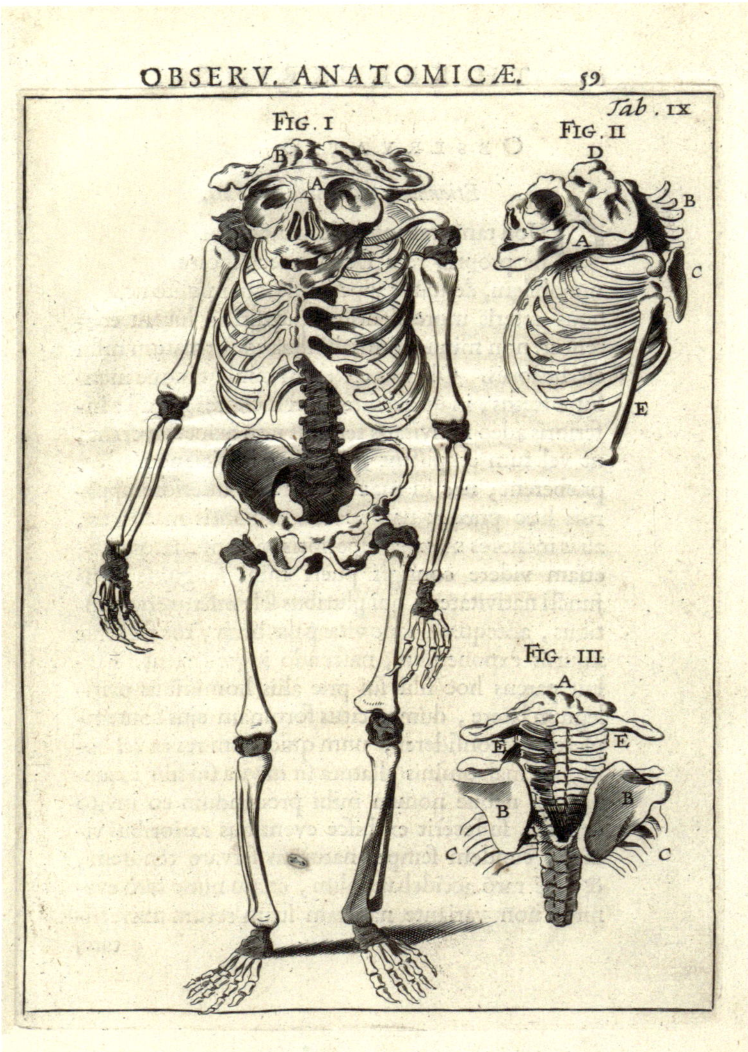

FIG. 7.8
ABRAHAM BLOOTELING
(Dutch, 1640–90).
The skeleton of an infant born without a brain, 1670, etching and engraving.
From Theodor Kerckring, *Spicilegium anatomicum* (Amsterdam: Sumptibus Andreae Frisii, 1670), 59, pl. 9.
Los Angeles, Getty Research Institute, 2017-B10.

prevents the uninformed reader from truly understanding the depicted specimen.

In contrast, the case of the intersex genitalia is without precedent in Ruysch's illustrated works, but this appears to be precisely the point of its depiction in this context. In his written description of this specimen, Ruysch explains, "It has never happened to me, to see a true hermaphrodite, and I also judge, that they have never been seen by others."[31] In the context of Ruysch's publications, the male reproductive organs are shown in two separate plates among his earlier works, and the genitalia of both sexes are featured in the images of infants included in the *Thesauri*, but the mature vulva is not illustrated in Ruysch's catalog. In this particular case, the intersex genitalia's origins from a sheep may have made the specimen permissible, in contrast to human samples.

Huyberts also used the practicalities of the page to moderate viewers' engagement with this image. In contrast to other figures within Ruysch's published oeuvre, the ovine figure is the sole represented specimen in Ruysch's *Thesauri* oriented so that the image is only rectified once the viewer physically turns the book from a vertical to a horizontal position. Whereas the other figures on the page are shown hanging from nails along the vertical axis, this specimen is suspended horizontally and consequently placed in a separate space of viewing. The particular mode of depiction reserved for this specimen makes it more challenging to interpret the black-and-white figure. The oddity of its portrayal is made all the more irregular when the viewer observes that, despite its gravity-defying suspension, the nail from which it hangs casts a shadow that is commensurate with the other figures that share the page. This detail suggests a common light source, unifying the objects on the page and confirming the purposeful positioning of the specimens. Through these subtle cues, Ruysch and Huyberts identify this specimen as unique, while informing the ways in which different viewers would approach and interpret the subject.

The viewer's access is not only restricted in the case of the reproductive organs; Ruysch places general limitations on his audience throughout the *Thesauri*. In fact, the majority of the objects listed are not represented, though this may have been due to the cost of images or a strategy to bring curious readers to the collection. Those selected for illustration are often included as proof of Ruysch's superior preparation technique or the rarity of the subject itself. When applicable, particularly if the preparation was the product of another anatomist's hand, Ruysch directed his reader to suitable illustrations available in other publications. This is the case with the skeleton of an infant born without a brain that was housed in Ruysch's cabinet but initially prepared by Theodor Kerckring (1638–93) and published in his *Spicilegium anatomicum* (1670) (fig. 7.8).

In similar cases, the author's written accounts suggest that access to these types of specimens may have also been curtailed within the cabinet. For example, in his eighth *Thesaurus*, Ruysch describes a prepared specimen of an infant with limb difference, which he kept behind a wet preparation of an intestine in the cabinet itself, explaining that it was "not to be seen by everyone."[32] A print of this figure is not included in Ruysch's catalog, and representations of infants with nonnormative bodies are rare in Ruysch's collected works.[33] This example suggests that within his cabinet, Ruysch could determine which visitors saw particular preparations. In his printed catalogs, the anatomist curated the experience of the viewer, often erring on the side of caution. Once in circulation, Ruysch no longer had control over readers' engagement with and interpretation of his specimens, and therefore he employs strategies in text and image to moderate the untrained eye.

Notes

1 Kroon 1911, 134–35; and Huisman 2009, 116, 126–27.
2 Witkam 1970, 23–24, no. 29; Molhuysen 1918, 265–66; Molhuysen 1920, 29; Huisman 2009, 89–90; and Kooijmans 2011, 159.
3 Scheurleer 1975, 217–77; Rupp 1990, 270–73; Huisman 2009, 48–59; Huisman 2015, 59–61; and Klestinec 2010, 35–37, 39.
4 Beukers 1989, 139–41.
5 Cook 2002, 237–47; and Mulder and Beukers 1991, 11.
6 The earliest reference to this technique is 21 January 1667. Swammerdam 1737, C, C2.
7 Cook 2007, 280–81.
8 Hansen 1996, 273; and Knoeff 2015, 32.
9 Kooijmans 2011, 88–89.
10 Hansen 1996, 669; and Van de Roemer 2010, 169, 173.
11 Margócsy 2014, 117–24.
12 Findlen 1994, 36–44; and Ogilvie 2006, 38–45.
13 Margócsy 2014, 127.
14 Van de Roemer 2010, 172–77.
15 Thieme, Becker, and Vollmer 1925, 195; and Groenendijk 2008, 429.
16 Groenendijk 2008, 115.
17 Ruysch 1744, 714–15.
18 Margócsy 2014, 129.
19 This interpretation draws on Ernst Gombrich's discussion of negative space in relation to the depicted object and the viewer in natural history images. Gombrich 1960, 193.
20 Working with Wandelaar's image in Ruysch's letter to Boerhaave, Robert Felfe interprets the projecting pins as locating the representation of the preparation between the viewer and the pictorial ground (Felfe 2013, 258).
21 References to specimens hanging from hairs are frequent throughout the catalog. In his eighth and ninth cabinets, Ruysch notes that several preparations hang on the doors of his cabinets. Ruysch 1744, 765, 810.
22 On the study of reproduction in the seventeenth-century Netherlands, see Cobb 2006.
23 Margócsy 2014, 127.
24 Van Leeuwenhoek 1677, 1042. Quoted in Cobb 2012, 5.
25 De Graaf 1686, *3v–*4r. (Translation mine.)
26 Bloemendal 2003, 20; and Fortgens 1958, 31–32.
27 Ruysch 1744, 227–35, 418–29. The reproductive organs are discussed and pictured in greater detail in these sources, but these texts were published in Latin and were not translated until after the anatomist's death.
28 This assessment does not take into consideration fetal or embryonic specimens. My thanks to Gijsbert van de Roemer for generously sharing his database of Ruysch's cabinets, which facilitated and expedited this research.
29 Knoeff 2012, 48.
30 Knoeff 2012, 48–49.
31 Ruysch 1744, 755–56. (Translation mine.)
32 Ruysch 1744, 754. (Translation mine.)
33 Kemp 2000, 25.

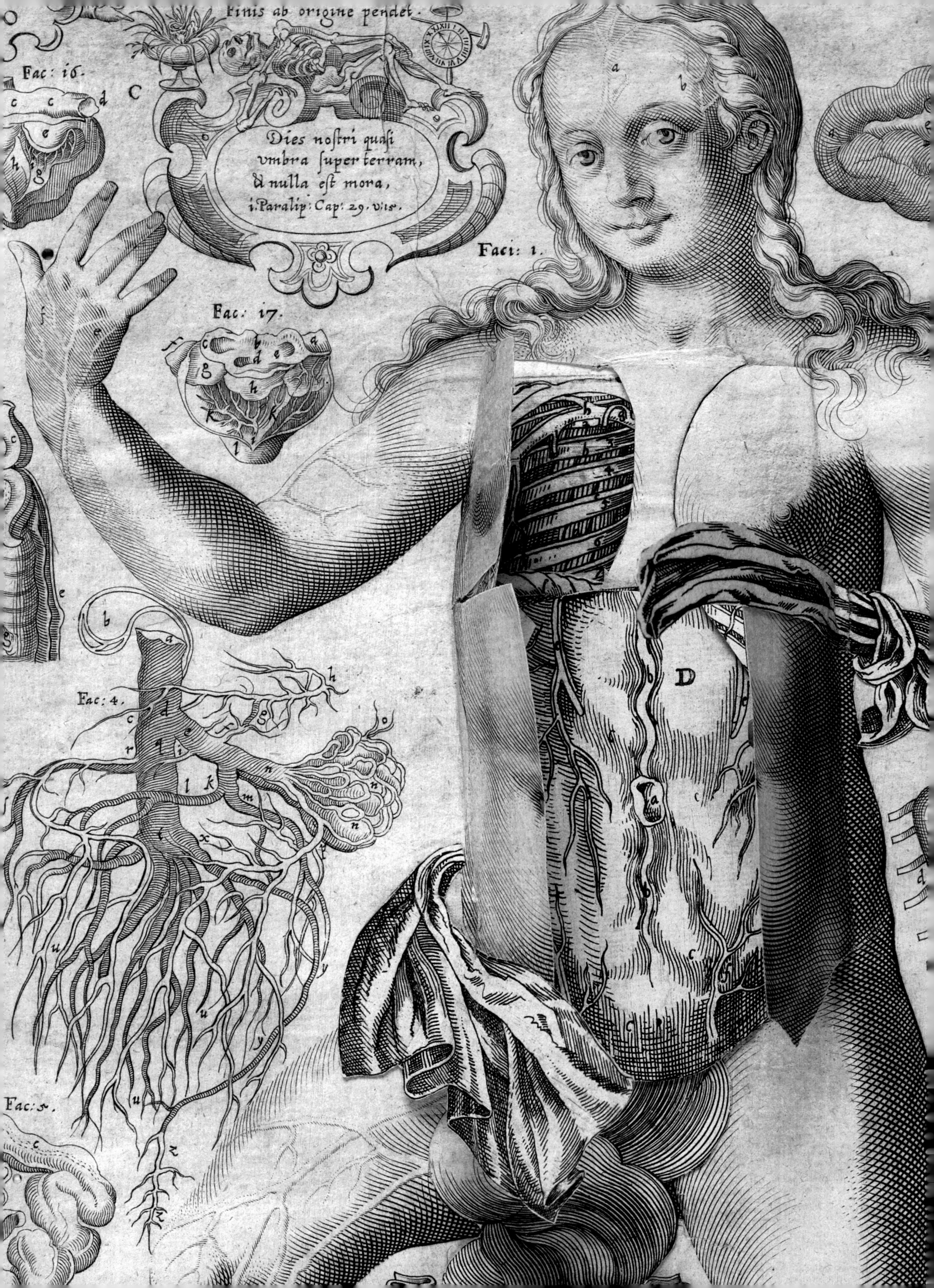
Finis ab origine pendet.
Dies nostri quasi
umbra super terram,
& nulla est mora,
1. Paralip: Cap: 29. v.15.
Fac: 16.
C
Faci: 1.
Fac: 17.
Fac: 4.
D
Fac: 5.

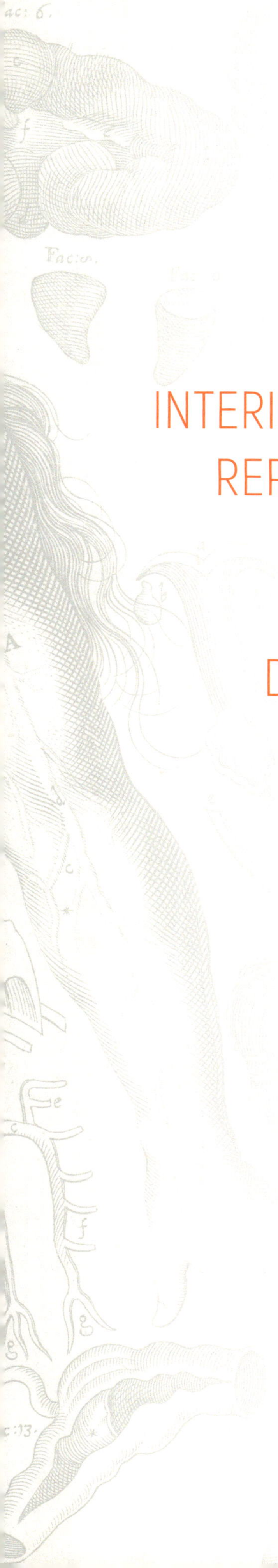

chapter eight

INTERIOR VISIONS: REPRESENTING THE BODY IN THREE DIMENSIONS

Thisbe Gensler

This my configuration of the human body will be demonstrated to you just as if you had the natural man before you. The reason is that if you want to know thoroughly the anatomical parts of man you must either turn him or your eye in order to examine him from different aspects, from below, from above, and from the sides, turning him round and investigating the origin of each part; and by such a method your knowledge of natural anatomy is satisfied.
—Leonardo da Vinci

In Leonardo da Vinci's "exploded view" of optical physiology, a brain and dangling nerves hover between two sides of a bisected skull (fig. 8.1). Schematic and figurative diagrams illustrating the cranial ventricles and the male reproductive system are illustrated on the same sheet, surrounded by expository notes in his distinctive mirrored script. These complex and manifold studies exemplify Leonardo's inventive illustrative vocabulary, as well as his astutely three-dimensional conception of the body. In his unpublished notebooks, Leonardo transcribed his observations in stark contrast to prevalent conventions of anatomical representation in the early sixteenth century. His innovative use of transparency, rotation, and topographical sectioning produced a multiperspectival vision of the body as a mechanical, vital structure that should be precisely examined from all sides and strata. Informed by his own dissections, as well as architectural drawing and engineering, Leonardo's studies articulate the corporeal density of the human form, his sketches serving as a synthetic act of knowledge acquisition.[1] Despite unfulfilled intentions to publish a treatise on anatomy, Leonardo's radical revisioning of the body could only anticipate the eventual adoption of these

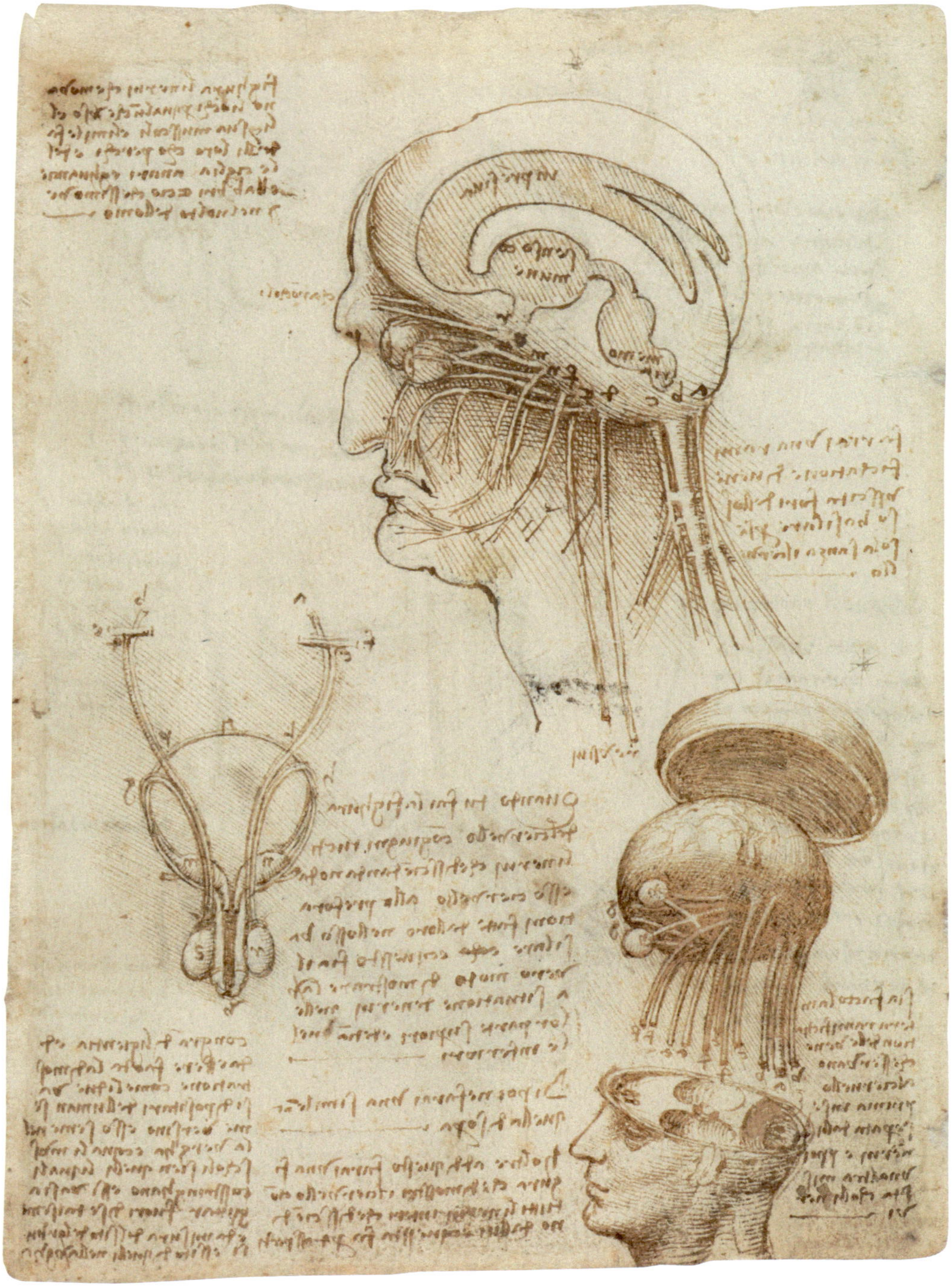

FIG. 8.1
LEONARDO DA VINCI (Italian, 1452–1519). Study of brain physiology, ca. 1508, pen and brown ink. Weimar, Kunstsammlungen zu Weimar Schlossmuseum, inv. KK 6287v.

graphic techniques, which remained unused and unknown for decades, if not centuries.[2]

In the intervening years, artists and anatomists approached the problem of three-dimensionality with a variety of visual strategies. As increasing emphasis on dissection opened up the body to observation, anatomists faced the challenges of communicating the depths of their discoveries, registering the dynamic networks of internal systems and the spatial relationships of parts, and contending with the limitations of the flat picture plane. In order to cultivate understanding of anatomy as a synthetic whole, artists had to render the body in three dimensions and did so according to the aesthetic conventions, scientific and epistemic values, medical ideologies, and technological possibilities of their times.

Artists sought to translate solid anatomy onto the two-dimensional page using a variety of graphic techniques, such as shading to round out forms or perspectival compositions to situate objects in space (cat. no. 15). Many anatomy books from the sixteenth century, beginning with those by Andreas Vesalius (cat. no. 30), employed sequential views of the body undergoing dissection, illustrating the flaying and dismemberment of écorchés and the progression to skeletal figures. At times, muscles appear to peel away from the body and fan out into space, as if to offer additional viewing angles within a single image (cat. nos. 14, 45). Some early modern anatomical figures even do so themselves, tugging at their own flesh and viscera to display the many sides of their earthly garments (cat. no. 8; see figs. 2.1, 2.9).

Printed fugitive sheets superimposed with anatomical paper flaps allowed a viewer to perform a dissection themselves, opening the body up like a book in multiple layers. So-called flap anatomies invited users to uncover the mysteries within by transposing the body's corporeal topography into flat, descriptive, and distinct planes, which could be physically manipulated to reveal concealed organs and demonstrate the physiological relationships of various parts. This tradition began as early as 1538 in Germany, where Heinrich Vogtherr and Jobst de Negker produced woodcut broadsides featuring male and female bodies, which opened at the torso to reveal rudimentary biological interiors. Printed cheaply and inscribed in vernacular, fugitive anatomy sheets also likely appealed to a popular lay audience.[3] In the *Epitome* (1543), Vesalius instructed his readers to cut out and assemble a flap anatomy figure—a three-dimensional, interactive pendant to the book.[4] As knowledge of the body advanced, flap anatomies became more scientifically and materially complex, featuring fine engraving and detailed, free-floating parts, such as Johann Remmelin's *Catoptrum microcosmicum* (1619) (cat. no. 51).

This simple, though effective, format persisted over centuries in various media and applications. Magnifying the minute anatomy of the eye, Georg Bartisch's ophthalmological manual, *Ophthalmodouleia: Das ist Augendienst* (1583), has several foldout woodcuts illustrating the anatomy of vision, alongside images of ocular disease and surgical techniques. Some three hundred years later, the French physician Gustave-Joseph Witkowski (1844–1923) used the flap method to animate the eye, updating the media in a stunning demonstration of its development over time. Advances in biological knowledge and improvements in printing technologies in the nineteenth century enabled the mass production of large-scale, colorful, and complex flap anatomies, which featured new materials and supports to accommodate their role as visual teaching aids. Witkowski's *Anatomie iconoclastique: Atlas complémentaire de tous les ouvrages traitant de l'anatomie et de la physiologie humaines, composé de planches découpées, coloriées et superposées* (1874–88) is a comprehensive atlas of anatomy in twelve fascicles, each separately devoted to a unique organ or physiological system. These elaborately hand-colored lithographic flaps activate the reader's participation in a virtual dissection through a process of progressive unveiling that traces the scalpel's passage through the body. Witkowski's section on the eye, "A Movable Atlas Showing the Mechanism of Vision" (in the English edition of 1878–88), is composed of fourteen layers made in a variety of materials—including fabric, cellulose acetate, and colored tissue papers—to differentiate the strata and imitate ocular matter (fig. 8.2). The flaps are printed on both sides in bold colors and open in various directions. Some sections even contain smaller internal foldouts or employ interlocking pieces so that the layers slot into one another and pop up, as if to defy the two-dimensionality of their paper supports.

In their kinesthetic simulation of a dissection, flap anatomies were useful pedagogical tools when cadavers were unavailable, whether due to scarcity, illegality, lack of refrigeration, or an absence of desire due to disgust or a sense of immorality. Likewise, in the eighteenth century, highly realistic wax models promised to replace the use of corpses by

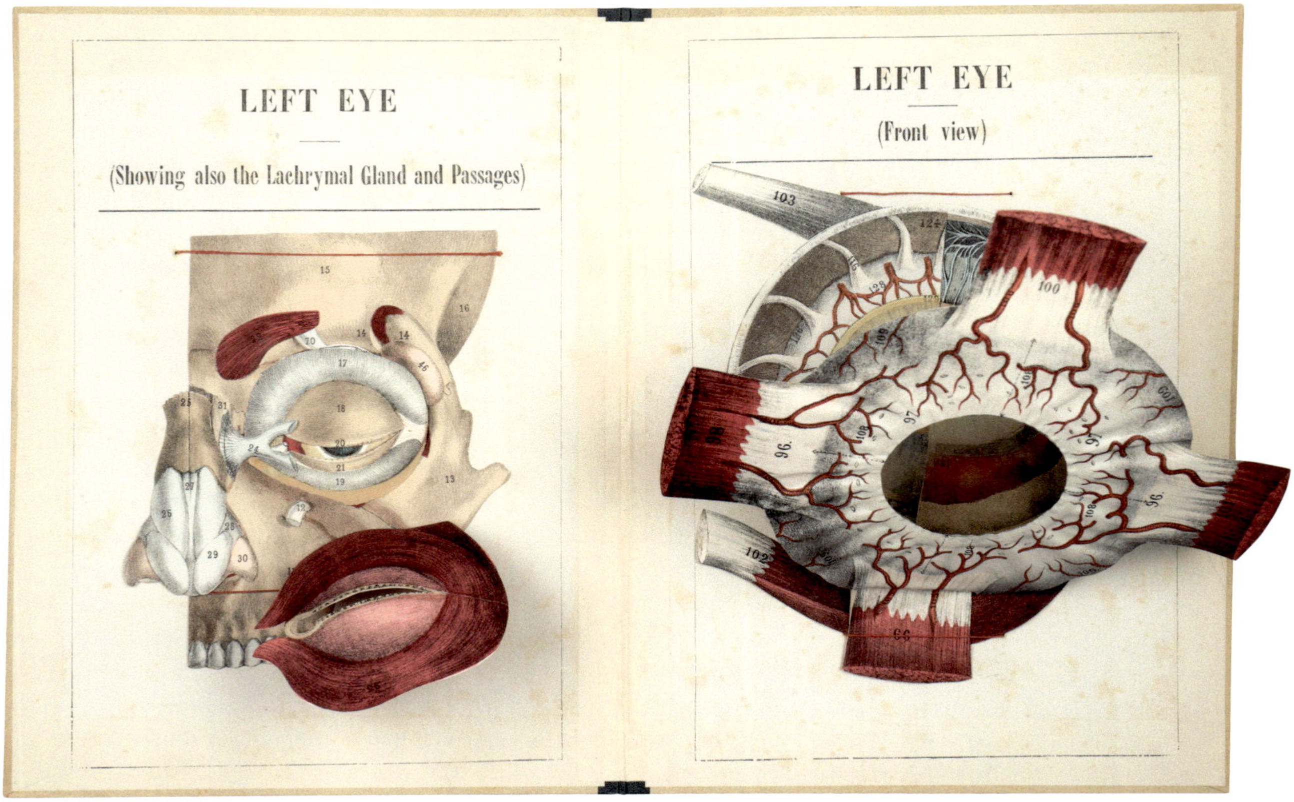

FIG. 8.2
Anatomy of the left eye, chromolithographs and various materials, including cellulose acetate, fabric, and colored tissue papers.
From Gustave-Joseph Witkowski, *Human Anatomy and Physiology. Part IV. A Movable Atlas Showing the Mechanism of Vision; The Eye; The Organs of Vision*, trans. and ed. Henry Power (London: Ballière, Tindall and Cox, 1878–88).
Los Angeles, Library Special Collections for Medicine and the Sciences, Louise M. Darling Biomedical Library, UCLA, BIOMED ** QS 17 W831a 1878.

providing scientifically accurate "artificial anatomies" that did not decompose.[5] Under the directorship of Felice Fontana, artists at the ceroplastics workshop of Florence's La Specola, such as Clemente Susini (1754–1805), produced life-size anatomical models in a viscerally naturalistic, though idealized, style. Modeled after real cadavers, they rendered minute anatomical details with great delicacy.[6] The series contained both topographical and regional sections of bodies,[7] as well as representations of complete, living humans, including some of beautiful female figures in various states of dissection. The famous *Anatomical Venus* (1780–82) lays supine in a confounding state of ecstatic half-sleep, her external bodily integrity belying her construction as a closed container of anatomical parts (fig. 8.3). Indeed, her chest covering can be removed to reveal seven subsequent layers of glossy musculature, organs, and veins, all the way down to a gestating fetus in her womb.

The technical accuracy applied to every texture, contour, and color of these body parts—exemplifying the Enlightenment ideals of realism and empirical observation—inspired Johann Wolfgang von Goethe to call such plastic anatomies "a worthy surrogate that, ideally, substitutes reality."[8] Fontana aspired to do just that with his incorruptible bodies. Highlighting the didactic utility of three-dimensional models for a medical and general public alike, as well as the shortcomings of dissection, he wrote eloquently on "the great usefulness of being able to consult the demountable anatomy whenever one wishes, in whatever season, with no need for a cadaver, without the risk of morbid infections, without the vexation of noxious odors, in short without distastefulness of any kind, a device which can be studied by even the frailest persons not destined for medicine, and in consequence the object of research at once useful and enjoyable."[9] And while he sought to transform anatomical education through these proxy cadavers, the wax proved to be fragile and malleable and thus impractical to be handled frequently. Fontana therefore attempted to fabricate a wooden version, on which a student

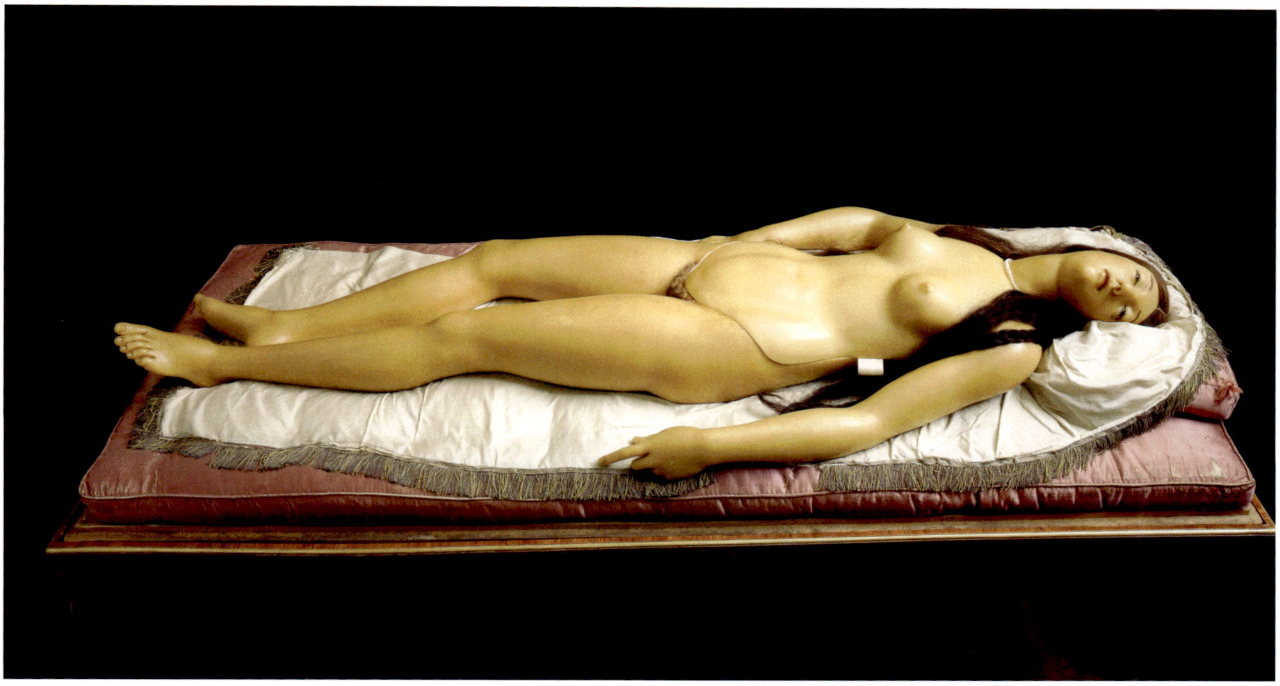

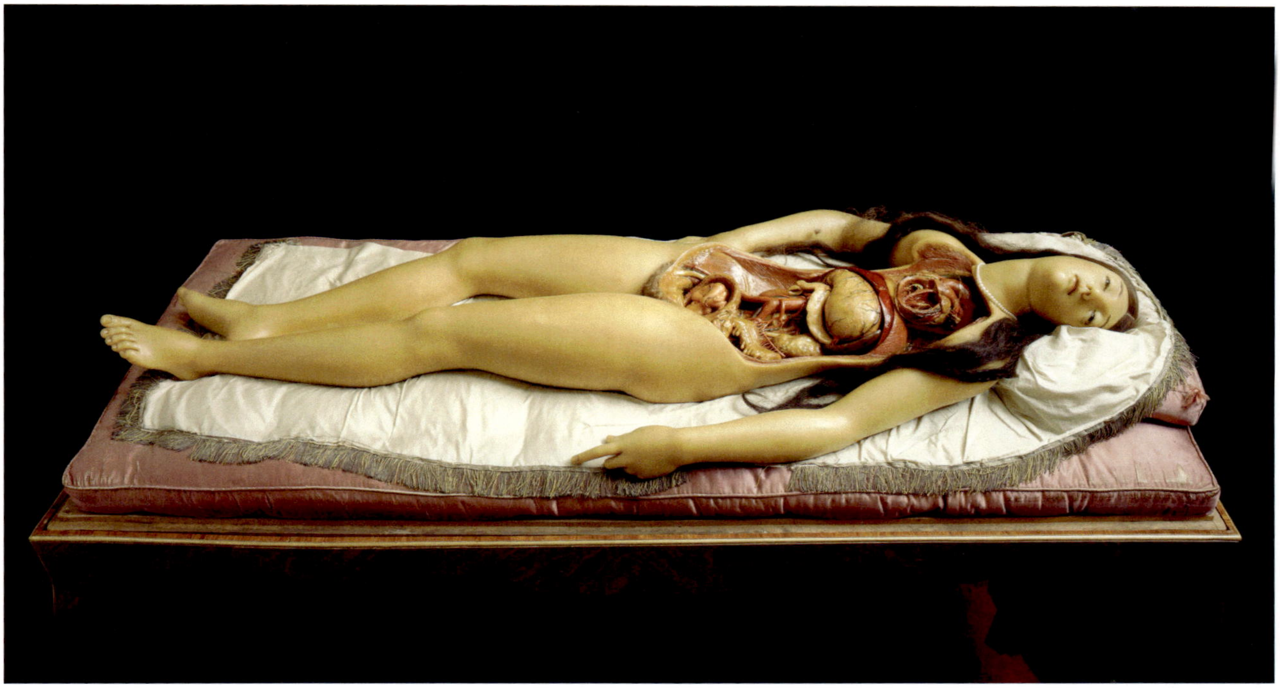

FIG. 8.3
WORKSHOP OF CLEMENTE SUSINI
(Italian, 1754–1805) and Giuseppe Ferrini
(Italian, 18th century).
Anatomical Venus, 1780–82, metal or
wood skeleton, transparent wax, and variously colored waxes with pigments,
180 × 80 cm.
Florence, Museo di Storia Naturale,
Sezione Zoologica "La Specola."

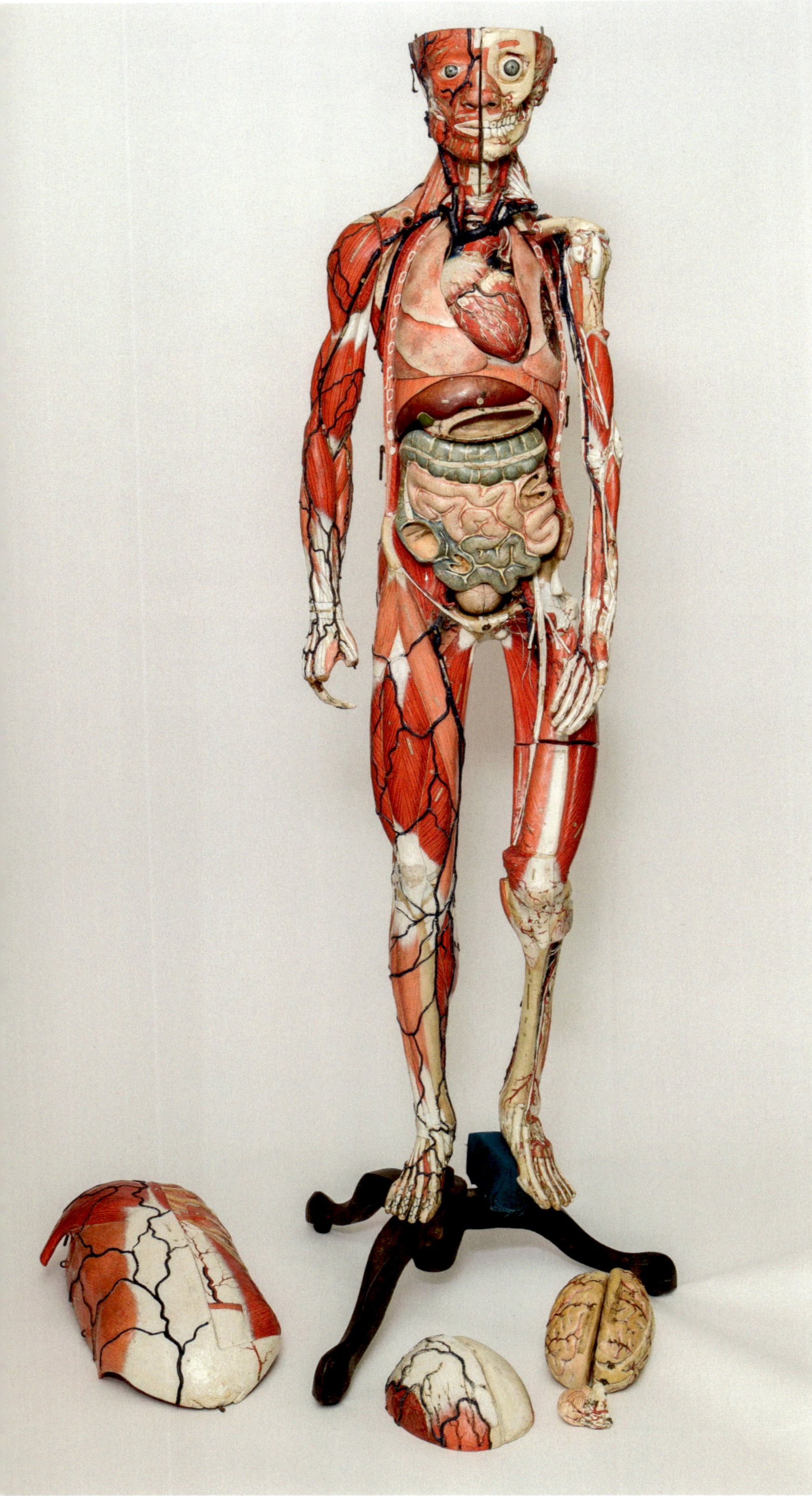

could rehearse the motions of dissection and grasp the relation of parts to the whole. However, he had to abandon the project with the discovery that the material was difficult to sculpt and warped in humidity.

Some decades later, the French physician Louis Auzoux (1797–1880) developed a papier-mâché-like paste that enabled him to realize this goal of a fully dissectible anatomical model (fig. 8.4). With this durable material and a serial production process, his factory mass-produced life-size human figures containing over two thousand removable parts designed for interactivity—including organs, arteries, muscles, veins, nerves, lymphatic vessels, and bones—which he considered a critical epistemological tenet. Advertised as inexpensive, mobile supplements, if not substitutions, for dissection, they were distributed to universities and hospitals around the world, as far as Egypt, where dissection was forbidden. Auzoux's marketing motto, *Nosce te ipsum* (Know thyself), indicates the models' broad reach beyond purely medical professionals, as part of the reform movement in public health he was aligned with.[10] Signaling its fragmentation, Auzoux referred to his multipart model as "anatomie clastique," from the Greek word *klao* (to break).[11]

Alternatively, other dismountable models promoted their durability—a reference to the superiority of the materials over the limited shelf life of a perishable human corpse (as well as flimsy paper). Smith's manikin of 1888 (fig. 8.5), a chromolithographic model standing at nearly four feet tall, was advertised as "The Indestructible Manikin" and came with thirty-three detachable steel organs. On flat metal sheets cut into the shapes of color-printed anatomical parts, these "unbreakable" components could be repeatedly removed from hanging prongs and handled by multiple students to learn the physiological composition of the body and the identification of its parts. An advertisement in the *American Journal of Clinical Medicine* touts the manikin as "indispensable to nurses' training schools" and "far superior to charts for practical teaching,"[12] indicative of the perceived instructional viability of movable parts.

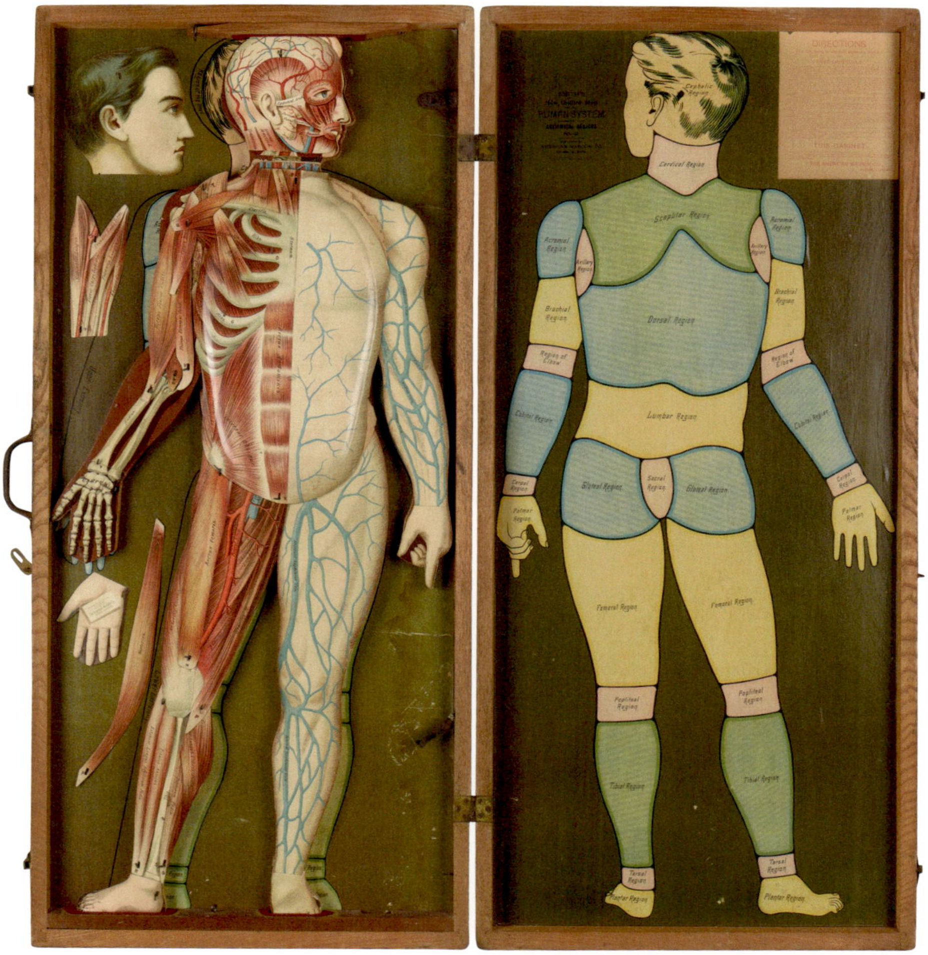

FIG. 8.4
LOUIS AUZOUX
(French, 1797–1880).
Anatomical model with some detachable parts, 1837–1920, hand-painted and varnished papier-mâché and plaster on metal tripod, 138 cm.
Cambridge, England, Whipple Museum of the History of Science, acc. no. 6361.

FIG. 8.5
AMERICAN MANIKIN CO.
Smith's New Outline Map of the Human System, 1888, chromolithography on paper and metal parts in wooden case, closed: 111 × 50 × 6.5 cm.
Los Angeles, Library Special Collections for Medicine and the Sciences, Louise M. Darling Biomedical Library, UCLA, BIOMED ** QS 17 S664 1888.

Designed to stimulate multisensorial interaction, and thus memory, these anatomical manikins instrumentalized the process of assembly as a pedagogical tool. Rather than passive observation, deconstructible models composed of anatomical organs and viscera simulated the process of dissection, encouraging manual exploration, as well as tactile and spatial understanding. The object's very form is testament to an epistemological culture that promoted haptic visuality and participation as an avenue to knowledge.

In the twentieth century, this interactive mode of learning was mobilized to educate the public in state-sponsored hygiene campaigns in Germany.[13] The *Gläserner Mensch*, or *Transparent Man* (fig. 8.6), was displayed at the Deutsches Hygiene-Museum in 1930 on the occasion of the Second International Hygiene Exhibition, and it was accompanied by an interactive audiovisual presentation in which various parts lit up to explain different anatomical features. This life-size model of a human skeleton built up with artificial organs, blood vessels, and a lymphatic system was encased in a clear plastic shell molded into a human form made using a newly developed material called Cellon. This vitreous silhouette provided a window onto the "invisible" anatomical parts in situ, demonstrating their structure and functions in a normal healthy body.

The *Transparent Man* was so popular that multiple editions were cast (including a *Transparent Woman*) and toured around the world, enlightening international audiences about the wonders of anatomy (and a healthy lifestyle).

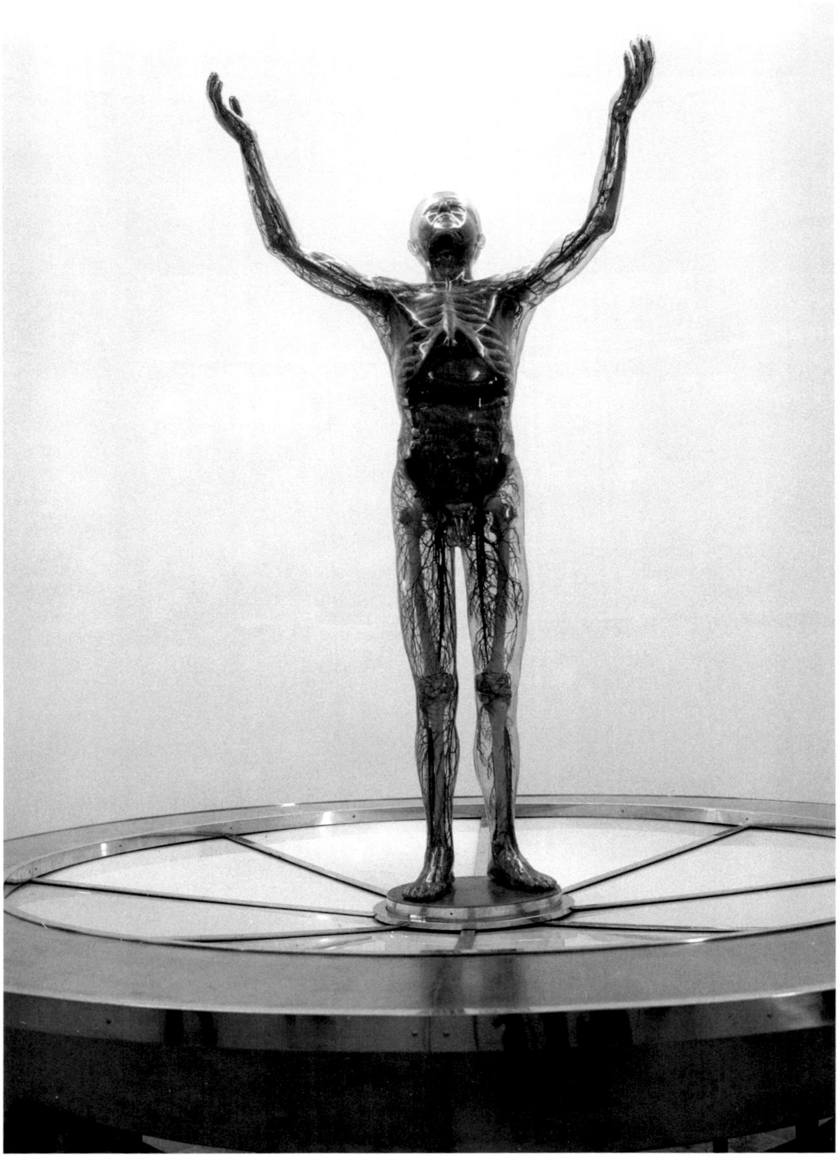

FIG. 8.6
GEORGE PAHL
(German, 1900–1963).
Transparent Man (Gläserner Mensch)
from the 1935 Berlin exhibition *Das Wunder des Lebens, 1935*.
Berlin, Bundesarchiv, Bild, 102–6746.

During the 1930s, it was featured at the Chicago and New York World's Fairs and at many museums across the United States. A few decades later, in the wake of this popularity, miniature do-it-yourself variations on the famous model were fabricated in the United States as toys called *The Visible Man* and *The Visible Woman* (fig. 8.7). Tiny plastic body parts could be painted at home and assembled to create a complete human figure encased in a see-through exterior skin. An accompanying illustrated guide featured information about the skeletal, respiratory, digestive, nervous, endocrine, urogenital, and circulatory systems—all referring to components of the figurine. Straddling the border between entertainment and education, these instructive toys were marketed as turning "science into hobbies . . . hobbies into science" and enabling "parents and children to share information which is otherwise so difficult to communicate."[14] *The Visible Woman* contained an optional feature: "The Miracle of Creation," which came with a fetus, an enlarged breast cover, and alternative viscera to fit the pregnant adaptation. This gestational model is reminiscent of a tradition of manikins that represent the pregnant female body, including diminutive figures made in ivory in the seventeenth and eighteenth centuries. While these types of models are not

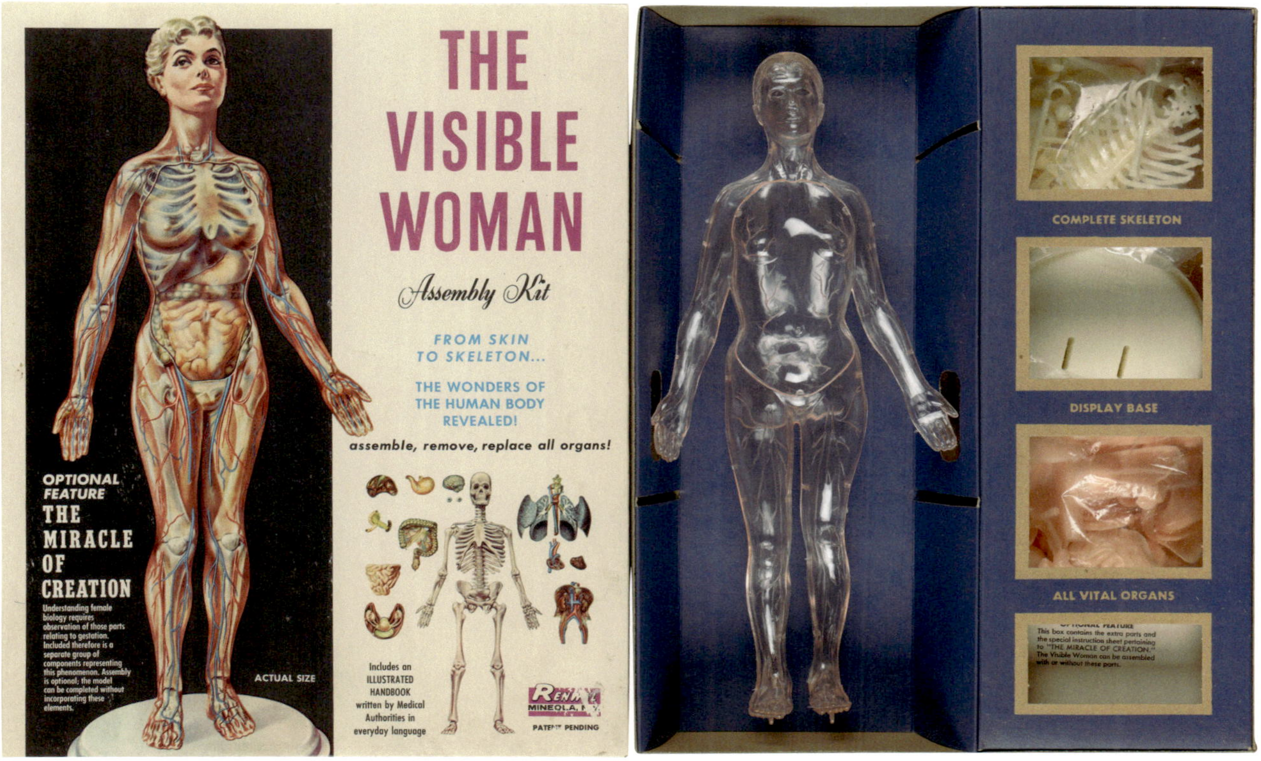

FIG. 8.7
RENWAL CO., MINEOLA, NEW YORK.
The Visible Woman,
ca. 1960, plastic model kit in cardboard box, 41 × 33.4 × 7.5 cm.
Los Angeles, Library Special Collections for Medicine and the Sciences, Louise M. Darling Biomedical Library, UCLA, BIOMED ** QS 26 V832 1960.

transparent, they provide entry into the interior of the body via a removable torso cover, under which finely carved (though anatomically crude) organs cradle a tiny fetus, often attached by a string umbilical cord.

While twentieth-century manufacturers exploited the framework of transparency by using clear plastics, the historical application of this conceptual apparatus is well rehearsed in graphic illustrations of anatomy dating back to the Renaissance.[15] Artists applied a diagrammatic approach to the concealed internal workings of the body, abstracting various layers superimposed upon one another. In these schematic illustrations, the body and its variously enveloped structures are reduced to simple outlines articulated in different colors or opacities, so that one could both "see through" the layers and locate them relative to one another. This logic of transparency solved several problems of dimensionality by adding spatial information and integrating multiple systems into one image. This modality was typically used in anatomy books for artists, since they concentrated mostly on the muscular and skeletal networks in order to better understand the construction and movement of the body. By overlaying the more visible myological landmarks onto the skeletal composition of a figure, an artist could visualize the alignment of anatomical elements to capture gestures or poses more accurately. This is exemplified by Jacques Philippe Bouchardon's anatomical figures (cat. no. 21), as well as Cornelis Ploos van Amstel's reinterpretation of Albinus's figures for the former's anatomy instruction for artists (see fig. 1.3). Like Ploos van Amstel, Jean-Galbert Salvage used red and black inks to differentiate overlapping bodily systems and to increase the clarity between the different layers (fig. 8.8). In the nineteenth century, Joseph Maclise gave the "location of viscera in vivo"[16] by tracing diagrammatic outlines of organs on top of fully rendered living humans, as well as by graphically tearing into the skin and subcutaneous tissues to unearth colorful segments of internal structures in hollow cavities (cat. no. 53).

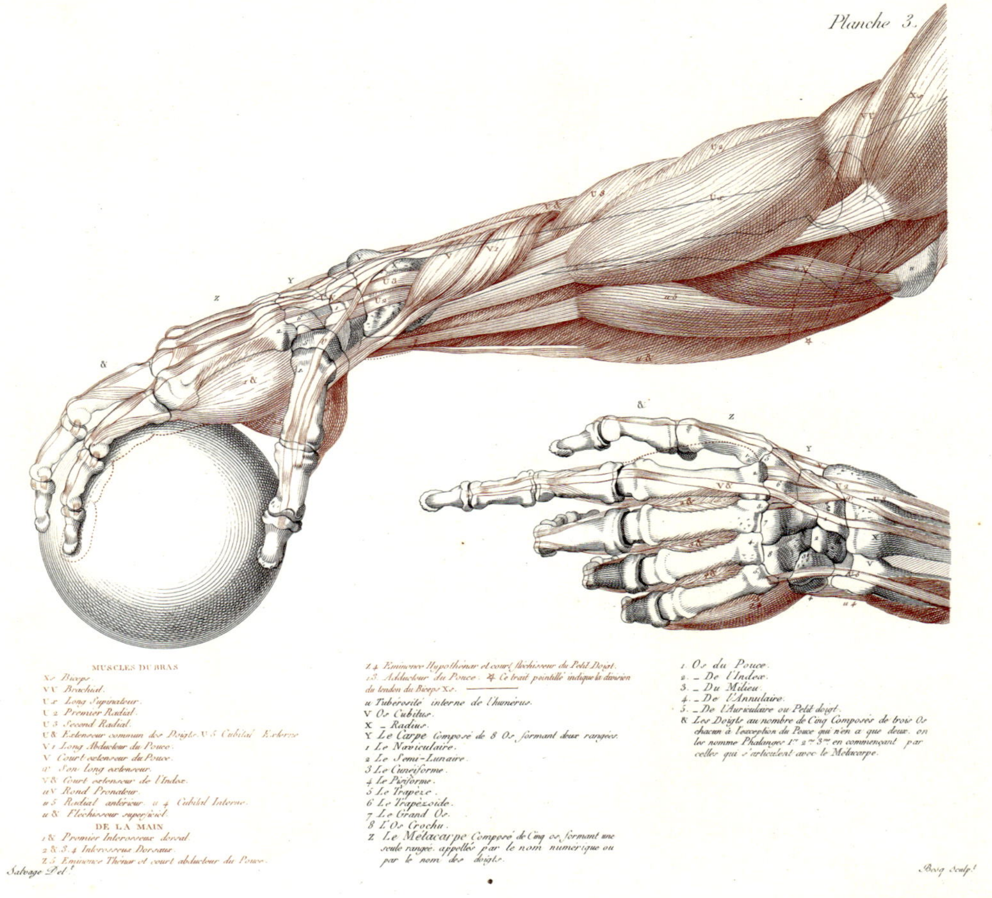

FIG. 8.8
JEAN BOSQ
(French, active 1801–40),
after Jean-Galbert Salvage
(French, 1770–1813).
Muscles of the Arm and of the Hand, etching and engraving from two plates printed in black and red ink.
From Jean-Galbert Salvage, *Anatomie du gladiateur combattant, applicable aux beaux arts; ou, Traité des os, des muscles, du mécanisme des mouvemens, des proportions et des caractères du corps humain* (Paris: Jean-Galbert Salvage and l'imprimerie de Mame, 1812), pl. 3.
Los Angeles, Getty Research Institute, 85-B12146.

This proto-X-ray vision removes the obstructing volume of interstitial blood, fat, and skin to make designated internal orders visible and decipherable. With the German physicist Wilhelm Röntgen's (1845–1923) development of true X-ray technology in 1895, one could access these deeply embedded structures without first subjecting the body to dissection (cat. no. 54). The surface dissolves as the rays virtually flay the subject and illuminate its skeletal interior. However, in doing so, the complex arrangement of organic parts is reduced to a single register, offering no real sense of depth or relief. Despite its enhanced threshold of vision, the disclosure of an X-ray remains flatly static in two dimensions.[17]

A different nineteenth-century imaging technology—namely, the stereograph—promised to fulfill the need for a fully rounded survey of the body's interior.[18] Transcending the perceptual limits of the flat pictorial plane, stereoscopic photography was used by anatomists to express three-dimensionality in a persuasive illusion of depth, solidity, and projection. The stereoscope, an optical device that mimics binocular perception, employs pairs of images taken from slightly different angles viewed through a lenticular apparatus to conjure the impression of three dimensions in the mind's eye. The striking effect of stereoscopic vision applied to the body lay not only in the virtual sensation of depth and projection but, moreover, in the exquisite quality of textural relief in emergent and recessing forms—a feat impossible to reproduce in graphic illustration. The incredible wealth of topographical details (often leveled by aesthetic representation) and the relative positioning of forms enabled the viewer to gain a visceral understanding of the interior of the body, to localize organs within cavities and systems, and to comprehend anatomical structures. Compared to flat illustrations, this binocular rounding offered a more accurate sense of scale, depth, and superficiality for the artist seeking to define the contours and coordinates of a body in space. The smoothness or roughness of a membrane, the wrinkles of the intestines, and the jagged edge of a bone were all physical characteristics that registered as haptic sensations, despite being generated by purely optical cues printed on two flat picture planes.

Stereographic photography was used to capture the anatomical body in medical illustration as early as 1865 and continued to be used into the twentieth century. *Stereoscopic Studies of Anatomy,* a comprehensive collection of over three hundred stereographs illustrating the body in progressive detail, uses topographical and sectional views to articulate an exhaustive vision of the human organism. The card featuring the thorax shows a heart in situ, seen through a ghostly rib cage (cat. no. 55). Composite photography is used to overlap two photographic impressions, thus enabling the viewer to situate the organ within the thoracic cavity. Reviewers hailed the stereoscope's immersive realism, accuracy, educational powers, and

> immense practical value in the modern method of studying and demonstrating medical science. . . . There is no form of illustration that affords such a satisfactory view of all dimensions of anatomical preparation or surgical region as the stereo-photograph, for here we find a most accurate relative position of structures which lie at different levels from the surface and in their natural relative proportion of size and angles. It is the nearest approach which has yet been found to the actual clinical or anatomical specimen.[19]

By imagining the dissection in three dimensions, these images harness the haptic, phenomenological experience of seeing a specimen in real life. Indeed, much of the critical response to stereoscopic atlases alluded to this idea of a hands-on, practical surrogacy—that the stereographic image came as close to a simulation of dissection as had ever been possible. In the preface to the *Edinburgh Stereoscopic Atlas of Obstetrics* (1908–9), the noted Scottish surgeon Sir J. Halliday Croom (1847–1923) extols the use of stereoscopy for creating images "so pronounced, clear and definite, that the student can now study these stereograms with as much accuracy as if he had the actual preparation in his hands."[20] The feeling that one could reach out to touch the body parts, protruding as they were into the viewer's field of vision with insistent solidity, defines the magic of the illusion as well as its heuristic power for a discipline rooted in the empirical act of dissection.

Just as stereoscopy represented the forefront of 3-D imaging technology in the nineteenth century, progressively new and radical modes of representation continued to overhaul traditional visions of the body throughout the twentieth century, and now into the twenty-first. Following X-rays, technologies including ultrasounds, magnetic resonance imaging (MRI), and computed technology (CT) scans scans demonstrate efforts to look inside the body without opening it up and to present three-dimensional records of such exploration. Virtual reality is increasingly present in medical education and diagnostics and is at the cutting edge of not only representational technology but also phenomenological simulation, allowing users to both visualize and manipulate the body in educational and clinical settings. The popularity of exhibitions such as Gunther von Hagens's *Body Worlds* (featuring a circus of "plastinated" cadavers) and the scale of the National Library of Medicine's Visible Human Project (a vast dataset of sectional anatomy)[21] point to the continued interest in representations of the body in three dimensions for scientific and popular audiences alike. Indeed, whether as pedagogical tool or source of entertainment, these models, manikins, and mock-ups of the dissected body articulate some of the various methods, frameworks, and conceptual principles with which artists made manifest the body's material and physiological complexity.

Epigraph

Keele and Pedretti 1978–80, 2:W 19061r, no. 154r, 594.

Notes

1 Gründler 2011, 134. Domenico Laurenza describes this synthetic quality as Leonardo's search for the Renaissance ideal of harmony through an integration of the distinctive, "dissective" anatomical parts into a complete, harmonious whole. He characterizes this sketch (see fig. 8.1) as a centripetal, imploded view, rather than a centrifugal and exploded vision, as it emphasizes the relationship of internal organs and external structures in a single composition (Laurenza 2003, 49). See also Landrus 2019, 72.

2 On Leonardo and anatomy, see Kemp 2006; Laurenza 2009; Nova and Laurenza 2011; Clayton and Philo 2012; and Bambach 2019, vol. 3.

3 For more on flap anatomies, see Carlino 1999a; Brown, M. 2013; and Buckley 2013.

4 Vesalius 1543b. See the colored example in the University of Cambridge Library, https://cudl.lib.cam.ac.uk/view/PR-CCF-00046-00036/39.

5 See Panzanelli 2008 on wax religious votives, effigies, museum curiosities, etc. For more on waxes, see Düring, Didi-Huberman, and Poggesi 1999; Panzanelli 2008; Ballestriero 2010; Maerker 2011; Maerker 2013; Maerker 2015; and Wagner 2017.

6 In addition to the use of cadavers, many of the waxes at La Specola were modeled after well-known anatomical engravings, such as William Smellie, William Hunter, Bernhard Siegfried Albinus, and Paolo Mascagni, which served as templates for the composition and content. See Hilloowala 1986; and Massey 2008.

7 On the gravid uteri models, see Massey 2008.

8 Quoted in Mazzolini 2004, 43.

9 Quoted in Mazzolini 2004, 61.

10 On the nineteenth-century efforts to improve sanitation and personal hygiene through public education and health reforms, see Maerker 2013, 544–55.

11 Louis Auzoux, *Anatomie clastique: Catalogue de 1869, du Docteur Auzoux* (Paris, 1869).

12 "Department of Progressive Advertisers," *American Journal of Clinical Medicine* 27, no. 6 (June 1920): 71; and Owen 2016, 44.

13 The *Transparent Man* engaged the popularity of hygienism, public health reform, and social engineering in the late nineteenth and early twentieth centuries, serving Nazi eugenics propaganda. See Vogel 1999; and Canadelli 2011.

14 The booklet is rife with euphemistic language, particularly surrounding reproduction. Notably, *The Visible Woman* does not even contain genitalia.

15 On the transparent body, see Kornell 1996, 62–64; Ciardi 2003; and Stoichita 2013.

16 See plate XXV in Maclise 1856.

17 See Gallay 2019 on stereoradiography, or 3-D X-rays.

18 For more on anatomy and stereoscopy, see Crary 1990, 14–19, 59–62, 119–132; Kemp 1997, 144–145; and Hallam 2016, 219–21.

19 M. A. G. 1910, 1185.

20 Simpson and Burnet 1908–9, 1:vii. For more on this stereoscopic atlas, see McGrath 2002, 130–44.

21 The Visible Human Project reconstructs the body in 3-D from cryosection slices of cadaver. On the history of frozen cross section "topographical" anatomy, see Al-Gailani 2016; and Sappol 2017.

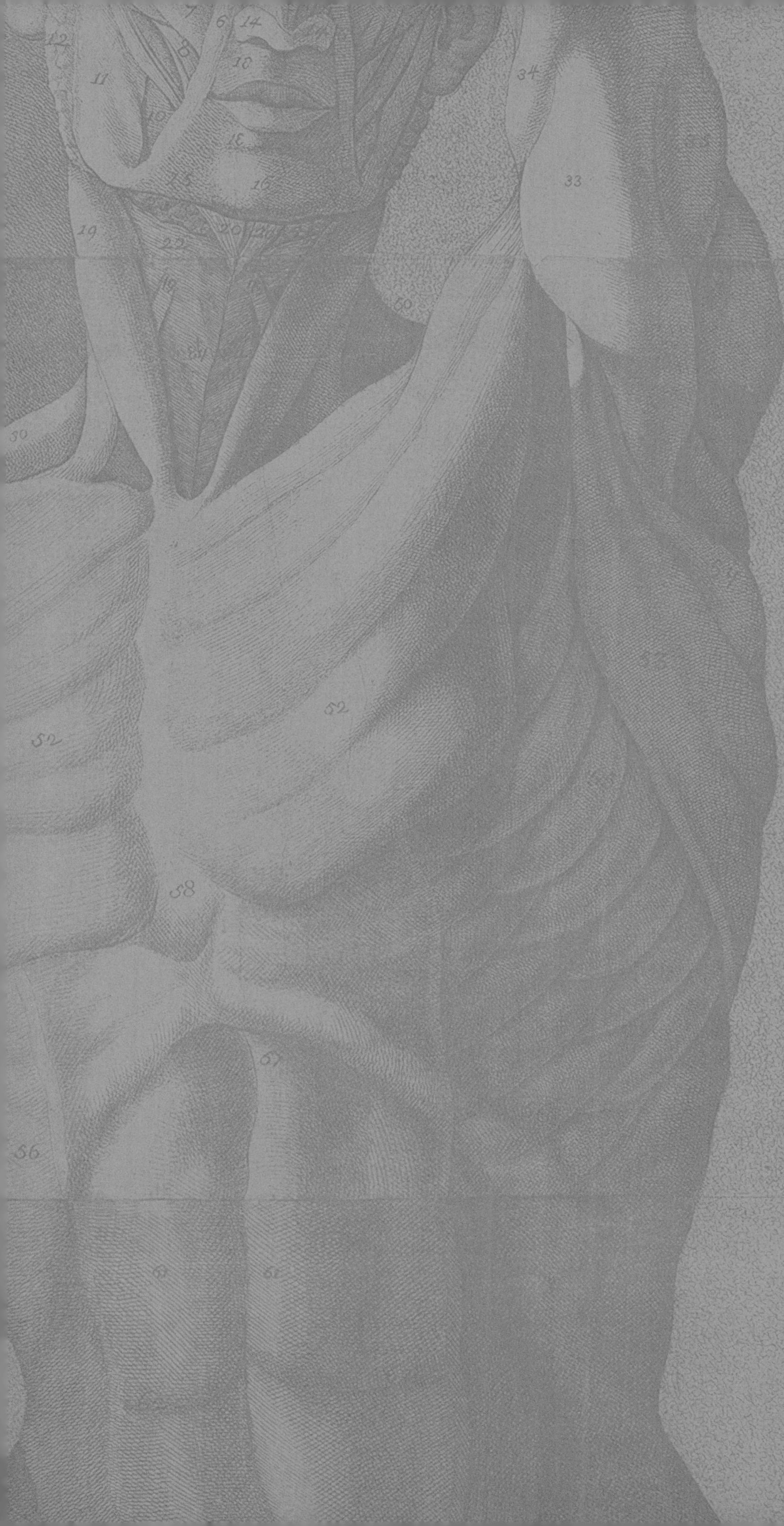

catalog

Thisbe Gensler

Monique Kornell

Naoko Takahatake

Erin Travers

1 **After Estienne de la Rivière**

(French, d. 1569)
Skeleton with nerves displayed
1530s
Woodcut
From Charles Estienne and Estienne de la Rivière, *De dissectione partium corporis humani libri tres, à Carolo Stephano, doctore Medico, editi: Unà cum figuris, & incisionum declarationibus, à Stephano Riverio Chirurgo compositis* (Paris: Simon de Colines, 1545), 59
Los Angeles, Getty Research Institute, 84-B31171

BIBLIOGRAPHY

Quesnay 1749, 228–29; Choulant 1945, 152–55; Russell 1952; Kellett 1964, 350; Burris 1966; Herrlinger 1970, 87–101; Cazort in Karp 1985, 156–57, nos. 5a–c; Kornell 1989b, 842–44; Hook and Norman 1991, 1:260–62, no. 728; Roberts and Tomlinson 1992, 168–87; Cazort in Cazort, Kornell, and Roberts 1996, 137–39, nos. 30–32; Carlino 1999b, 23–26; Talvacchia 1999, 161–87; Laurenza 2003, 97; Carlino 2004, 83–84; Cazes 2004, 38–40; Comar 2008b, 140–41, no. 5; Pantin 2014, 40–43; Canalis 2018, 145.

This skeleton displaying the nerves has a sense of monumentality gained from being depicted from a low viewpoint against a landscape that drops away to a distant port with sailing vessels. According to the explanation of the plate, the figure holds its detached jaw aloft so that the course of a nerve can be seen as it enters along the inside and exits through a foramen below the teeth (Estienne and La Rivière 1545, 60; Roberts and Tomlinson 1992, 176). The springy ends of the nerves give the figure a busy energy that is heightened by the coiled clouds behind him and the pointer lines that dart in and across one another, making it challenging to trace them back to the letters keyed to the text identifying the parts. This labeling method predated the *Fabrica,* for which Vesalius placed the identifying marks directly on his figures, a practice that sometimes had its own issues of visibility, particularly in the shaded areas of the first edition, as Vesalius himself acknowledged.

This richly illustrated folio was the result of a collaboration between Charles Estienne (1504–64), a physician and classicist from a famous family of French printers, and Estienne de la Rivière, a close friend of fellow surgeon Ambroise Paré. Although the earliest dated block is from 1530, the book wasn't published until 1545, first in Latin, with a French translation the following year. By then, the impact of being the largest and most profusely illustrated atlas of anatomy to date had gone instead to Andreas Vesalius's *Fabrica,* published in Basel in 1543 (cat. no. 30).

A court case had dramatically halted printing of *De dissectione* in 1539. Apparently, a disagreement had arisen between the two authors when Estienne had attempted to publish the book as his work alone (Quesnay 1749, 228–29; Kellet 1964, 350). Signaling the compromise that was needed to resolve the case, the title and preface define the contributions of the authors. La Rivière is credited as responsible for the illustrations and their explanations, and for carrying out the dissections. The initials of the Latin version of his name, Stephanus Riverius, appear on a tablet next to a skeleton in the first illustration of the book (Herrlinger 1970, 89). Several blocks bear the marks and signatures of the cutter "Mercure" Jollat and the printer Geofroy Tory (Russell 1952, 147). A curious feature is that the majority of the woodblocks—excepting the full-figure skeleton, muscle, nerve, and vein figures—have been altered by an insert into the block showing the area of dissection (Talvacchia 1999, 164–65). Overall, there is great variation in style. The most sophisticated figures can be found in book three and are adapted from Giovanni Jacopo Caraglio's print series *Loves of the Gods.* —MK

CORP. HVMANI LIB. I. 59

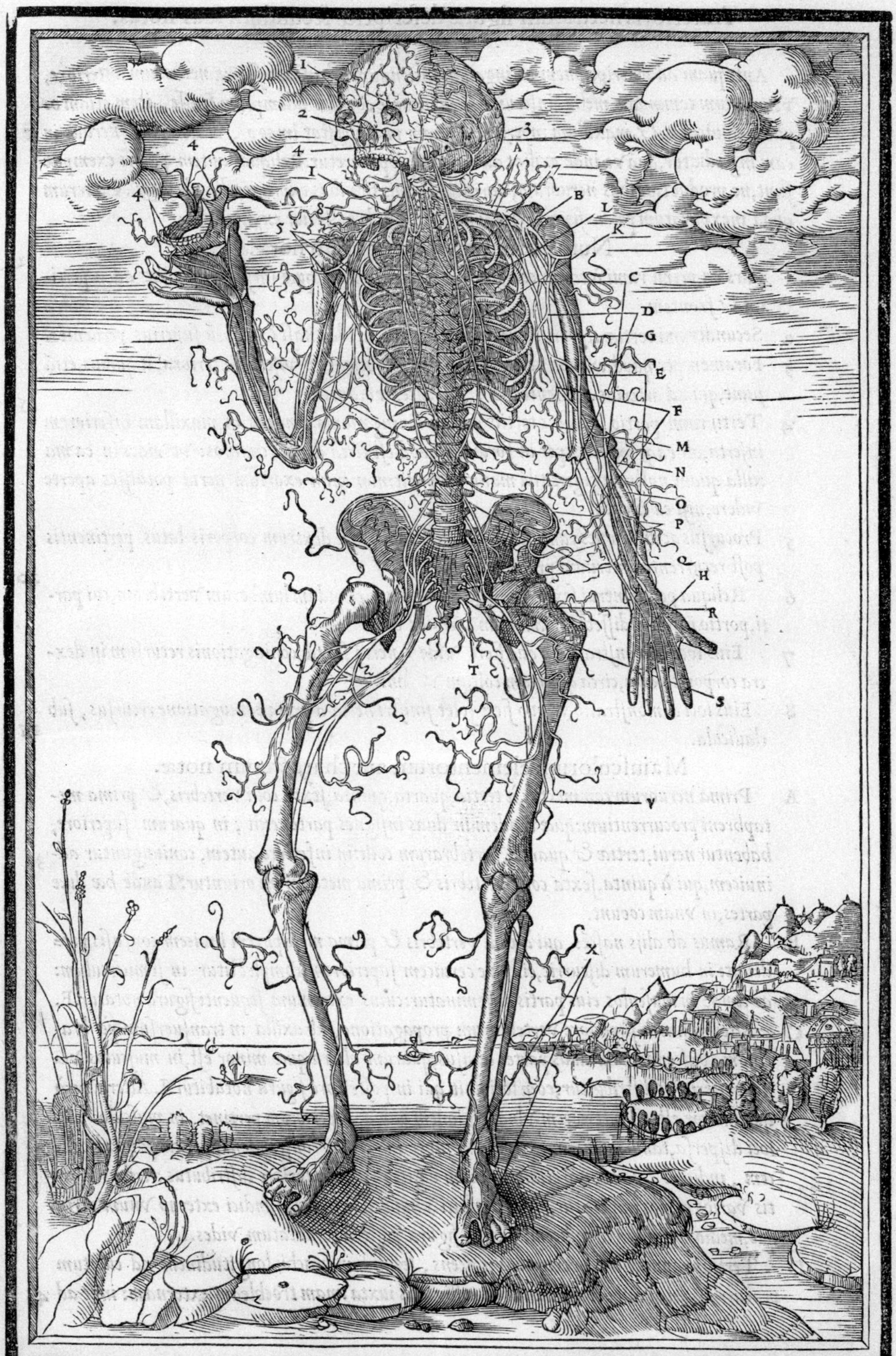

2 Domenico Bonaveri

(Italian, 1653–1731), after Jan Steven van Calcar
(North Netherlandish, ca. 1515–ca. 1546)
Lamenting or praying skeleton
Etching and engraving
From Domenico Bonaveri, *Notomie di Titiano*
([Bologna]: n.p., ca. 1685–90), pl. 2
Los Angeles, Getty Research Institute, 84-B18369

BIBLIOGRAPHY

Choulant 1945, 196; Cushing 1962, 134, no. VI.D.-9; Simons and Kornell 2008; Kornell 2016a; Margócsy 2019, 321.

Three skeleton and fourteen muscle figures after Andreas Vesalius's *Fabrica* (1543) (cat. no. 30) were "reengraved for the use of those who draw" by the Bolognese printmaker Domenico Bonaveri and were probably intended for the students of an art academy that met in the late 1680s at the Bolognese palazzo of Senator Francesco Ghisilieri (1650–1712), the dedicatee of this book (Kornell 2016a). The title, which translates as "Anatomies of Titian," credits the *Fabrica*'s figures to Titian (ca. 1488–1576), as does an inscription on the first plate. However, they were likely by Jan Steven van Calcar, a Northern artist active in Venice in Titian's circle whom Giorgio Vasari (1511–74) identified as Vesalius's artist (Kornell 2018). The remaining plates bear initials indicating Titian as the designer and Bonaveri as the etcher.

This skeleton, whose pose of prayer or lamentation grants it a sense of inner life, depicts the third skeleton from the *Fabrica* in reverse direction. It wonderfully demonstrates Vesalius's observation, paraphrasing Galen, that the spine is made up of several bones rather than just one in order to allow for movement, such as bending over or standing erect (Vesalius 1543a, 58). In the instructions he gives on articulating a skeleton, Vesalius favors cleaning the bones by immersing them in boiling water and—before they are allowed to dry too much and harden—piercing them with an awl and threading them with copper wire that is heated so as to bend more easily (Vesalius 1543a, bk. 1, ch. 39; Kornell 2000). The vertebrae were threaded with an iron rod.

Because they appeared without an explanatory text, Bonaveri's copies omit the lettering seen on the figures in Vesalius's original woodcuts (Margócsy 2019, 321). One consequence of separating the image from text is that the elements of ape and canine anatomy, introduced by Vesalius into the fifth and sixth muscle figures to demonstrate Galen's errors, remain unexplained (Saunders and O'Malley 1950, 100–102). Bonaveri preserved the nonhuman anatomy in his copies of those figures, whereas Juan Valverde de Amusco, when he adapted the *Fabrica* woodcuts for his anatomy book of 1556, corrected them (cat. no. 31). —MK

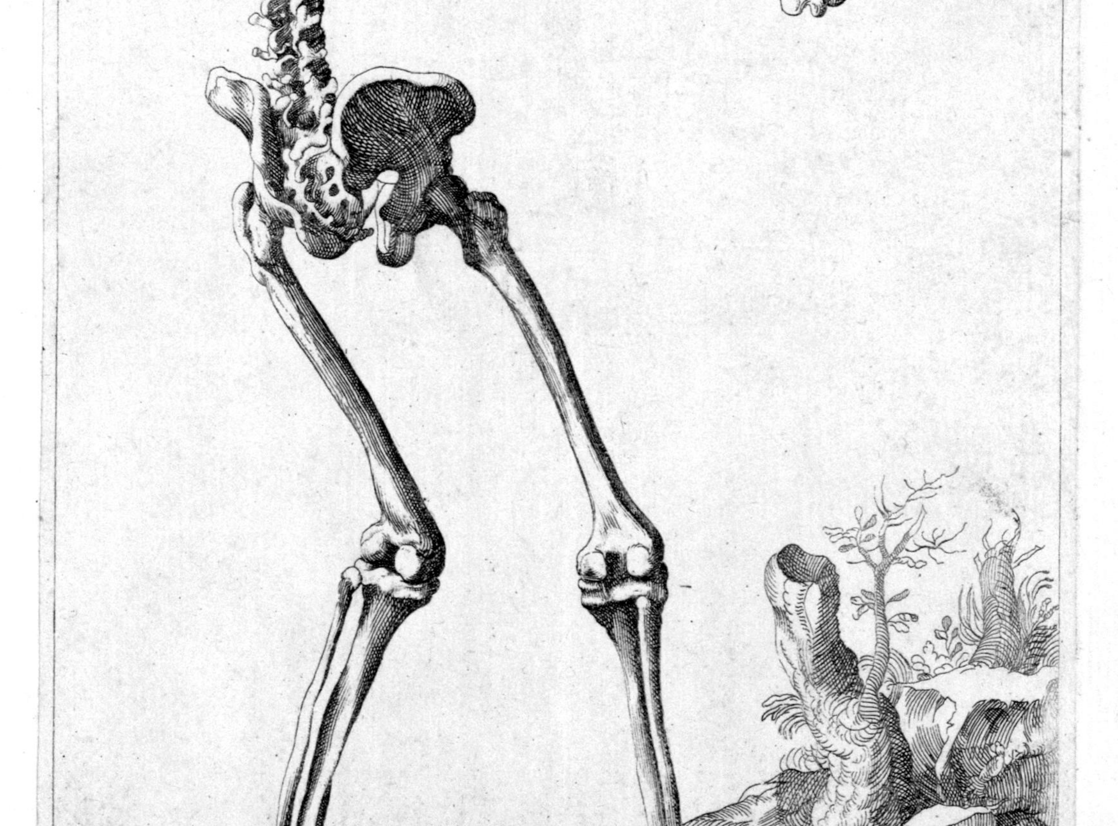
2
T.I.D.
D.B.S.

3 Stefano della Bella

(Italian, 1610–64), after Alfonso Parigi II
(Italian, 1606–56)
Funeral of Prince Francesco de' Medici in the church of San Lorenzo, Florence, on 30 August 1634
Circa 1634–37
Etching with engraving, trimmed within image at top: 29 × 21 cm; sheet: 32.5 × 41.1 cm
Los Angeles, Getty Research Institute, 2554-292; P910002** (FF. 247)
Inscriptions: lower left: "Al. Parigi In."; lower right: "Stef. della bella Fe."; monogram "SDB" partly hatched out

BIBLIOGRAPHY

Cavalcanti 1634; Berendsen 1961, no. 56; Gaeta Bertelà and Petrioli Tofani 1969; De Vesme and Massar 1971, no. 74.3; Forlani Tempesti 1973, no. 6; Schraven 2014.

Elaborate funeral decorations for grand personages first arose in Italy in the late sixteenth century and often included skeletons in animated poses. Stefano della Bella's etching depicts the interior of the church of San Lorenzo on the occasion of Francesco de' Medici's (1614–34) funeral, held on 30 August 1634. This is the third state of the print. A second state was included as a foldout illustration in Andrea Cavalcanti's account of the funeral, *Esequie del serenissimo principe Francesco . . .* (Cavalcanti 1634). The Tuscan prince had died of the plague at age nineteen in the previous month while serving in the Imperial army at the siege of Regensburg, in southeastern Germany. His catafalque is set on a raised platform in the crossing, which is flanked by lance-bearing equestrian skeletons in armor, befitting a military death. Above the catafalque, Medici coats of arms are suspended.

Along the nave, enormous skeletons on pedestals tower over the mourners. The skeletons are as elegant in gesture and pose as the funeral guests and have a jaunty demeanor, despite the occasion. One of the skeletons on the right of the nave raises a hand in greeting, welcoming the mourners. Della Bella's delicate touch as an etcher captures the airy plumes of the headdress of this skeleton and those of his companions. A plumed headdress similarly adorns a skeleton on horseback riding into battle in a print by Della Bella dated to the late 1640s (De Vesme and Massar 1971, no. 93). Oversize skeletons had earlier been employed in the decoration of the Medici funerals in San Lorenzo for Duke Cosimo in 1574 and Duke Francesco in 1587 (Schraven 2014, 90–93, fig. 3.5).

The funeral decorations of 1634 were planned by Alfonso Parigi II, who brought his skills as a stage designer to the project. Parigi also designed the obsequies for Emperor Ferdinand II, which was held in the church of San Lorenzo in 1637. In etchings by Della Bella, similar skeletons are seen flanking the nave and equestrian skeletons are on the facade (De Vesme and Massar 1971, nos. 75, 76), suggesting that Parigi was following an established Medici custom of recycling funerary decorations (Schraven 2014, 95–98). In a similar fashion, the third state of Della Bella's view of the nave has been noted as appearing in the published account of the emperor's funeral of 1637 (De Vesme and Massar 1971, no. 74.3). —MK

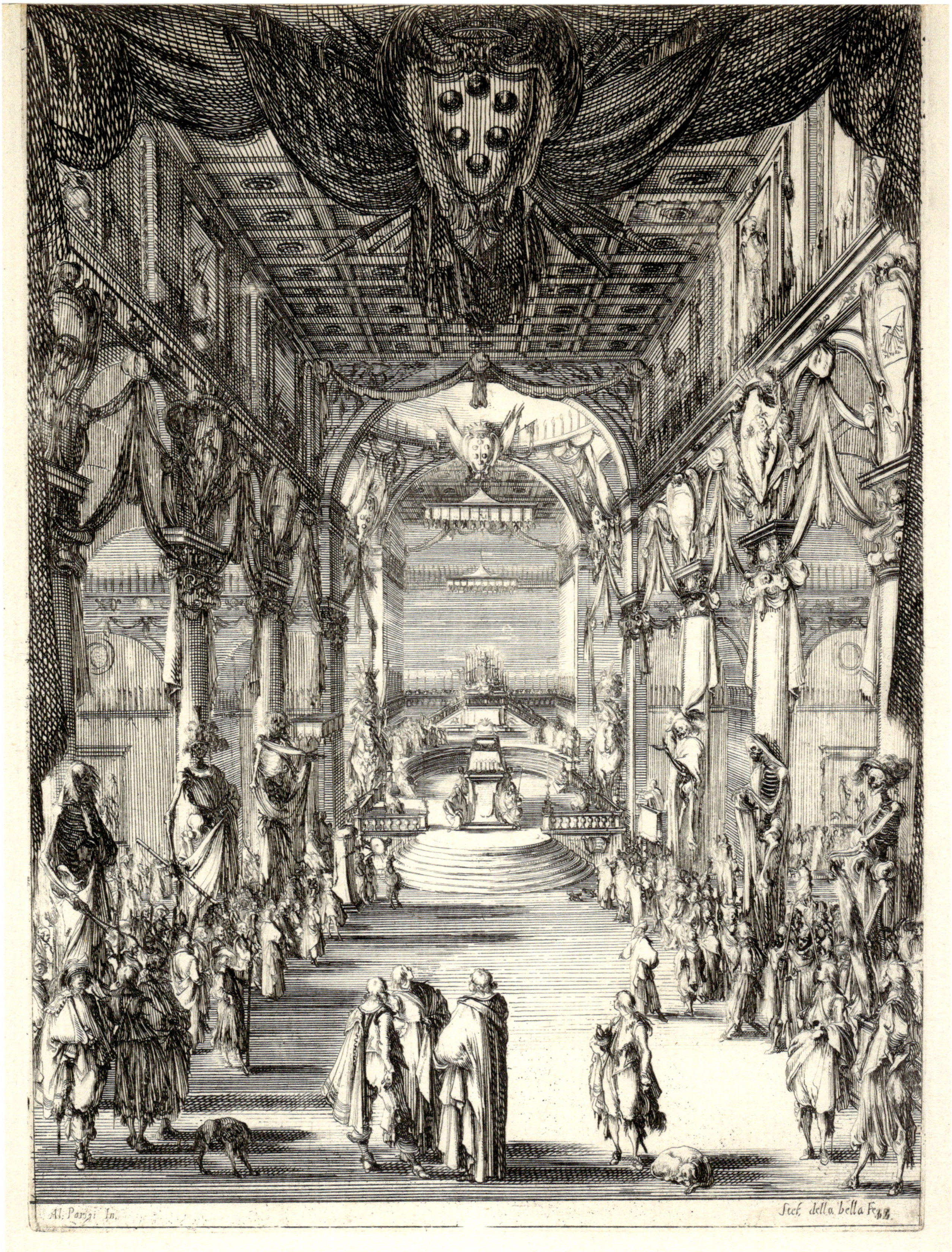
Al. Parigi In.
Stef. della bella Fe.

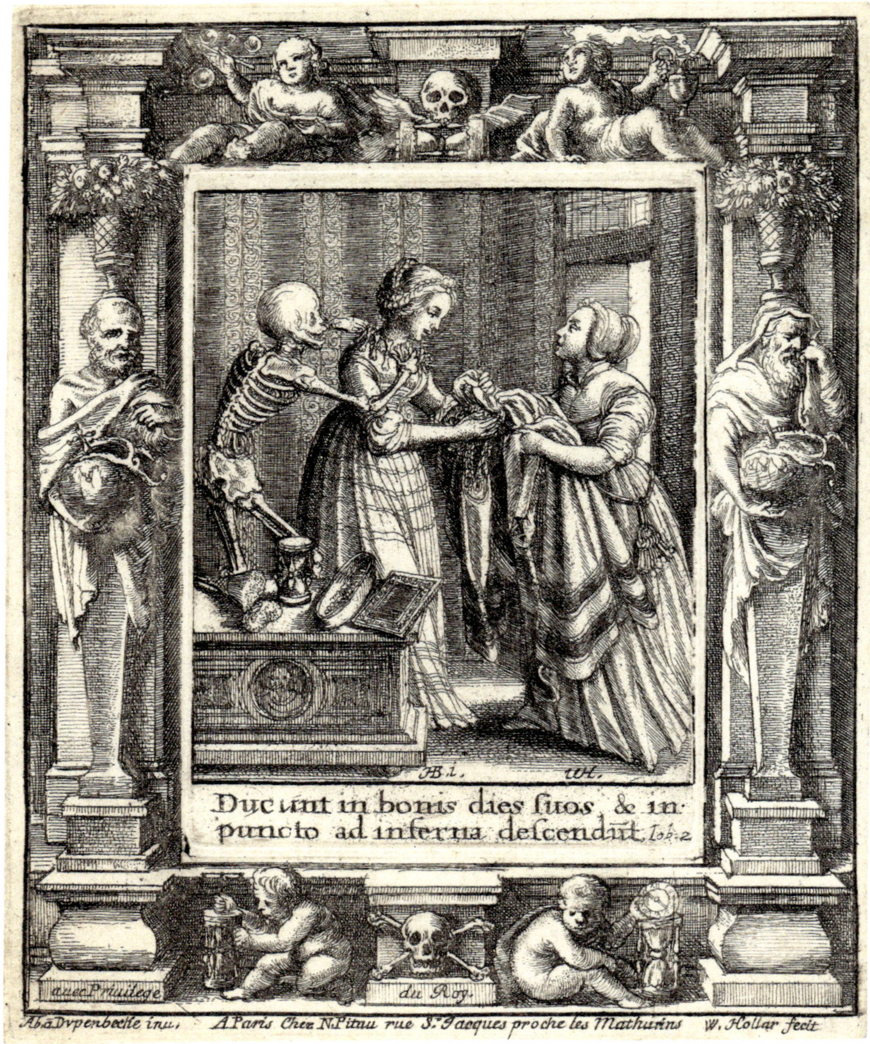

4 **Wenceslaus Hollar**

(Bohemian, 1607–77), after Hans Lützelberger (German, fl. 1517–26), after Hans Holbein (German, 1497/98–1543) (main scene); Wenceslaus Hollar (Bohemian, 1607–77), after Abraham Jansz. van Diepenbeeck (Flemish, 1596–1675) (border)
a) *The Countess* (Death gives a young woman being dressed a necklace of bones)
b) *The Physician* (Death gives a urine sample to a doctor)
Etchings, plate mark ca. 7.5 × 5.5 cm, enclosed within etched border with a platemark ca. 11.5 × 9.3 cm (for both)
From [Dance of Death] (Paris: Nicolas Pitau, ca. 1680)
Los Angeles, Getty Research Institute, 84-B30676

BIBLIOGRAPHY
Collins 1978; Warthin 1931, 43–56; Pennington 1982, nos. 249, 253; Hollstein 1988, vol. 14a, no. 99 (26), no. 99 (34); Hollstein 2010, nos. 1143, 1147.

In the "Dance of Death" or "Danse macabre," exuberant and lively skeletons or cadaverous figures interact with people from every level of society and all ages to lead them in a dance that was first depicted on walls of cemeteries and churches in fifteenth-century northern Europe. *The Countess* and *The Physician* were part of a set of Hans Holbein's designs for woodcuts first printed in Lyon in 1538, in which scenes of daily life are interrupted by Death, whose victims are either oblivious, alarmed, or occasionally welcoming.

Wenceslaus Hollar's etchings after Holbein for the main scenes, and after Abraham Jansz. van Diepenbeeck's design for the decorative borders, were first published in Antwerp in 1651. The initials "HB.i" and "WH" were added to the etching to credit Holbein as inventor and Hollar as etcher. The Getty Research Institute's edition is Nicolas Pitau's reprinting of about 1680 in Paris.

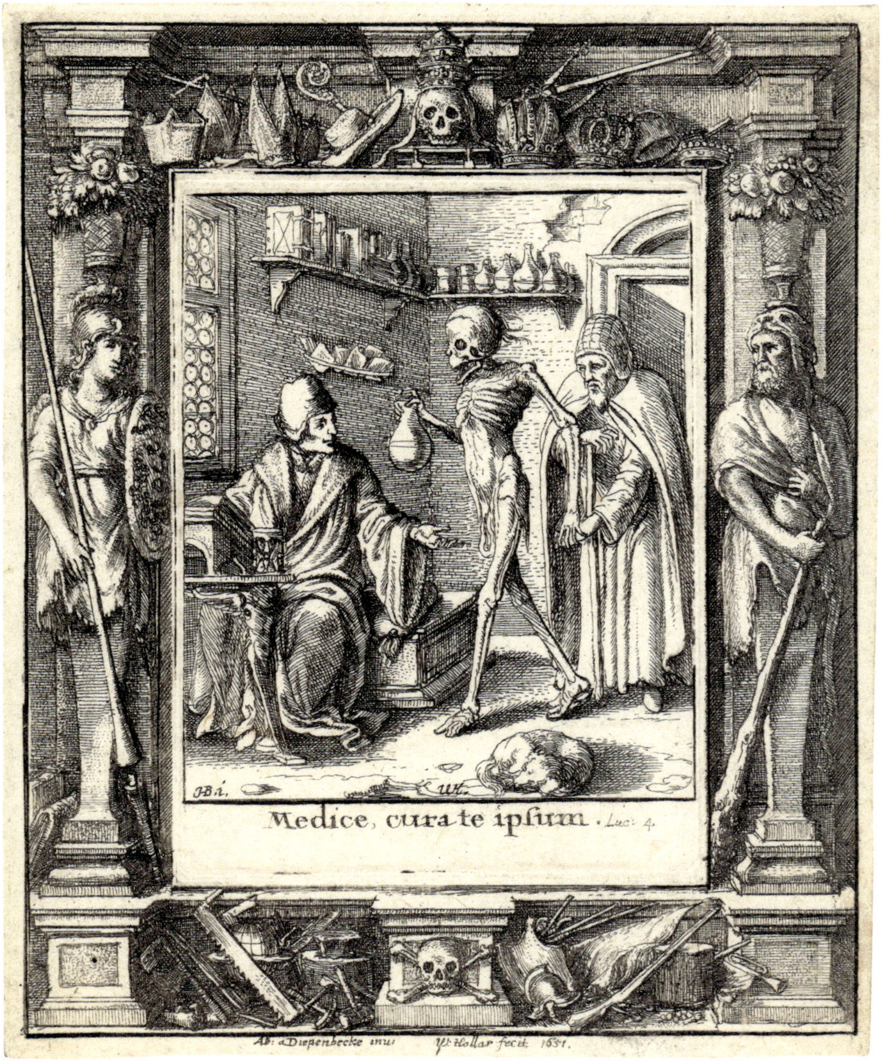

In these two prints, Death inserts himself into the activities of the living. At left, he delicately places a necklace of bones around the neck of a young countess who is being handed a garment and a gold chain. Compared to Holbein's original design, Hollar has simplified the necklace, making it easier to see that she is being adorned by bones, not jewels. Below is a verse from the book of Job: "They spend their days in wealth, and in a moment go down to the grave" (Job 21:13).

On the right, Death strides into a physician's room. Despite his decaying form, he is far sprightlier than the elderly patient leaning on a cane. Death carries a urine sample for the physician, who sits at his desk, on which rests an hourglass: a reminder of the passing of time that is present in many of the scenes. The phrase below, "Physician, heal thyself" (Luke 4:23), indicates that Death has come for both doctor and patient.

Some of the earliest prints after Leonardo da Vinci's anatomical studies, now in the Royal Collection at Windsor Castle, are by Hollar, who had access to them while they were in the collection of the great English collector Thomas Howard, the Earl of Arundel (1585–1646). —MK

5 Samuel van Hoogstraten

(Dutch, 1627–78)
Skeleton and flayed figure with laurel crown
Etching
From Samuel van Hoogstraten, *Inleyding tot de Hooge Schoole der Schilderkonst, anders de Zichtbaere Werelt* (Rotterdam: François van Hoogstraten, 1678), pl. A
Los Angeles, Getty Research Institute, 86-B12032

BIBLIOGRAPHY
Blanc 2008, 68–72; Van de Roemer 2011, 196; Van Hoogstraten 2021, 104–10.

A pupil of Rembrandt van Rijn (1606–69), Samuel van Hoogstraten is best known today for his illusionistic paintings and his art-theoretical treatise, *Inleyding tot de Hooge Schoole der Schilderkonst, anders de Zichtbaere Werelt* (*Introduction to the Academy of Painting; or, The Visible World*). Designed to provide a program of universal study for both painters and interested individuals, the publication is divided into nine classrooms (*leerwinkels*), each of which is dedicated to a Greek Muse and addresses subjects ranging from perspective and composition to proper coloring and the use of light and shadow (Brusati 1995, 5; Weststeijn 2008, 61, 84).

The second classroom is where the Muse Polyhymnia (the Rhetorician) leads the discussion on muscles and bones. She instructs the reader in skills needed for the proper depiction of the human figure, which Van Hoogstraten identifies as the most important subject for artists (Van Hoogstraten 2001, 93). Van Hoogstraten's primary concern was the study of proportion, for which he had devised a new system of measurements and ratios based on the works of Vitruvius (first century BC), Albrecht Dürer (1471–1528), and Karel van Mander (1548–1606) (Bolten 1985, 212–14). The broader contents of Van Hoogstraten's text can be situated within a body of period literature on rhetoric and art theory, but his publication is notable for the degree of anatomical instruction provided and the position it is awarded within the art of painting.

To instruct his reader, Van Hoogstraten provides two labeled anatomical plates featuring four figures that present naturalistic rather than classical bodies (Van Hoogstraten 2021, 10). Seen from the front, side, and rear, they are accompanied by a register containing the names for the muscles and bones, following the example of anatomical atlases (see fig. 2.5). He advises young artists to actively engage with his images and, for best results, to "copy the engraving and look up the names of the accompanying letters" (Van Hoogstraten 2021, 105). In this specific example, graphite annotations of a skull, "-4" at the pelvis, and the outline of the ribcage have been added to the etching, recording a past reader's interaction with the image. Notably, this method of study privileged knowledge of the form and names of the muscles and bones over their function, distinguishing it from the more in-depth studies undertaken by physicians and surgeons in the period. The first etching, shown here, supports Van Hoogstraten's larger aims within Polyhymnia—principally, the study of proportion, as seen in the grid that surrounds the skeletal figure and that corresponds to the illustrations of proportionate figures found later in the same classroom. Van Hoogstraten's concern with providing a credible model is recorded in his testimony stating that he had "measured this skeleton from life [*nae 't leven*]" (Van Hoogstraten 2021, 105). Breaking through the grid, the animated skeleton reaches out to place a laurel crown on the écorché, an attribute that is referenced throughout Van Hoogstraten's text in association with fame and honor (Van Hoogstraten 1678, 69). Thus, this image visually reinforces the author's claim that the study of this subject will elevate and perpetuate the work of the young painter, even after death. —ET

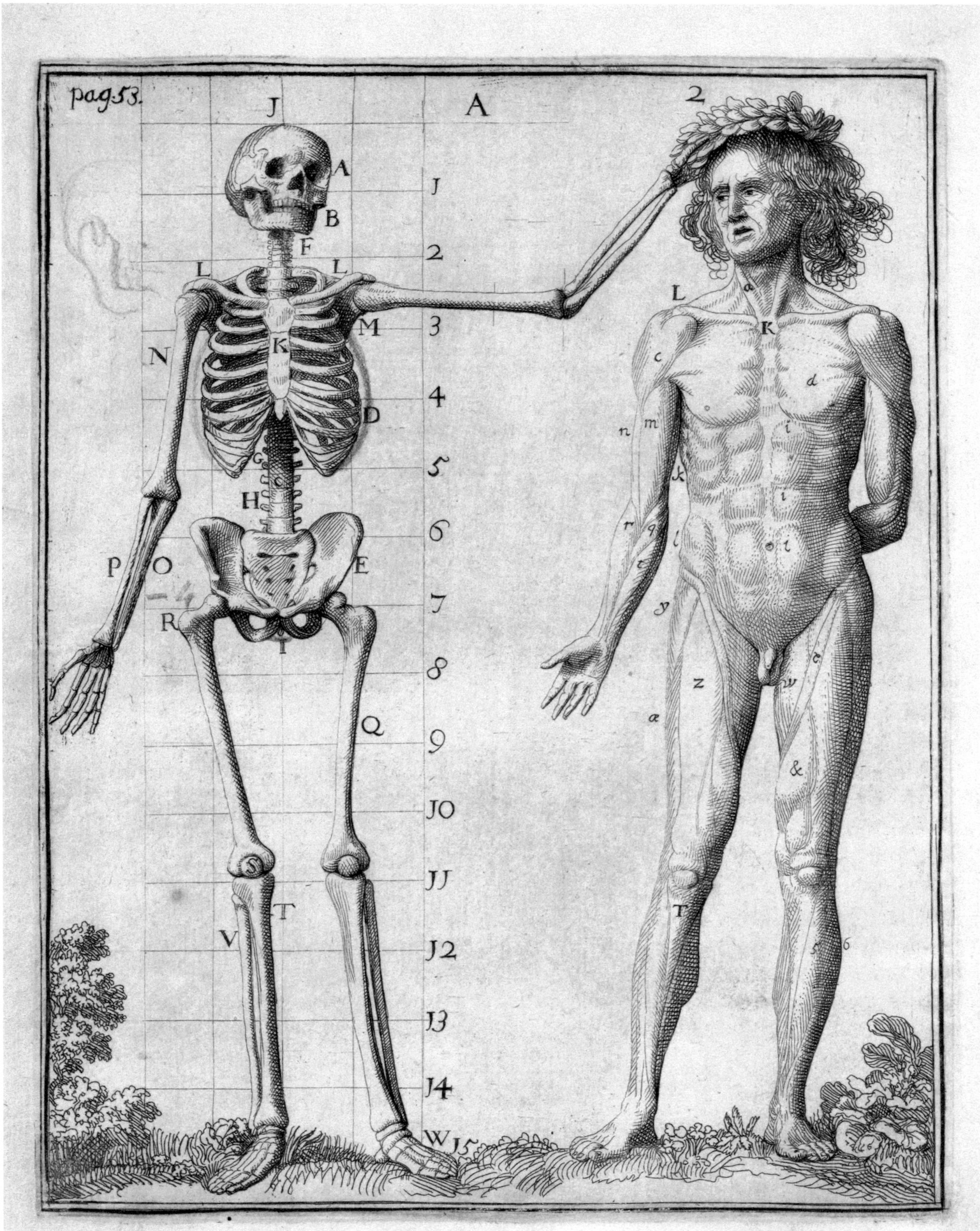
pag.53.

6 Cornelis Huyberts

(Dutch, 1669/70–ca. 1712)
Fetal skeletons in an anatomical assemblage
Etching and engraving
From Frederik Ruysch, *Thesaurus anatomicus tertius* (Amsterdam: Johannes Wolters, 1703), vol. 2, pl. 1
Los Angeles, Getty Research Institute, 94-B12267

BIBLIOGRAPHY
Cook 2002, 241–42; Cook 2007, 283; Rifkin, Ackerman, and Folkenberg 2006, 39–43; Van de Roemer 2010, 171.

A skilled physician, anatomist, and botanist, Frederik Ruysch (Dutch, 1638–1731) gained notoriety and fame for his preparation technique and collection of specimens in Amsterdam (Hansen 1996, 669). At the time of his collection's sale in 1717 to Peter the Great (1672–1725), the czar of Russia, it contained more than 1,125 preparations, which were displayed in fourteen cabinets within the anatomist's home. Several of these objects are recorded in a series of ten catalogs, or *Thesauri,* that Ruysch published to advertise his skill, collection, and findings (Margócsy 2014, 127). After the sale, Ruysch quickly began work on a new collection, which numbered at least 383 objects at the time of his final catalog's publication in 1728 (Van de Roemer 2010, 169, 173).

Ruysch's specimens assumed a range of forms during the anatomist's long life. In its earliest phase, marked by the publication of Ruysch's first catalog, the *Museum anatomicum Ruyschianam* (1691), the cabinet was composed primarily of solitary, dried specimens interspersed with arranged fetal skeletons that held attributes or written inscriptions conveying the fleetingness of life. With his shift to a wet preparation technique in the early eighteenth century, Ruysch changed the appearance and organization of his collection, and several of his dried specimens were repurposed to construct small tableaux vivants laden with *vanitas* imagery (Van de Roemer 2010, 173–74). Though none survive today, eight of these assemblages are recorded in Ruysch's catalogs, and three have been preserved as prints executed by Cornelis Huyberts.

The engraved tableau included in Ruysch's catalog for his third cabinet, which was originally published in 1703, shows five fetal skeletons standing around and on top of a "rocky" outcrop composed of stones removed from the body. Dried and injected vascular specimens mimic vines and trees to evoke a kind of ruinous landscape. The skeletons assume the guises of musicians playing stringed instruments made of tissue and bone. Though not included in the image, the accompanying description records that the infants were displayed with annotations. The first, at the top of the structure, was marked by the passage "Ah Fate. Ah, bitter fate." The figure to the left reinforces the air of dejection and states, "Oh, how miserable is the condition of mankind in this life!" The larger fetus on the right informs the viewer: "The first hour that gave me life, took it away again." At the edge of the wooden pedestal below, a reclining skeleton utters, "Like a flower of the field, quickly I emerged, and again I am snatched away." The fly grasped in its limp hand reinforces the short span of the child's life, as Ruysch explains to his reader that the insect matures and dies within a day (Ruysch 1703, 1–9). —ET

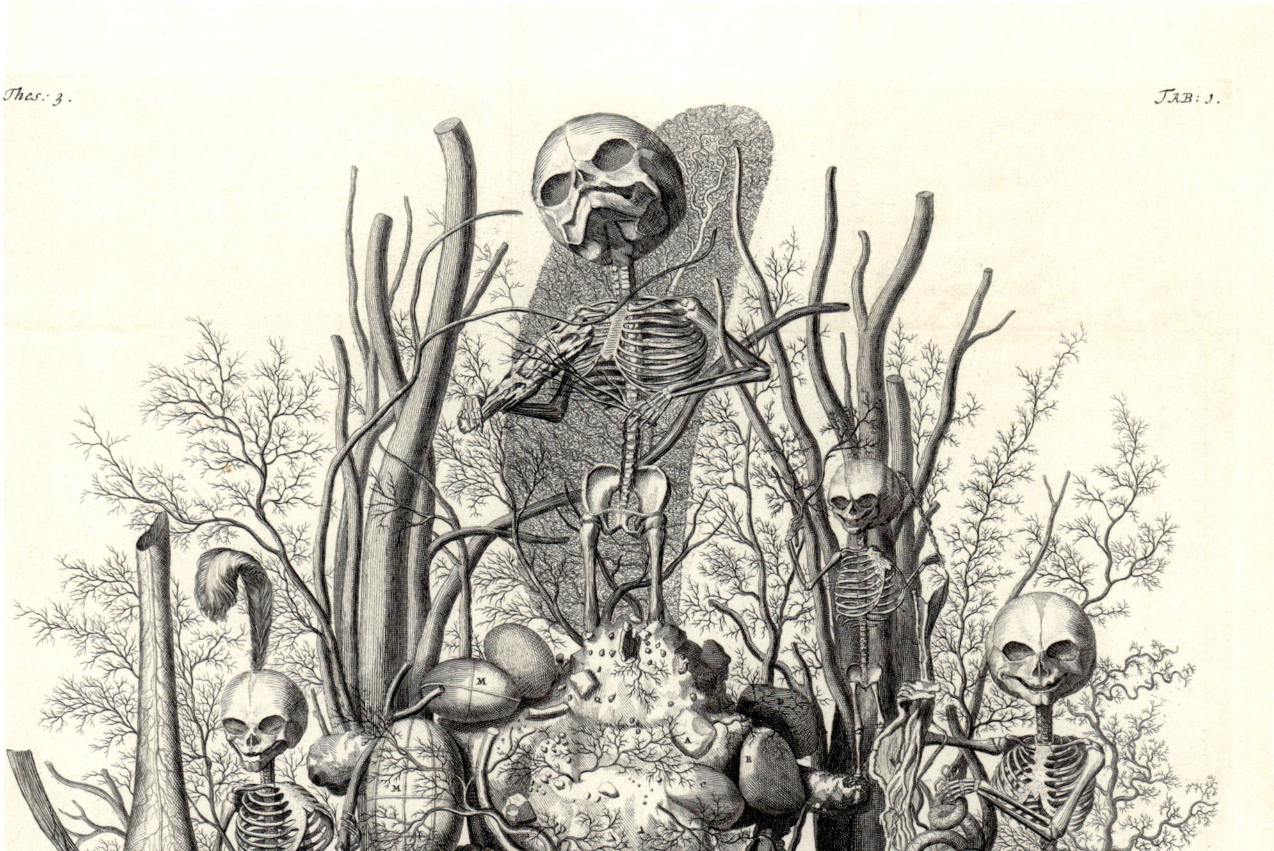
Thes: 3.
TAB: 1.
C. Hüÿberts.
ad. vivum Sculpsit

7 THE LEIDEN ANATOMY THEATER

Etching

From Pieter van der Aa (Dutch, 1659–1733), *Les delices de Leide, une des célébres villes de l'Europe* (Leiden: Pieter van der Aa, 1712), pl. 8

Los Angeles, Getty Research Institute, 88-B29804

BIBLIOGRAPHY

Gogelein 1975, 105–6; Witkam 1980, v; Cazort in Cazort, Kornell, and Roberts 1996, 206–9; Knoeff 2011, 166; Vrolijk 2017, 16–17.

Constructed between 1591 and 1594, under the direction of Universiteit Leiden's first professor of anatomy, Pieter Pauw (1564–1617), the Theatrum Anatomicum served as a site of medical instruction and collection until 1821. Over the course of the seventeenth century, the university grew and attracted students and visitors from throughout Europe. Published in the early eighteenth century, this guidebook includes the university among the city's main attractions and features two unsigned prints of the theater. The first print shows the space "as it was," and is based on a 1610 engraving after Jan Cornelis Woudanus (ca. 1570–1615). This print quickly became one of the most iconic images of the space and was copied in Jan Janszn Orlers's city description for Leiden, *Beschrijvinge der stad Leyden* (1614). In his preface, Van der Aa explains that he had purchased Orlers's plates and decided to publish a French guidebook to make this information more available to non-Dutch speakers.

Seen here, a second, new print by an unknown artist depicts a contemporary view of the room (*L'Anatomie d'apresent*). The majority of the image is dedicated to the six-tiered, elliptical theater, which was designed after the example of Padua and provided space for two hundred visitors at a time. Medical students and university professors were seated in the lower levels, closest to the cadaver, followed by city magistrates and other lay individuals who occupied the outer tiers (Huisman 2009, 33–34). At the center of this structure, the professor of anatomy conducted his annual demonstrations over a period of three to five days during the cold winter months, which offered favorable conditions for warding off decay.

In the early days of the theater, the skeletons of several anatomical subjects were preserved and put on display during the summer, when demonstrations were not possible. These figures were arranged to hold flags that displayed Latin phrases concerning the transience of life, while plaques detailed their former crimes for curious visitors. Several of these figures are depicted in the print, such as "The Skeleton of a Man who stole Cows and sits on the back of an Ox," seen to the right, and "Two Deserters, stuffed and covered in their own skins" (Van der Aa 1712, 86). Under successive professors of anatomy, the collection expanded to include paintings, prints, natural specimens—some of which can be seen hanging from the ceiling—and objects acquired through trade networks. For example, a "great mummy" brought from Egypt is shown on top of the case in the back right-hand corner and can still be found in Leiden today (Gogelein 1975, 106; Vrolijk 2017, 16). Together, this collection of art, natural history, instruments, and specimens formed a kind of microcosm of the known world, inviting visitors to consider their own place within it (Huisman 2009, 48–59). —ET

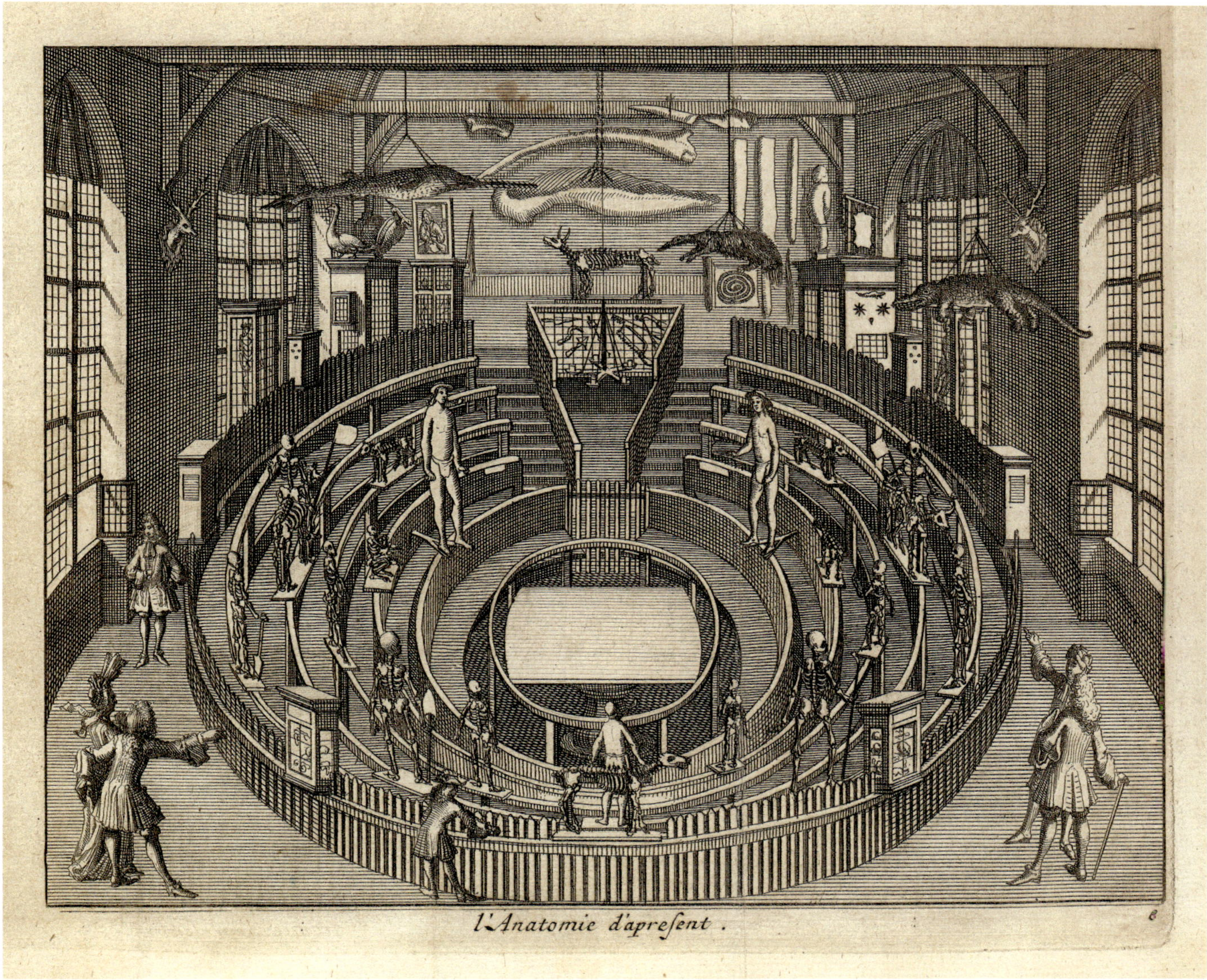
l'Anatomie d'apresent.

8 MUSCLES OF THE ABDOMEN

Woodcut

From Jacopo Berengario da Carpi, *Isagogae breves perlucidae ac uberrimae in anatomiam humani corporis a communi medicorum academia usitatam a Carpo in almo Bononiensi gymnasio ordinariam chirurgiae docente, ad suorum scholasticorum preces in lucem datae* (Bologna: B. Hectoris, 1523), fols. 6v, 7r

Los Angeles, Getty Research Institute, 84-B28214

BIBLIOGRAPHY

Roth 1889; Putti 1937; Choulant 1945, 136–42; Berengario 1959; Mayor 1984; French 1985, 61–62; Park 1994, 23; Carlino 1999a, 95–96; Roberts and Tomlinson 1992, 70–83; Kemp 1993, 94–96; Sawday 1995, 118; Cazort 1996, 38–40; Nutton 2001, 71; Gamberini 2005, 73–74, 76; Laurenza 2012, 19; Pantin 2014, 29–32.

Jacopo Berengario da Carpi (ca. 1460–ca. 1530) lectured in Bologna on surgery, and his medical services were sought after by nobility, particularly the Medici of Florence. Berengario's lengthy illustrated commentary on the anatomical treatise of Mondino de' Luzzi, *Carpi commentaria . . . super anatomia Mundini,* was published in 1521. Many of the woodcuts were reused for *Isagoge breves,* a compendium of the *Commentaria,* first published at the end of 1522. Just seven and a half months later, on 15 July 1523, this second edition was printed. Other editions of this "short introduction to anatomy" quickly followed (Carlino 1999a, 24), demonstrating the popularity of the handy textbook that was, according to the title, published at the request of Berengario's students. A series of abdominal figures assist in their own anatomies by holding back dissected layers, an early example of a method of display that was to become popular (see figs. 2.1, 2.9). The arm that crosses the chest of the dramatically backlit figure on folio 6v displays its principal veins, which were useful for the practice of bloodletting, a common medical treatment in the period (cf. fols. 66r–v).

Renaissance artists Benvenuto Cellini and Giorgio Vasari both remarked on Berengario's connoisseurship and art collecting (Cellini 1901, 55–56; Vasari 1568, part 3, 1:83). The pose of the "rope man" écorché (fig. A) derives from Michelangelo Buonarroti's *David* (Mayor 1984, 91), which Berengario no doubt admired on his trips to Florence, one of them in 1513 to treat Alessandro Soderini at Pope Leo X's behest. David's slingshot is transformed here into a noose and rope, a reference to the gallows as an official source of many bodies for dissection in this period. Berengario's interest in artistic matters is further demonstrated by the numerous changes he made to the illustrations between the *Commentaria* and the two editions of *Isagogae breves* (1522 and 1523), which he oversaw. Rather than relying on the existing woodblocks, he kept tinkering and refining. Each edition that appeared while he was alive had some new illustrations—some replacing earlier ones of the same subject, like the vertebrae. This process was made easier because the images primarily stood alone with their captions and were not extensively cross-referenced in the text. In the 1523 edition, there are new elements including woodcut illustrations of the heart, brain, uterus, spinal column (with details of a vertebra and pelvic bones), a profile of an écorché figure, and a floral border (Putti 1937, 165–94). Also new is a dialogue in verse by Berengario's friend Bernardino Partenio, in which a thief who had been executed and dissected asks the Greek god of the underworld, Pluto, for the return of his body's missing parts (see Kornell, "The Living Dead," this volume, p. 18). —MK

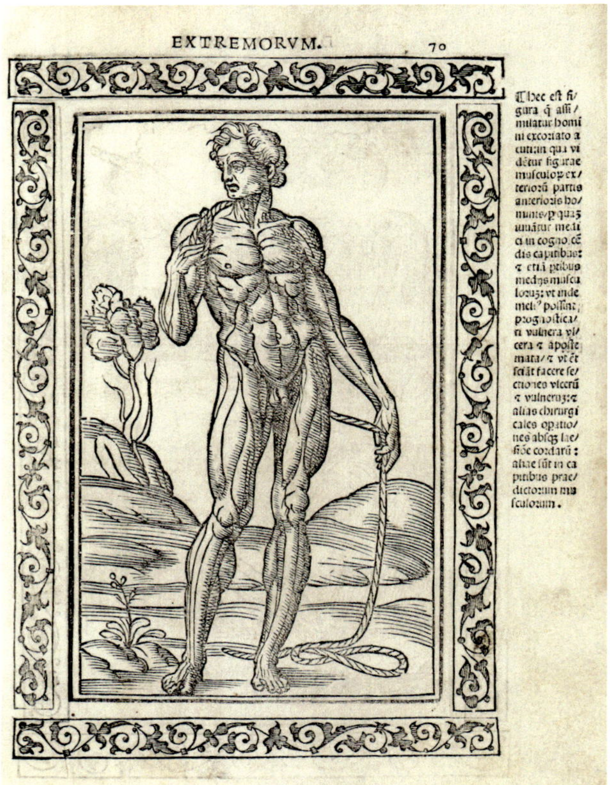

Fig. A
Rope man, woodcut.
From Jacopo Berengario da Carpi, *Isagogae breves* (Bologna: B. Hectoris, 1523), fol. 70r.
Los Angeles, Getty Research Institute, 84-B28214.

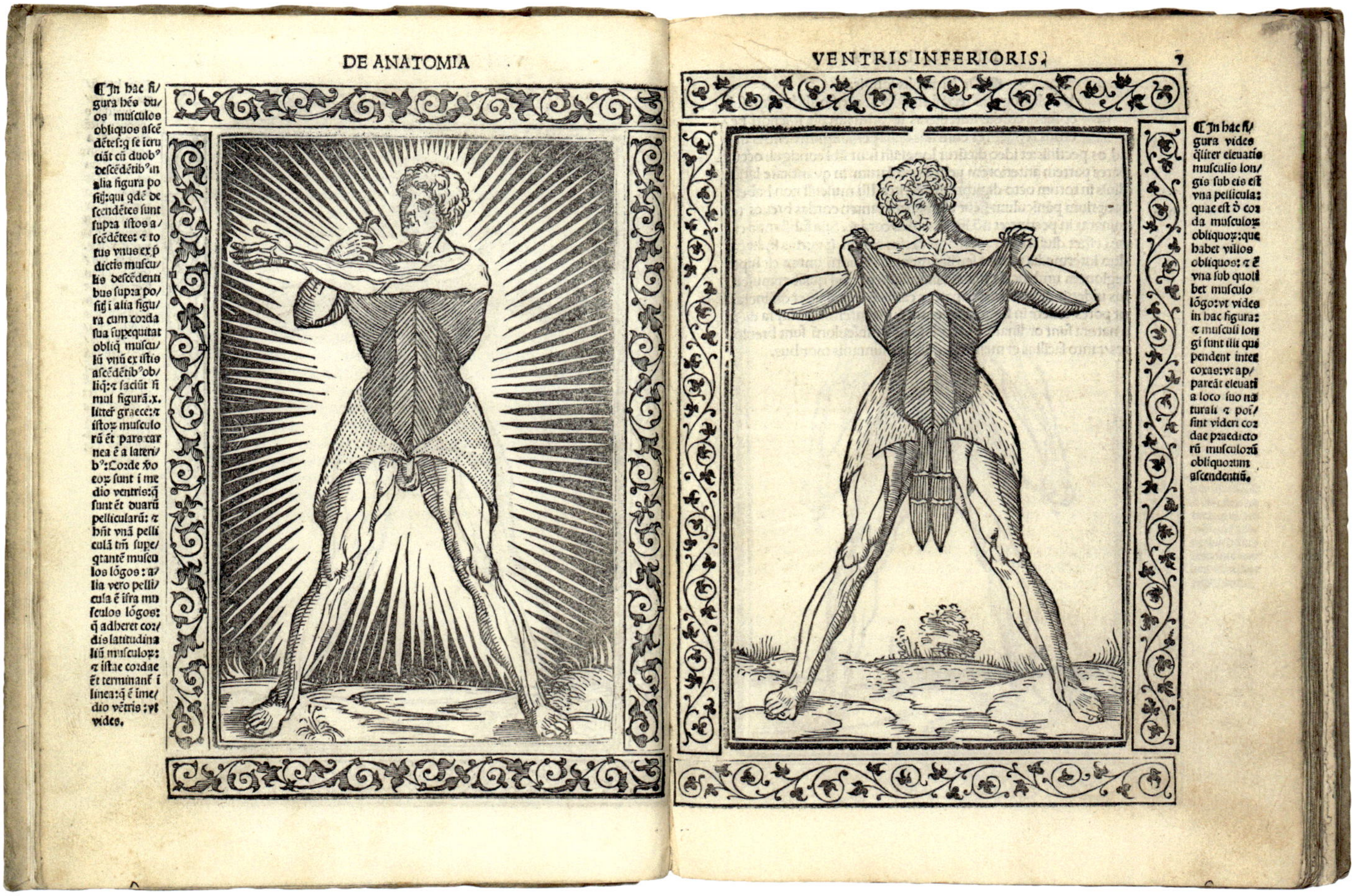
DE ANATOMIA
VENTRIS INFERIORIS.

9 After Nicolas Beatrizet

(French, 1507 or 1515–ca. 1565), after Gaspar Becerra (Spanish, ca. 1520–ca. 1570), after Jan Steven van Calcar (North Netherlandish, ca. 1515–ca. 1546)
Walking muscle figure in profile with two views of the anatomy of the eye
Engraving
From Jacques Guillemeau, *Tables anatomiques, avec les pourtraicts et declaration d'iceulx: Ensemble un denombrement de cinq cens maladies diverses* (Paris: Jean Charron, 1586), p. 84, pl. 2
Los Angeles, Getty Research Institute, 84-B28033

BIBLIOGRAPHY
Portal 1770–73, 2:180–88; Turner 1878, 129–30; Choulant 1945, 213; Mortimer 1964, no. 267; Herrlinger 1970, 167–68.

Jacques Guillemeau (1550–1613) was a student of the surgeon Ambroise Paré and served as a surgeon to French royalty. He published on diseases of the eye (1585), surgery (1593), and obstetrics (1609), and he brought Parè to a wider scholarly audience by translating his collected works into Latin (1582). Guillemeau himself wrote in French—in emulation of Paré, whose book on anatomy is cited in the address to the reader—in order to share medical knowledge with readers without an understanding of Greek and Latin. The *Tables anatomiques* is composed of synoptic tables of anatomy and a listing, according to the title, of five hundred different diseases. The illustrations are adapted from Juan Valverde de Amusco's engraved copies, first published in his *Historia de la composicion del cuerpo humano* (1556), after Andreas Vesalius's *Fabrica* (cat. nos. 30, 31), but it is Vesalius who is acknowledged as the source for them in the address to the reader. In his willingness to select and rearrange the Vesalian illustrations, Guillemeau was following the example set by Valverde and Paré.

On the title page (fig. B), an allegorical figure of surgery is depicted sitting among an array of surgical instruments fanning decoratively out behind her. Nested in a ribbon at the bottom is the monogram of the engraver Léonard Gaultier (ca. 1561–ca. 1630 or 1641) (Mortimer 1964, I, no. 267). Facing the first synoptic table in this copy is an engraved portrait of Guillemeau at age thirty-five that is dated 1585 and signed by Alexandre Vallée (active 1583–1610). The nineteen anatomical plates, however, are unsigned.

The muscle figure shown here is taken from Valverde after Vesalius's second muscle figure (see fig. 2.2) and is one of six (out of the fourteen) Vesalian muscle men that Guillemeau illustrates. Selected body parts culled from the Valverde plates fill the space around them. Floating at the upper right are two figures depicting muscles of the eye adapted and reversed from illustrations original to Valverde (see figs. XX, XXIII in fig. C). Guillemeau improves upon Valverde's fig. XXIII of plate 15 by showing all six muscles of the eye instead of four, and the trochlea, a fibrocartilaginous pulley associated with the superior oblique muscle (Guillemeau 1586, pl. 2, fig. II, no. 7). A later example of filling up available space with anatomical elements is found in Petrioli's *Tabulae anatomicae* of 1741 (cat. no. 14). —MK

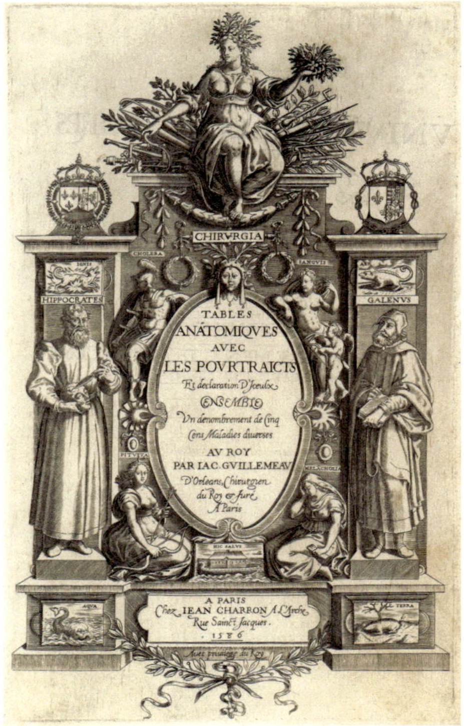

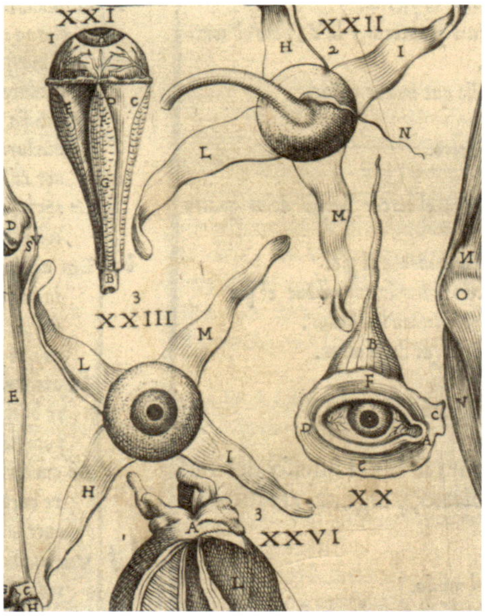

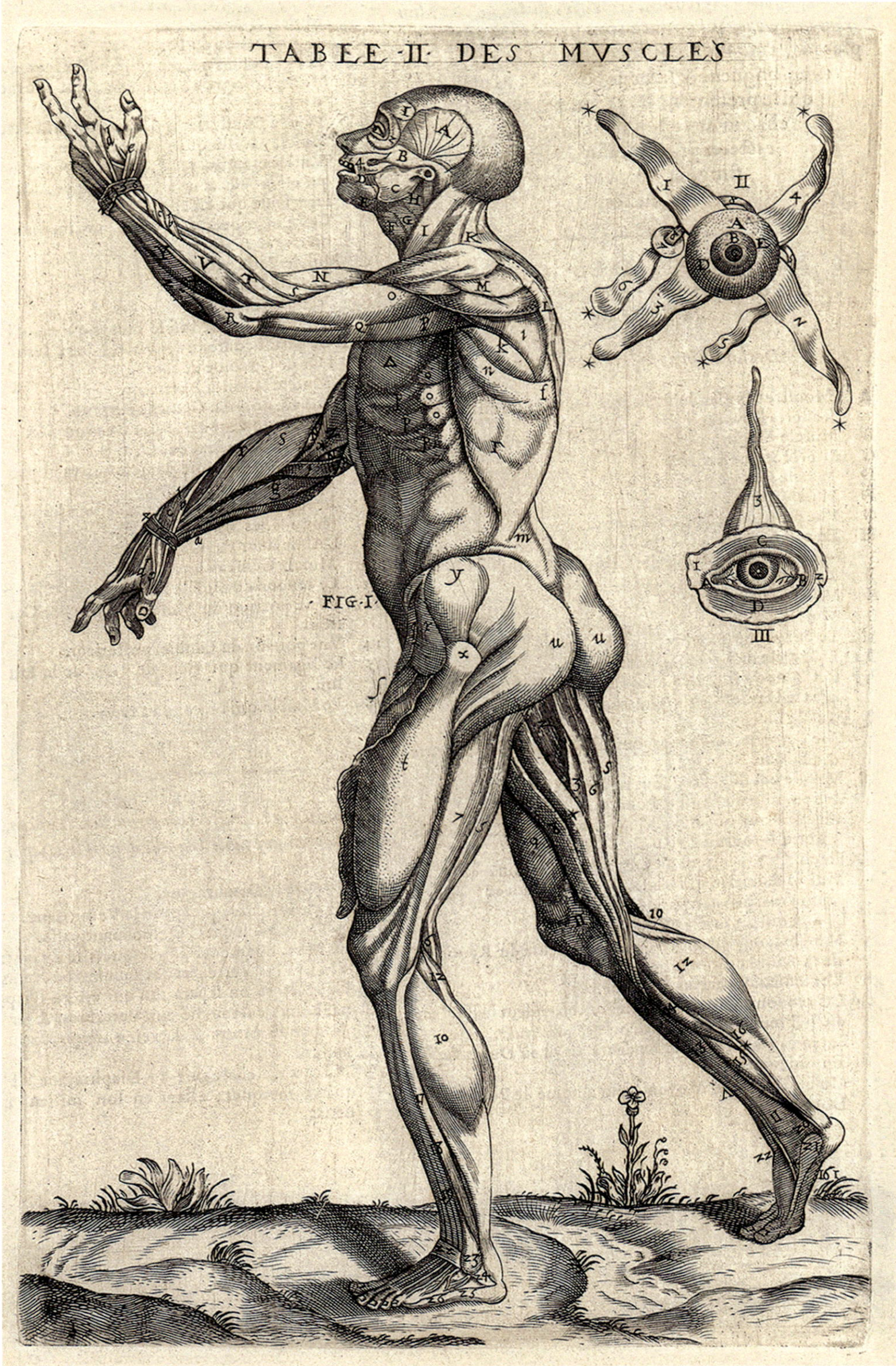

Fig. B
Léonard Gaultier (French, ca. 1561–ca. 1630 or 1641).
Title page, with an allegory of surgery, Hippocrates, Galen, the four humors, and the four elements, engraving.
From Jacques Guillemeau, *Tables anatomiques, avec les pourtraicts et declaration d'iceulx: ensemble un denombrement de cinq cens maladies diverses* (Paris: Jean Charron, 1586).
Los Angeles, Getty Research Institute, 84-B28033.

Fig. C
Nicolas Beatrizet (French, 1507 or 1515–ca. 1565), after Gaspar Becerra (Spanish, ca. 1520–ca. 1570), after Jan Steven van Calcar (North Netherlandish, ca. 1515–ca. 1546).
The eye and its muscles (detail), engraving.
From Juan Valverde de Amusco, *Historia de la composicion del cuerpo humano* (Rome: Antonio Salamanca and Antonio Lafrerij, 1566), bk. 2, pl. 15.
Bethesda, National Library of Medicine, 2294023R.

10 ÉCORCHÉ HORSE IN A LANDSCAPE, SEEN FROM THE FRONT

Woodcut
From Carlo Ruini, *Anatomia del cavallo, infermità, et suoi rimedii* (Venice: Gasparo Bindoni the Younger, 1599), 1:295 [293]
Los Angeles, Getty Research Institute, 87-B1750

BIBLIOGRAPHY
Fantuzzi 1781–94, 7:240, 9:180; Ercolani 1873; Bayon 1935; Cole 1949; Fanti and Chiossi 1984; Sørensen 2002; Laurenza 2012, 28.

In the history of veterinary medicine, Carlo Ruini's *Anatomia del cavallo* is considered the first book devoted to a single animal (Cole 1949, 83). The book is divided into two volumes, the first being on anatomy and the second on diseases of horses and their treatments. It is unknown what medical training Ruini (1530–98) had, if any. His love of and admiration for horses, however, is clear. In the preface to the section on anatomy, he discusses the worthy nature of the horse, judging that his book would be pleasing and profitable for those who wished to know the anatomy of such a generous and useful animal and how to treat its ailments. The Ruini, a noble Bolognese family, owned several horses—an inventory from 1553 lists twenty-three at the family's farmhouse, each by its own name (Fanti and Chiossi 1984, 39). Just a month after penning the dedication of the first edition to Cardinal Pietro Aldobrandini on 7 January 1598, both Ruini and his wife are recorded as deceased (Ercolani 1873, 85–86).

Work on the illustrations had begun by 1590, the date appearing on a woodcut illustration of the skeleton profile in the first volume. In their presentation, they were modeled on those of Andreas Vesalius's *Fabrica* (1543) (see fig. 2.2; cat. no. 30). Ruini's animated écorché of a horse, caught in midstride, is placed in a landscape setting. The walled city in the distance is probably meant to represent Bologna, where the first edition was published in 1598. The horse's strong shadow oddly extends toward instead of away from the bright sun seen just over the trees on the right of the image.

Ruini's book was the *Fabrica* of equine anatomy in that its illustrations proved enduringly influential. Ruini's book went through multiple editions and was the source for copies through to the nineteenth century (Ercolani 1873, 88–93; Bayon 1935, 140–43; Cole 1949, 88–89; Sørensen 2002), with a particularly impressive exception being the English artist George Stubbs's *Anatomy of the Horse* (1766). The Getty Research Institute's copy of *Anatomia del cavallo* is the second edition, which was published in Venice by Gasparo Bindoni the Younger and reused the original woodblocks. The new dedication by Bindoni was to César de Bourbon, Duke of Vendome (1594–1665), the illegitimate first son of Henry IV of France (1553–1610), who at the time was only four or five. Bindoni, seeking the favor of the duke's father, mentions that Ottavio Ruini, one of Ruini's sons, served Henry IV in the siege of Amiens in 1597. Bindoni describes the late Carlo Ruini as having spent the greater part of his life observing "all that one could consider concerning the beauty and the goodness, as well as the health of horses, of which he had the noblest and most exquisite breeds that one could find." —MK

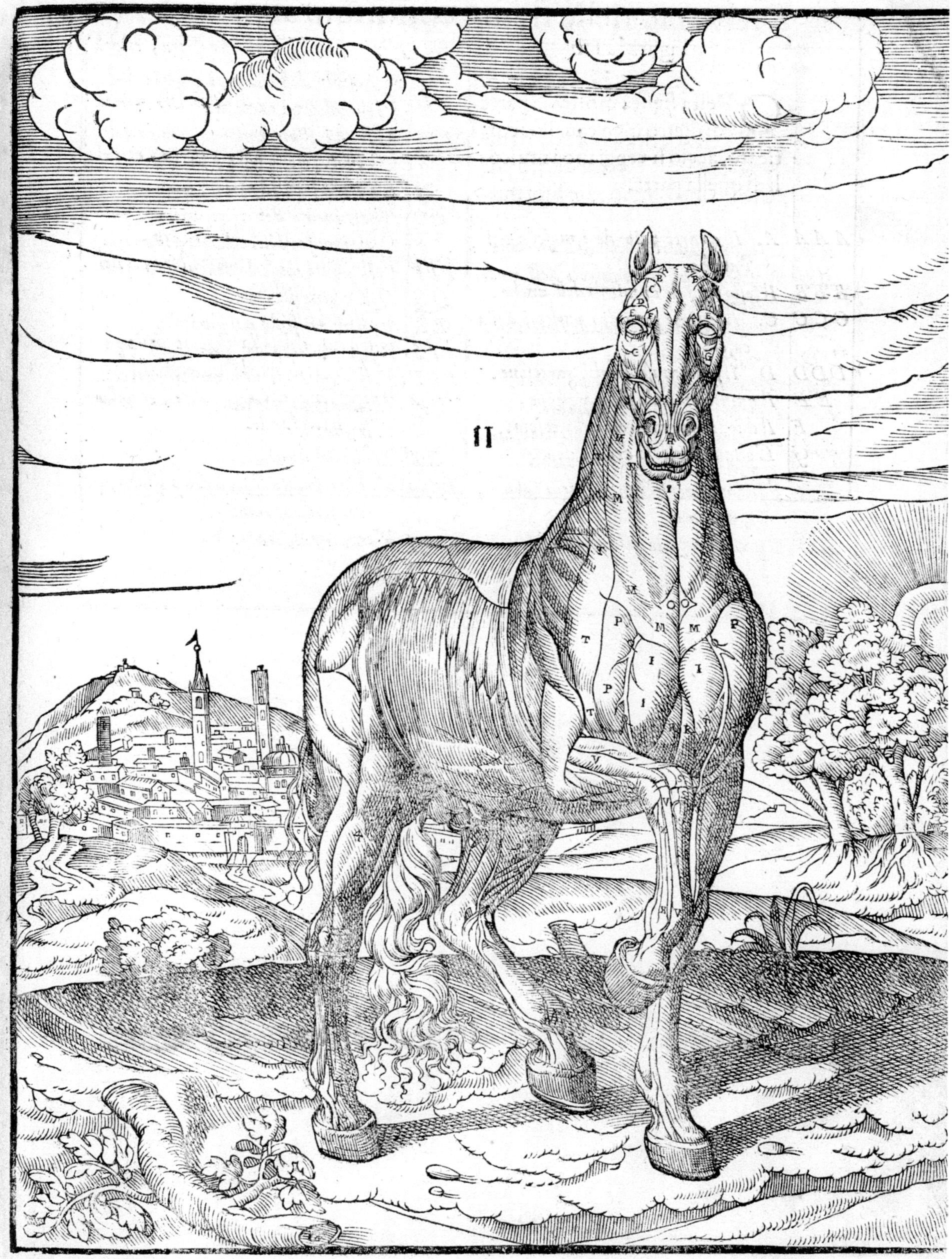
II

11 ***THE MUSCLES OF THE EYE IN ITS NATURAL POSITION AND THE MUSCLE OF THE EYELID, SHOWN SEPARATELY***

1609
Engraving
From Thomas Bartholin, *Anatome ex omnium veterum recentiorumque observationibus*, 5th rev. ed. (Leiden: Jacobum Hackium, 1686), bk. III, p. 507, pl. IX
Los Angeles, Getty Research Institute, 93-B10050

BIBLIOGRAPHY

Sterzi 1910, 38–42; Choulant 1945, 223–24, 245–46; Cole 1949, 112–14; Cunsolo 2005, 98–102; O'Malley 2008.

Thomas Bartholin (1616–80), a member of a renowned family of Danish physicians, was responsible for an early description of the lymphatic system, which is included in this edition. The book is one of several revisions to the textbook of his father, Caspar Bartholin the Elder (1585–1629), *Institutiones anatomicae corporis humani*—first published without illustrations in 1611. These figures of the eye are copied from Giulio Casseri's book on the five senses, *Pentaesthesеion*, from the section devoted to sight (Casseri 1609, 259, fig. 7; 265, fig. 2). Caspar Bartholin's friendship with Casseri (1561–1616) is attested by his laudatory verses that appear at the beginning of *Pentaestheseion;* according to Thomas, they had dissected many bodies together in Padua (Bartholin 1651, *5v).

In figure I, the sockets of the eye have been exposed to display the muscles of the eye. The connecting glance of the right eye, resting in its socket and turned toward the viewer, imparts a vivid impression of life to the face, an effect seen in illustrations of both human and animal in Casseri's book (see fig. 2.10). At the same time, the drooping left eye, suspended by its muscles, adds a note of pensiveness. The lower half of the face remains intact, and a pleasing sense of volume is evoked by the change of pattern in the engraving of the chin against the neck and by the areas left free of hatching. Figure II is a view of the muscles of the eye and the eyelid (see "H"), removed from the head.

The unsigned illustrations in *Pentaestheseion* may have been drawn by the Swiss artist Josias Murer II (1564–1630), who was living in Casseri's house in 1593 for the purpose of "painting anatomical figures," as the anatomist comments in his *De vocis auditusque organis historia anatomica* (Casseri 1600–1601, pt. 2, *De auris auditus organi historia anatomica*, bk. 1, ch. 13, p. 79). The Bolognese-born artist Odoardo Fialetti (1573–1626/27) might also have been responsible for some of them, as he had been working for Casseri on illustrations for an anatomy book, for which he completed 150 copperplates by 1613—a project that likely took several years to prepare (cat. no. 12).

In 1678, Thomas Bartholin's son, Caspar the Younger (1655–1738), continued the tradition of publishing parental works by issuing a second edition of his father's book on unicorns and other horned creatures, *De unicornu observationes novae*, first published in 1645, to which he added illustrations (Bartholin 1678) (GRI, 93-B13194). —MK

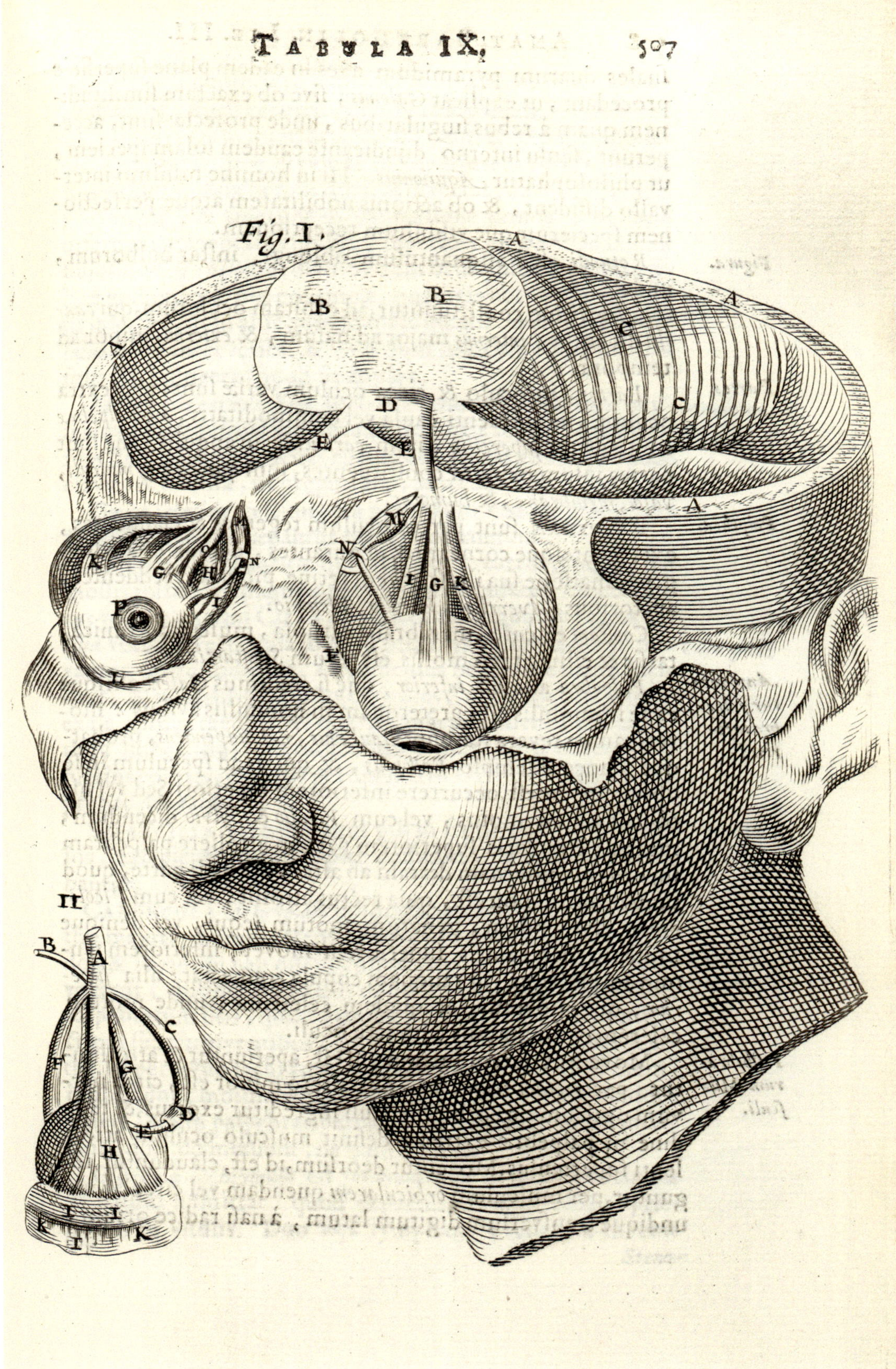
TABULA IX.
507
Fig. I.
II.

12 Francesco Valesio

(Italian, active 1598–1624), after Odoardo Fialetti (Italian, 1573–1626/27)
Dissected legs walking in a landscape
By 1616
Engraving
From Adriaan van den Spiegel, *Opera quae extant, omnia,* edited by Johannes Antonides van der Linden (Amsterdam: Joan Blaeu, 1645), bk. 4, p. 91, pl. 31
Los Angeles, Getty Research Institute, 84-B2833

BIBLIOGRAPHY

Sterzi 1910, 44–45, 100; Choulant 1945, 223–28; Roberts and Tomlinson 1992, 266–67; Cazort in Cazort, Kornell, and Roberts 1996, 168; Maugeri 1997; Riva et al. 2001; Cunsolo 2005, 107; Walters 2009, 131–41, 145; Houtzager 2014, 213; Moretti 2014, 19n15.

In his application for the professorship of anatomy in Padua in 1613, Giulio Casseri (1561–1616) states that there are 150 figures in his house ready to be engraved for a work on the anatomy of the human body—and on which he had expended three thousand ducats (Sterzi 1910, 45, 100, doc. 24). Casseri had earlier announced a forthcoming illustrated anatomy book in his *De vocis auditusque organis historia anatomica* (1600–1601) and again in his *Pentaestheseion* (1609) (see fig. 2.10), yet when he died in 1616 it still had not appeared. Instead, the illustrations prepared for Casseri first appeared in Venice in the *Tabulae anatomicae* (1627), published with the posthumous works of Adriaan van den Spiegel (1578–1625), professor of anatomy and surgery in Padua (Casseri 1627). This edition was overseen by Daniel Rindfleish, known as Bucretius (ca. 1600–1631), whose preface identifies Odoardo Fialetti and Francesco Valesio as Casseri's original artistic team, from whom Bucretius also commissioned additional plates copied after Vesalius and other authors.

Hieronymus Fabricius ab' Acquapendente, Casseri's onetime master and then rival, also had plans for a large work of anatomy; related paintings and prints of human and comparative anatomy in color for the unpublished volume survive in Venice (Rippa Bonati and Pardo-Tomás 2004). But whereas Fabricius's illustrations are records of inanimate dissected bodies, Casseri's figures are restlessly alive like these walking legs, which are both eerie and amusing at the same time.

This 1645 edition of Van den Spiegel's complete works includes his *De formato foetu,* which was first published by his son-in-law Liberalis Crema in about 1626, with nine illustrations on gestation and the fetus that had also been prepared for Casseri by Fialetti and Valesio. Fialetti may not have seen any of his illustrations for Casseri in press, since he is described as deceased in his son's baptismal record of 6 March 1627 (Moretti 2014, 19n5). The Bolognese-born Fialetti was the son and brother of physicians. His brother Tiberio, who studied medicine at Padua, brought Odoardo with him to the Veneto, where Odoardo went on to have a successful career as a painter and printmaker. Cesare Carlo Malvasia relates how Tiberio encouraged Odoardo to draw the individual bones, skeleton, and muscles (Malvasia 1678, 1:313–14). Odoardo illustrated several books and published a drawing manual in two variant issues in Venice in 1608. Correspondences may be seen between Casseri's dissected male figures and the foreshortened heads in *Il vero modo et ordine per dissegnar tutte le parti et membra del corpo humano* (GRI, 86-B10115), which would have been drawn within a few years of each other.

Among the other illustrated works included in this edition are William Harvey's *De motu cordis* (1628) and Gaspare Aselli's *De lactibus* (1627). The chiaroscuro woodcuts of the latter are an early example of the use of color in anatomical illustration but are rendered in this edition as uncolored engravings. —MK

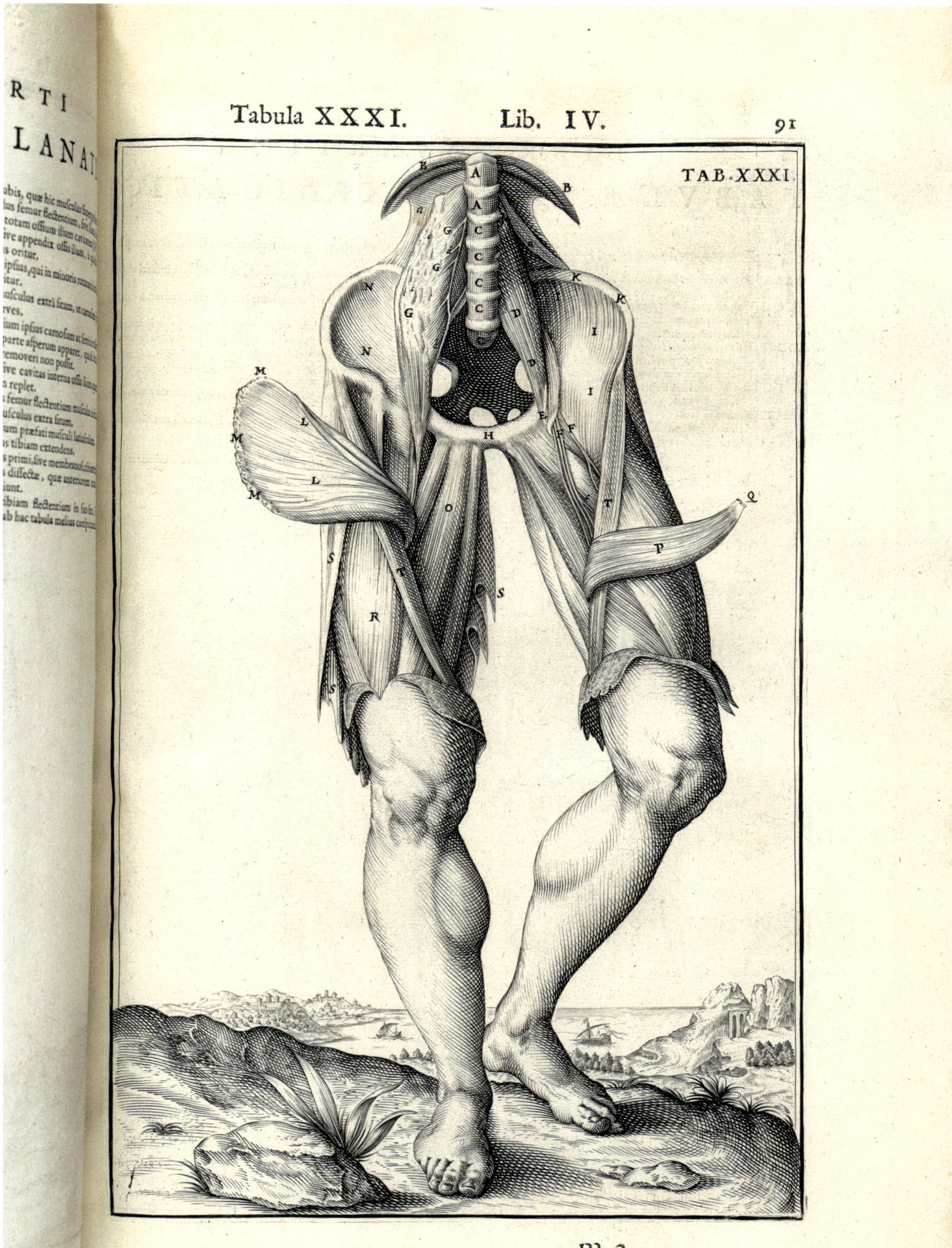
Tabula XXXI.
Lib. IV.
91
TAB. XXXI.
m 2

13 William Cowper

(English, 1666/67–1710) or Michael Vandergucht (Flemish, 1660–1725, active England), after William Cowper
Muscle figures in profile view
Etching
From William Cowper, *Myotomia reformata; or, An Anatomical Treatise on the Muscles of the Human Body: Illustrated with Figures after the Life*, 2nd ed. (London: Printed for Robert Knaplock, William and John Innys, and Jacob Tonson, 1724), pl. 5
San Marino, California, The Huntington Library, 409834

BIBLIOGRAPHY
Roberts and Tomlinson 1992, 416–17, 420–21; Kornell in Cazort, Kornell, and Roberts 1996, 186–90, nos. 72–74.4; Petherbridge and Jordanova 1997, 14–15, no. 54; Sanders 2005; Black with Dulau in Black 2007a, 163, nos. 117–20; Hanson 2012, 159–61; Flis in Campbell and Flis 2018, 216, no. 31; Kornell 2019.

The anatomist James Douglas described the English surgeon and anatomist William Cowper as "equally Famous for his wonderful Dexterity in Dissecting, and great Skill in Designing" (Douglas 1707, xii–xiii). Although Cowper is better known today for what has been considered his scandalous reuse of Gérard de Lairesse's illustrations for *The Anatomy of Humane Bodies* (1698), in which Cowper's commentary replaced the often-cursory one by Govard Bidloo (cat. no. 40), Cowper made many original contributions to the history of anatomical illustration, supplying the drawings for his own publications, such as the first edition of the *Myotomia reformata* (1694), and for those of others. Cowper also provided the drawings as well as some of the etchings for the second edition of his *Myotomia reformata* (Kornell 2019, 493), the publication of which was overseen after Cowper's death by the physician, collector, and fellow member of the Royal Society Richard Mead (1673–1754). It is illustrated with sixty-six new plates, to which Mead added a lengthy introduction on muscular motion by Henry Pemberton (1694–1771). The copy in the Huntington Library bears a presentation inscription on the flyleaf from Mead. The name of the original recipient has been crossed out but to the left has been written the suggestion of "Dr Littlejohn." Many of the illustrations, to discourage their removal, bear the stamp of the Edinburgh Medical Society—the library of which was partially dispersed in sales in 1969.

In this fifth plate, Cowper follows the standard set by Andreas Vesalius of gradually stripping the whole body to the bone, but Cowper innovates by placing his figures in the same pose, side by side, allowing for a comparison of anatomical layers while their matched steps and gestures give them a choreographed air that would not be out of place in a well-rehearsed stage show. Cowper's aim was to provide illustrations of every muscle described in the text. In addition to full figures, Cowper provided depictions of individual muscles on a more ambitious scale than had previously been attempted, with Charles Estienne and Estienne de la Rivière's *De dissectione partium corporis humani* being a notable early example (Estienne and La Rivière 1545, bk. 3, 311–35). Cowper's text is embellished with head- and tailpieces of écorché figures set in dramatic poses as well as decorated initial letters inspired by Vesalius's *Fabrica* that reflect the theme of the chapters they adorn. The University of Glasgow Library has many of Cowper's preparatory drawings, which were purchased by the anatomist and physician William Hunter at Mead's posthumous sale. Among these are red chalk drawings by Cowper for the figures in plate 5 in MS. Hunterian Ay.2.5, fol. 43 (left figure) and fol. 45 (right figure). —MK

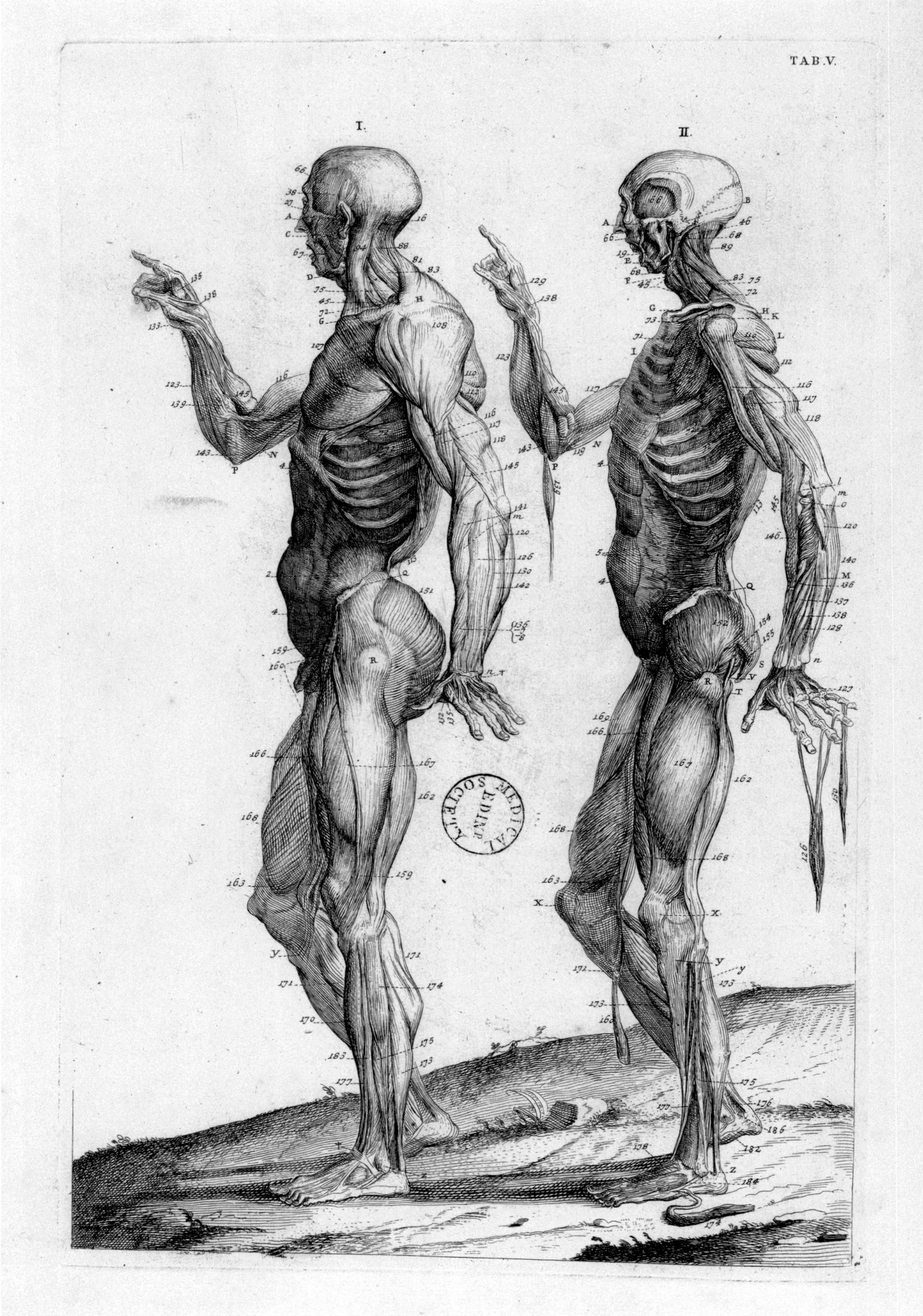
TAB.V.
I.
II.

14 Luca Ciamberlano

(Italian, b. ca. 1570), after Pietro da Cortona
(Italian, 1597–1669)
Nerves and muscles of the back and legs; various bones
Engraving
From Gaetano Petrioli, *Tabulae anatomicae a celeberrimo pictore Petro Berrettino Cortonensi delineatae, & egregiè aeri incisae nunc primum prodeunt, et a Cajetano Petrioli Romano doctore, regis Sardiniae chirurgo, publico anatomico, & inter arcades Erasistrato Coo notis illustratae* (Rome: Printed at the expense of Fausto Amidei by Antonio de' Rossi, 1741), pl. 19
Los Angeles, Getty Research Institute, 84-B4213

BIBLIOGRAPHY
Petrioli 1750, xlii, cviii, xcii [cxii]; Moehsen 1771, 99–103; Haller 1774–77, 1:579; Petraglia 1788; Choulant 1945, 235–39; Savio 1968, 266–67; Putscher 1972, 15; Kemp 1976; Duhme 1980; Norman 1986; Merz 1991, 25–35; Roberts and Tomlinson 1992, 272–79; Cazort in Cazort, Kornell, and Roberts 1996, 181–83, no. 67; Merz 1997, 57–59; Black 2007a, 161–62, no. 114; Măgureanu 2013–14; Campbell and Flis 2018, 214, no. 29.

The young Pietro da Cortona, who went on to be a leading painter and architect in seventeenth-century Rome, designed most of the plates for this book by 1618, the date visible on the first plate in some copies, along with Cortona's signature (Duhme 1980, 7). The monogram of Luca Ciamberlano, the engraver, appears on tables 1 and 4. Cortona was sufficiently famous to be mentioned in the title as the artist when the illustrations were first published in 1741 with anatomical explanations supplied by Gaetano Petrioli, a physician and surgeon to the king of Sardinia, Charles Emmanuel III. Petrioli also provided commentary for an edition of the illustrations of the sixteenth-century anatomist Bartolomeo Eustachi (see fig. 1.4) (Petrioli 1740).

Cortona's drawings for the first nineteen tables as well as the twenty-seventh table of female anatomy survive in Glasgow and were once owned by the anatomist, physician, and collector William Hunter (Kemp 1976). It is not known for whom they were originally made. An inscription with the drawings states that Cortona was aided by the surgeon "Nicolo Lache." This name has been associated with Nicholas Larche (1602–65), a French surgeon working in Rome who had instructed the French painter Nicolas Poussin (1594–1665) in anatomy; however, Larche would have been only about sixteen years of age and therefore too young (Merz 1997, 32–33).

The illustrations highlight the location of the nerves within the body. In the nineteenth plate, the leg muscles have been fanned out in a decorative fashion to expose the nerves and their attachments. An innovative feature of the illustrations is the use of roundels and tablets to display secondary views of anatomical structures, often on a larger scale. The subsidiary images that fill up available space around the figures and on architectural elements—not present in the original drawings—are based on Andreas Vesalius, Giulio Casseri, and other sources, as are plates 20–26 (Duhme 1980, 29; Măgureanu 2013–14, 260–61). Hunter considered the supplemental figures to the plates related to the Glasgow drawings "injudiciously put in" (Kemp 1976, 148), and the Roman professor of philosophy and medicine Francesco Petraglia had them wiped from the plates (Petraglia 1788). This anatomical horror vacui has been attributed to Gaetano Petrioli (Kemp 1976, 147); however, his description of all "25 [*sic*] plates in folio, adorned with 221 figures" that he dates to 1620 and says were drawn by the "celebrated Pietro da Cortona" implies that he considered all the figures original to the artist (Petrioli 1750, xlii, no. 12). —MK

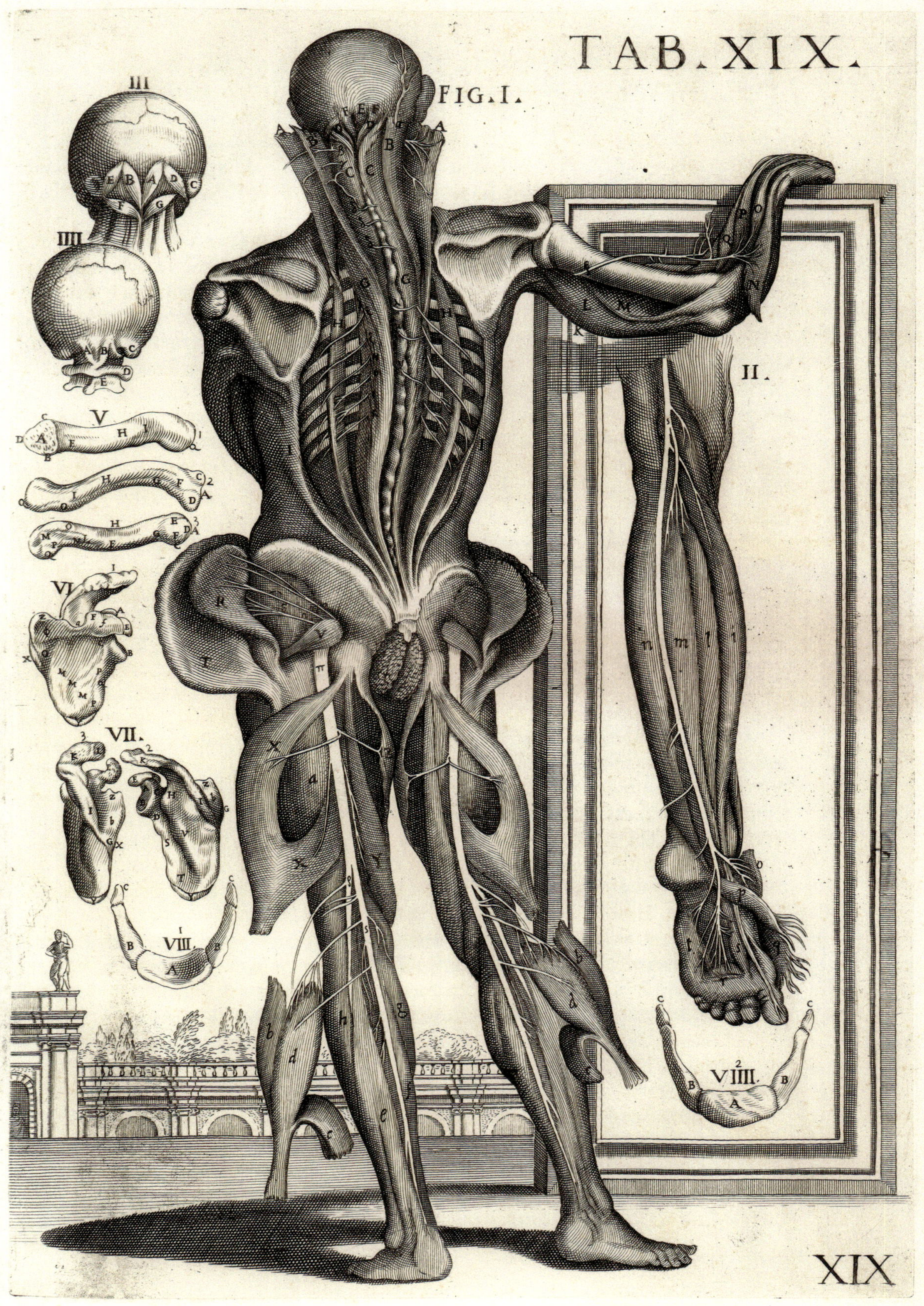
TAB. XIX.
FIG. I.
III
IIII
V
VI
VII.
VIII.
II.
VIIII.
XIX

15 Jan Wandelaar

(Dutch, 1690–1759)
First muscle figure
1739
Etching and engraving, plate mark: 56.1 × 40 cm;
sheet: 76 × 54 cm
From Bernhard Siegfried Albinus, *Tabulae sceleti et musculorum corporis humani* (Leiden: Johannes and Herman Verbeek, 1747), unbound sheet
Los Angeles, Library Special Collections for Medicine and the Sciences, Louise M. Darling Biomedical Library, UCLA, BIOMED ** WZ 260 A337t 1747 plates

BIBLIOGRAPHY

Choulant 1945, 276–83; Punt 1983; Croke 1992, 37–55; Huisman 1992; Roberts and Tomlinson 1992, 320–29; Kemp 1993, 107–12; Cazort in Cazort, Kornell, and Roberts 1996, 194–200, nos. 77–88.

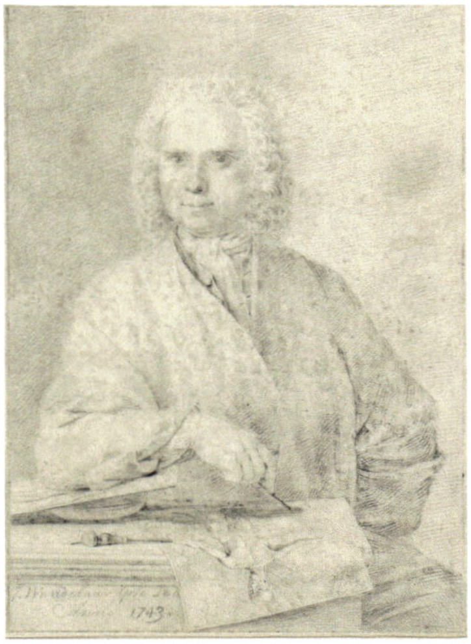

Fig. D
Jan Wandelaar (Dutch, 1690–1759).
Self-portrait, 1743, black chalk on parchment, 21.5 × 15.5 cm.
Amsterdam, Rijksmuseum, RP-T-1940–288.

The German-born anatomist Bernhard Siegfried Albinus (1697–1770), professor of medicine at Universiteit Leiden, first worked with the draftsman and engraver Jan Wandelaar on the *Opera omnia* of Andreas Vesalius (1725), which Albinus edited with his former teacher Herman Boerhaave. Thus began a lifelong collaboration with the artist, who provided plates for several works by Albinus, and for his editions of Hieronymus Fabricius ab' Acquapendente (1737) and Bartolomeo Eustachi (1744).

Upon publication, the *Tabulae sceleti et musculorum corporis humani* was met with great success, and its illustrations soon replaced Vesalius's (cat. no. 30) as the standard images of the muscles and bones and the preferred source of copyists (cat. no. 22; see fig. 1.9). Arnauld Éloi Gautier Dagoty was of the opinion that they "surpass all that has hitherto appeared, for precision, sharpness, and accuracy" (Gautier Dagoty, A. É. 1773, *Plan de l'ouvrage*). Following the series of skeletons and muscle figures in three views, each muscle is depicted individually, showing their insertions on the bones. Separate outline plates carry lettering that is keyed to an anatomical explanation, leaving the main figures unobscured. Albinus's goals were accuracy and perfection, and his figures are the result of a careful selection from multiple bodies over several years. In this respect, Albinus is the anatomical counterpart to the ancient Greek artist Zeuxis, who chose the most beautiful features from the maids of Croton to create his portrait of Helen of Troy.

Albinus's detailed account of the preparation of the plates reveals his desire to control not only the type of body represented but also the working methods of his artist, Wandelaar, as well as the printing process, in order to ensure the accuracy of the plates. He oversaw Wandelaar's work so closely that he remarked it was as if "I made the figures myself" (Albinus 1754, xxii). Albinus also valued that Wandelaar both drew and engraved the plates, thereby lessening the possibility of errors and misinterpretations. Like those for William Cheselden's *Osteographia* of 1733 (cat. no. 33), the first drawings for the *Tabulae* were abandoned in favor of those that were mechanically guided. To avoid any distortion of perspective or scale, two grids of rope were employed at different distances through which Wandelaar viewed an articulated skeleton using a stationary eyepiece. The closer grid was used to capture the details of the bones. The resulting drawings served as a base on which selected muscles were drawn. Wandelaar's preparatory drawings, including some life-size figures, which were then scaled down for publication, survive in the library at Universiteit Leiden.

The lush backgrounds, brimming with vegetation, architecture, classical tombs, rocks, streams, and, famously, a rhinoceros (see fig. 1.8) were suggested by Wandelaar as a way to enhance the three-dimensionality of the figures. This effect is heightened when the viewer looks at them, as Albinus suggests, through a hand "placed before the eye in the manner of a spy-glass," so as to cut out the surrounding light (Albinus 1754, xxii). The evident pride Wandelaar took from the project is seen in his drawn self-portrait, signed and dated 1743, four years before the publication of the *Tabulae*, in which he depicts himself at work on the first muscle figure (fig. D). —MK

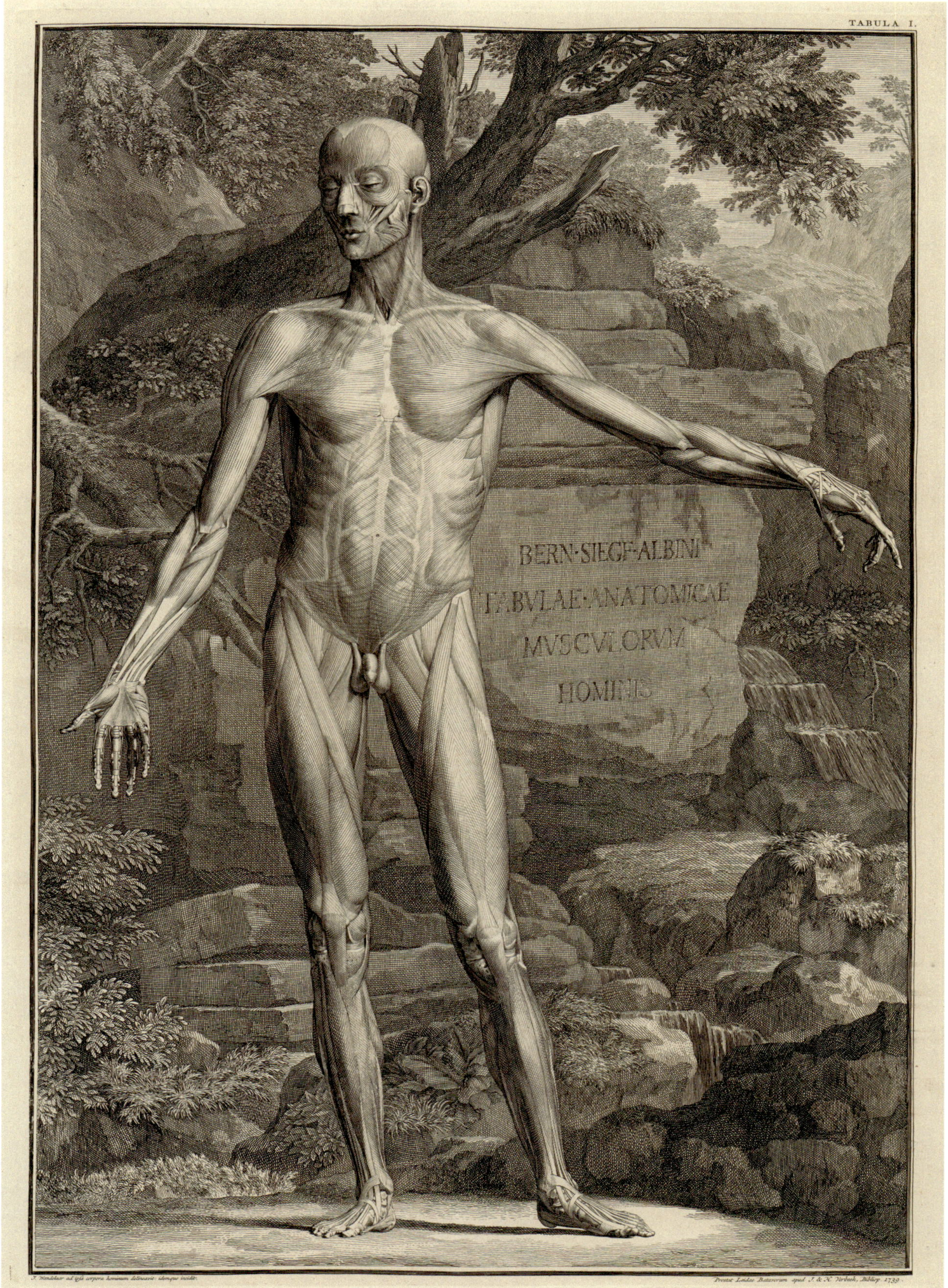
TABULA I.
BERN·SIEGF·ALBINI
TABVLAE·ANATOMICAE
MVSCVLORVM
HOMINIS
J. Wandelaar ad ipsa corpora hominum delineavit: idemque incidit.
Prostat Leidae Batavorum apud J. & H. Verbeek, Bibliop. 1739.

16 TITLE PAGE SHOWING DISSECTION SCENE

Woodcut
From Realdo Colombo, *De re anatomica libri XV*
(Venice: Nicolò Bevilacqua, 1559)
Los Angeles, Library Special Collections for Medicine and the Sciences, Louise M. Darling Biomedical Library, UCLA, BIOMED BENJ * WZ 240 C717dr 1559

BIBLIOGRAPHY

Russell 1953; Coppola 1957; Steinberg 1982; Ciardi 1984; Carlino 1999a, 59–66; Ciardi 2003, 82–83; Hillard 2007; Bylebyl 2008; Cunningham, A. 1997, 148–50; Colombo 2014, xlvii–li; Resta 2014.

A rival of Andreas Vesalius and teacher of Juan Valverde de Amusco, the Cremona-born surgeon and anatomist Realdo Colombo (ca. 1515–59) is known for his description of the pulmonary circulation of the blood. He also claimed the first description and explanation of the clitoris. On the woodcut title page to his *De re anatomica*—the book's only illustration—he is shown with scalpel in hand standing before a cadaver's opened abdomen and surrounded by onlookers. As in the title page to Vesalius's *Fabrica* (see fig. 1.1), the anatomist carries out the dissection himself rather than relying on a prosector. Colombo looks out to his audience while he gently supports the cadaver's left hand, which is draped over his wrist. Below him, in the left-hand corner, a young man—likely an artist rather than an amanuensis, given the position of his pen—sits with one leg tucked under the other, ready to record the scene.

An illustrated book is held by a man to the left of the body, and it serves as an evocation of the plans that Colombo originally had for *De re anatomica.* Writing to Duke Cosimo de' Medici on 17 April 1548, Colombo informs his employer that he is working on a book of anatomy and requests leave from his post teaching anatomy in Pisa to stay in Rome. The reasons are twofold: He wants to take advantage of the good supply of bodies to dissect in that city, and "the best artist in the world" is assisting him (Coppola 1957, 45–46, 55–56). He also mentions a desire to "supervise the painters," indicating that he was working with a team of artists.

Who was the unnamed artist Colombo refers to in his letter to Duke Cosimo? It is usually assumed, with good reason, to be Michelangelo Buonarroti, since he was both a friend and patient of Colombo's. According to Michelangelo's student Ascanio Condivi, the two conferred on anatomy (Condivi 1553, between fols. 42, 43). It has been

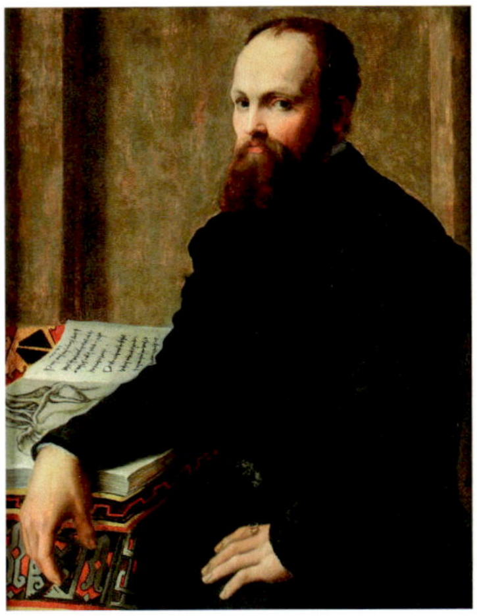

Fig. E
Attributed to Francesco Salviati (Italian, 1510–63). *Portrait of Realdo Colombo*, oil on panel, ca. 1548, 89 × 71 cm.
Knutsford, Cheshire, England, Tatton Park, NT 1298173.

suggested that Michelangelo is the man whose fingers are held by the putto in the foreground (Ciardi 1984). However, his features do not accord with contemporary portraits and descriptions of Michelangelo, who in his later years had a deeply furrowed brow and a sparse, bifurcated beard. Nor is it a depiction of a man of eighty-four, Michelangelo's age at the time.

Some evidence that engraved illustrations for the book were prepared is found in a history of Colombo's birthplace, *Cremona fedelissima città* (1585), written by the painter, architect, and engraver Antonio Campi (1523–87). Campi mentions Colombo's "great and very learned volume of anatomy in which with the purest Latin style he taught all that one could know on the subject, and to further facilitate the way of learning it, he also printed in copper infinite figures of anatomy" (Campi 1585, bk. 3, xlij). Another trace of the existence of the illustrations may be found in a portrait in Tatton Park, outside of Manchester; the sitter may be identified as a younger Colombo at an age that corresponds with the period during which he was working on the illustrations in the late 1540s (fig. E). It shows him resting an arm on a manuscript that is open to a page of illegible script and what appears to be an unclear drawing of neck muscles and the clavicles.

Despite his promises to the duke, Colombo never returned to Pisa, remaining in Rome and teaching at the Sapienza Università. The copy shown here bears a dedication to Pope Paul IV, who died in August of 1559. By the time a second issue was published later the same year and dedicated by Colombo's sons to the new pope, Pius IV, the anatomist was dead. The woodblock for the title page illustration survives in the Princeton University Library (GA 2011.01417). The British Museum, London, holds a poorly printed impression of the woodcut that lacks the title lettering in the cartouche (inv. 1874,0808.1372). —MK

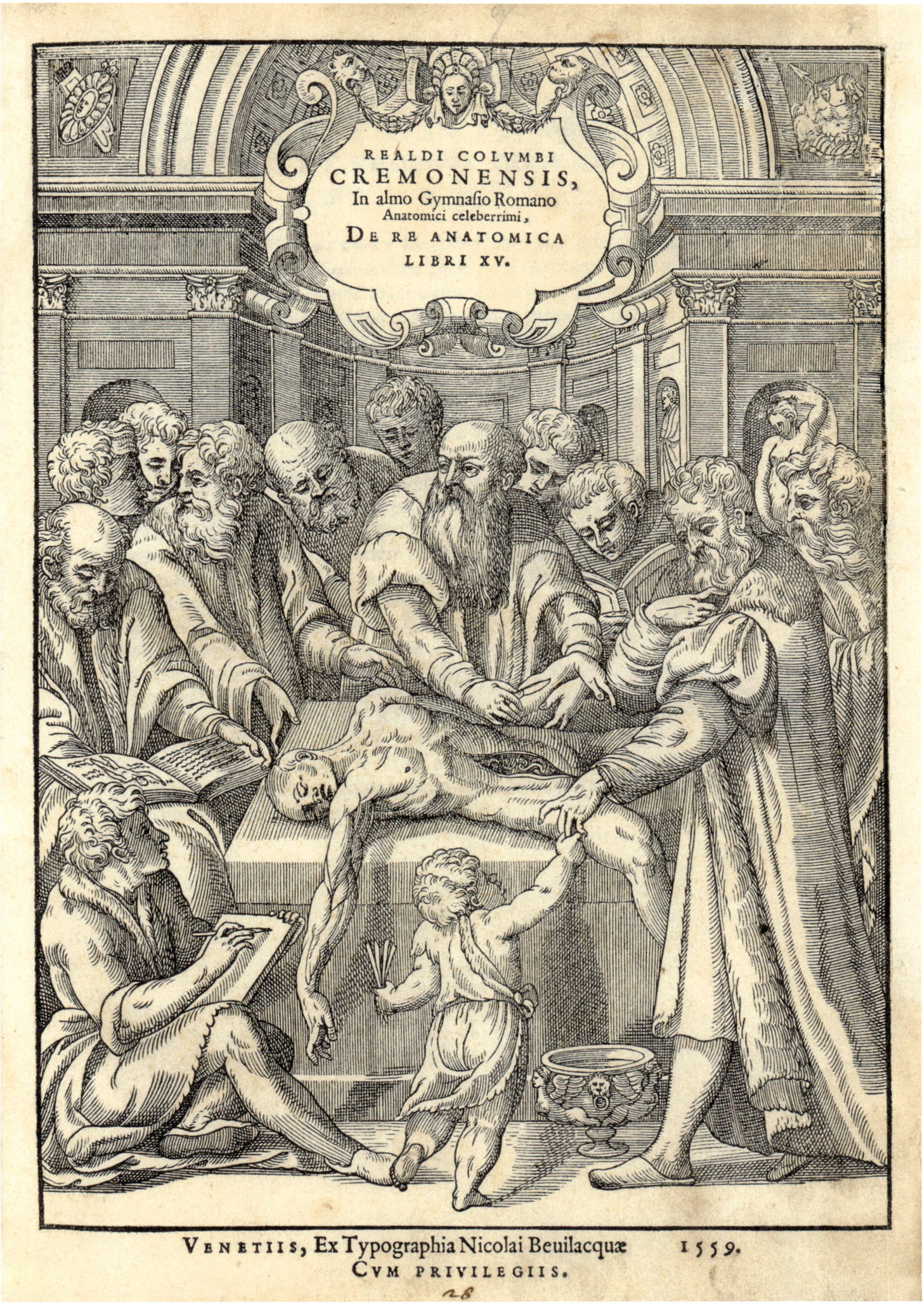
REALDI COLVMBI
CREMONENSIS,
In almo Gymnasio Romano
Anatomici celeberrimi,
DE RE ANATOMICA
LIBRI XV.
VENETIIS, Ex Typographia Nicolai Beuilacquæ 1559.
CVM PRIVILEGIIS.

17 Juan de Arfe y Villafañe

(Spanish, 1535–1603)
Muscles of the face, neck, and chest
Woodcut
From Juan de Arfe y Villafañe, *Varia conmensuración para la escultura, y arquitectura* (Madrid: Francisco Sanz, 1675), bk. 2, fol. 27r
Los Angeles, Getty Research Institute, 87-B24799

BIBLIOGRAPHY

Röhrl 2000, 82–100; Portmann 2014; Skaarup 2015, 72–74, 228–29, 246–56; Nanobashvili 2018, 69, 100.

Several Italian artists in the sixteenth century attempted to write about anatomy for artists: Alessandro Allori, Michelangelo Buonarroti, Benvenuto Cellini, Vincenzo Danti, Leonardo da Vinci, Rosso Fiorentino, and Bartolomeo Passarotti. Varying traces of these texts remain today; however, it was the Spanish sculptor and goldsmith Juan de Arfe (Arphe) y Villafañe who first published a substantial consideration of the subject for artists, albeit as part of a more general work. First published in Seville in 1585–87, his book is divided into four illustrated parts: geometry; proportions, anatomy, and perspective; animals; and architecture.

Arfe y Villafañe, who owned a copy of Juan Valverde de Amusco's book on anatomy (cat. no. 31) (Portmann 2014, 129), followed up his study of bones by attending a dissection led by Cosme de Medina at the Universidad de Salamanca (Arfe y Villafañe 1585–87, bk. 2, fol. 25v). Aside from considering the anatomization of the executed and the poor to be "horrible and cruel," he felt that it was not useful for artistic purposes and that the muscles of the arms and legs could be observed in living models. This critical view of the relative worth of dissection for artists was fairly rare for the period.

Although Arfe y Villafañe proclaims in the prologue that without an understanding of "the bones and muscles of a figure, you will not be able to do anything except with a thousand errors," in the commentary to his woodcut illustrations of anatomy he prioritizes the appearance of the surface of the body. Arfe y Villafañe states that the muscles seen in a flayed face were useful for doctors and surgeons rather than for sculpture and painting; accordingly, his illustration of the anatomy of the face is rendered as rounded lumps (*bultos redondos*) rather than as muscles and their fibers (Arfe y Villafañe 1585–87, bk. 2, fol. 25v).

In this second edition, of 1675, the woodcut portrait—which depicts the author in profile and wearing pince-nez and is set in a roundel—has moved from the verso (where it appeared in the first edition) to the title page. In the Getty Research Institute's copy of the second book, leaf 31rv and the woodcut illustrations of the posterior and left profile torso in écorché are missing. —MK

1

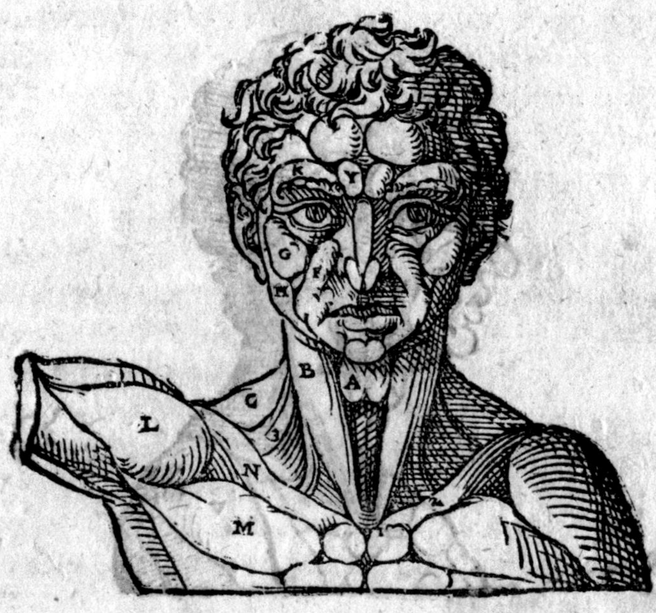

2

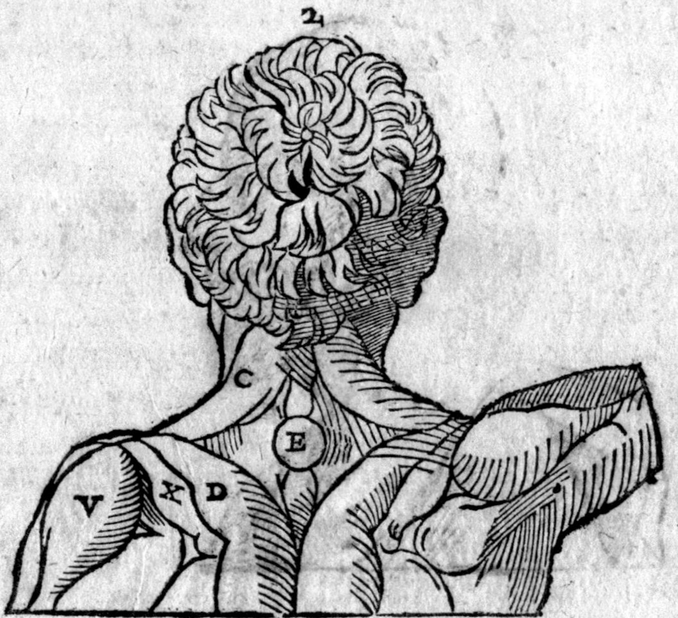

L 3

18 After Jean Cousin the Younger

(French, ca. 1536/37–ca. 1594)
Details of the Legs Seen from Inside and Outside with Their Proportions and Measures
Before 1595
Woodcut
From Jean Cousin, *Livre de pourtraicture; La vraye science de la pourtraicture descrite et demontrée par maistre Jean Cousin, peintre & geometrien tres-excellent* (Paris: Guillaume le Bé, 1671), fol. 20r
Los Angeles, Getty Research Institute, 86-B8575

BIBLIOGRAPHY
Cousin 1595; Firmin Didot 1872, 119–21; Bolten 1985, 179–86; Grodecki 1986, 170–71, no. 759; Auclair 2011; Grivel 2013, 256–59; Leproux 2013, 19–21; Engelskirchen 2014; Nanobashvili 2018, 69–70.

Printed drawing manuals, produced for artists as well as amateurs, functioned as pattern books of the body and sometimes included anatomy. Jean Cousin's manual, first published in 1595, shows the body whole and in parts, with several foreshortened views. While a significant number of its illustrations are devoted to anatomy—six out of the thirty-six in this edition—no real anatomical explanation is given, and they are instead analyzed for their proportions and measures relative to other body parts. For example, the leg is divided into four head lengths, and a width of three noses is assigned to the upper thigh, while the knee is given the width of a forefoot.

The plans for a book on figures and landscapes by Jean Cousin the Elder (ca. 1503–ca. 1560–62), mentioned in his *Livre de perspective* of 1560, were carried out instead, at least in part, by his son Jean the Younger, who signed a contract with the publisher Jean Leclerc in 1589 for a book with figures of the human body in different views, including foreshortening, but with no mention of anatomy (Grodecki 1986, 170–71, no. 759). The earliest surviving copy of the younger Cousin's *Livre de pourtraicture* has a privilege of 13 July 1593 and a printing date of 10 March 1595, by which time the artist was dead, since the preface mentions "the late Jean Cousin" (Leproux 2013, 21). The 1671 copy in the Getty Research Institute appears to reuse the same woodblocks from the first edition. Versions of Cousin's book were published through to the nineteenth century (Firmin Didot 1872, 118–24). In his anatomy book for artists of 1812, the surgeon Jean-Galbert Salvage cites Cousin's drawing manual and lists him among artists in France who have devoted themselves to the study of the body (Salvage 1812, i).

One of the changes to the revised edition published by François Jollain in circa 1685 (fig. F) is indicated on the title page, which advertises "an exact description of the bones and muscles of the human body, and their functions and uses" (Firmin Didot 1872, 122). To facilitate this, lettering was added to the illustrations and keyed to a new explanatory text. Also new is the final woodcut signed by "Le Sieur L'Aimé" of three skeletons, which are after Andreas Vesalius's *Tabulae anatomicae sex* (1538). —MK

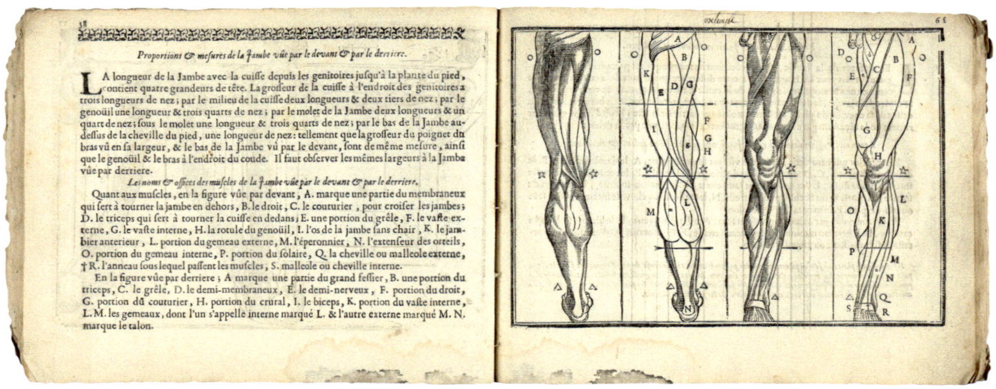

Fig. F
After Jean Cousin the Younger (French, ca. 1536/37–ca. 1594). Views of an écorché leg, with proportional measures and labeled muscles, woodcut. From Jean Cousin, *L'art de dessiner* (Paris: François Jollain, ca. 1685), 39.
Los Angeles, Getty Research Institute, 94-B5375.

Particularitez des jambes veuës par dedans & dehors, avec leurs proportions & uresures. 20

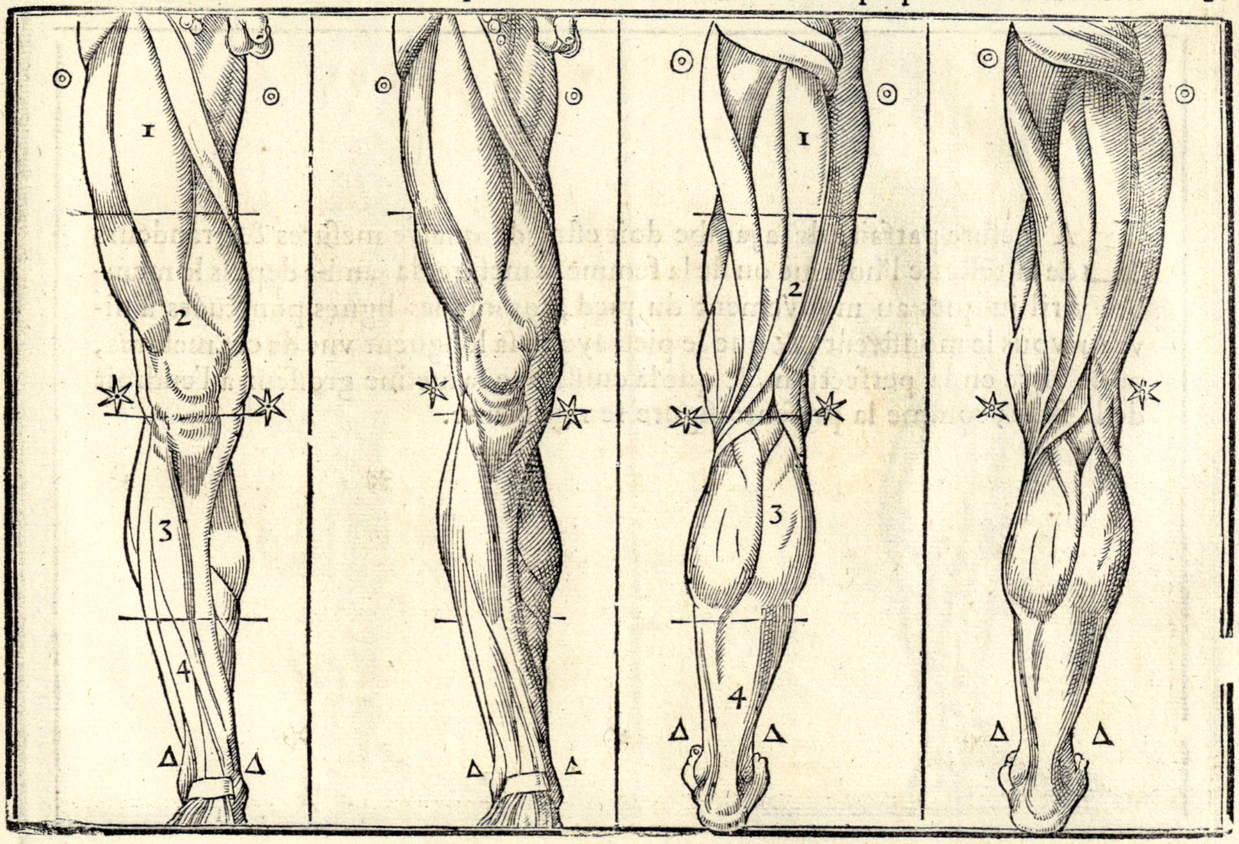

19 Jacob van der Gracht

(Dutch, 1593–1652)
Frontispiece
Etching
From Jacob van der Gracht, *Anatomie der wtterlicke deelen van het menschelick lichaem* (The Hague: Jacob van der Gracht, 1634)
Los Angeles, Getty Research Institute, 84-B4212

BIBLIOGRAPHY
De Lairesse 1701, 57; Kornell 1996, 47; Houtzager 2014, 214; Travers 2019, 255.

A practicing painter and printmaker in The Hague, Jacob van der Gracht professed to have undertaken anatomical study during his travels abroad, likely through parts of Spain and Naples (Van der Gracht 1634, A2v; López Martínez 1928, 56). Van der Gracht's claims to practical study lent credibility to his anatomical manual, for which the author produced illustrations and then published at his own expense. However, the contents of this book do not present the physical cadaver as a source of knowledge; instead, they place an emphasis on learning through images, which Van der Gracht augmented from medical publications already in circulation. The most notable and famous among these is Andreas Vesalius's *De humani corporis fabrica libri septem* (1543), to the extent that *Anatomie* has often been interpreted as one of many copies after the sixteenth-century anatomical atlas (Choulant 1945, 242; De Feyfer 1914, 36–39; Cushing 1962, 138). Yet, in preparing this manual for "painters, sculptors, printmakers, and also surgeons" (Van der Gracht 1634, frontispiece), Van der Gracht drew on at least five distinct publications, demonstrating the artist's familiarity with early modern anatomical literature and his ability to select figures that he deemed most useful for members of his profession (Travers 2019, 263–72). These images were then simplified and displayed alongside a register of the muscles and bones in order to aid artists in learning the different components of the body. Throughout his preface, Van der Gracht encourages his reader to treat anatomical study as a foundation for working after sculptures and live models—the proper applications of which could result in more convincing figures.

In his frontispiece, Van der Gracht blends the realms of art and anatomy. The viewer is presented with a scene that resembles an anatomy theater, in which an anatomist instructs a group of onlookers who have assembled to witness the demonstration. The suspended cadaver reinforces this association to the theater, while referencing scenes of artists' academies (see fig. 3.2). In the foreground, personifications of Painting and Sculpture recline while pointing toward the activities above them. Their inclusion signals both Van der Gracht's primary concern with providing instruction for members of his profession and the porous boundaries between these fields in the early modern period.

Though little known today, *Anatomie* was printed in two editions in the seventeenth century and is often referenced in the period's art literature, including *Grondlegginge ter Teekenkonst* (1701) by Gérard de Lairesse, who also designed images for Govard Bidloo's anatomical atlas (cat. no. 40). Later artists' books—such as John Singleton Copley's manuscript known as the *Anatomy Book* (ca. 1756) or Edward Cooper's *An Abridgment of Anatomy Taken from Titian & Other [of] the Best Italian Masters* (1715–20) (Russell 1987, 52, no. 204)—also include copies made after Van der Gracht's plates and testify to the continued use of these images even as their origin was slowly overshadowed by their association with Vesalius (Travers 2019, 272–77). —ET

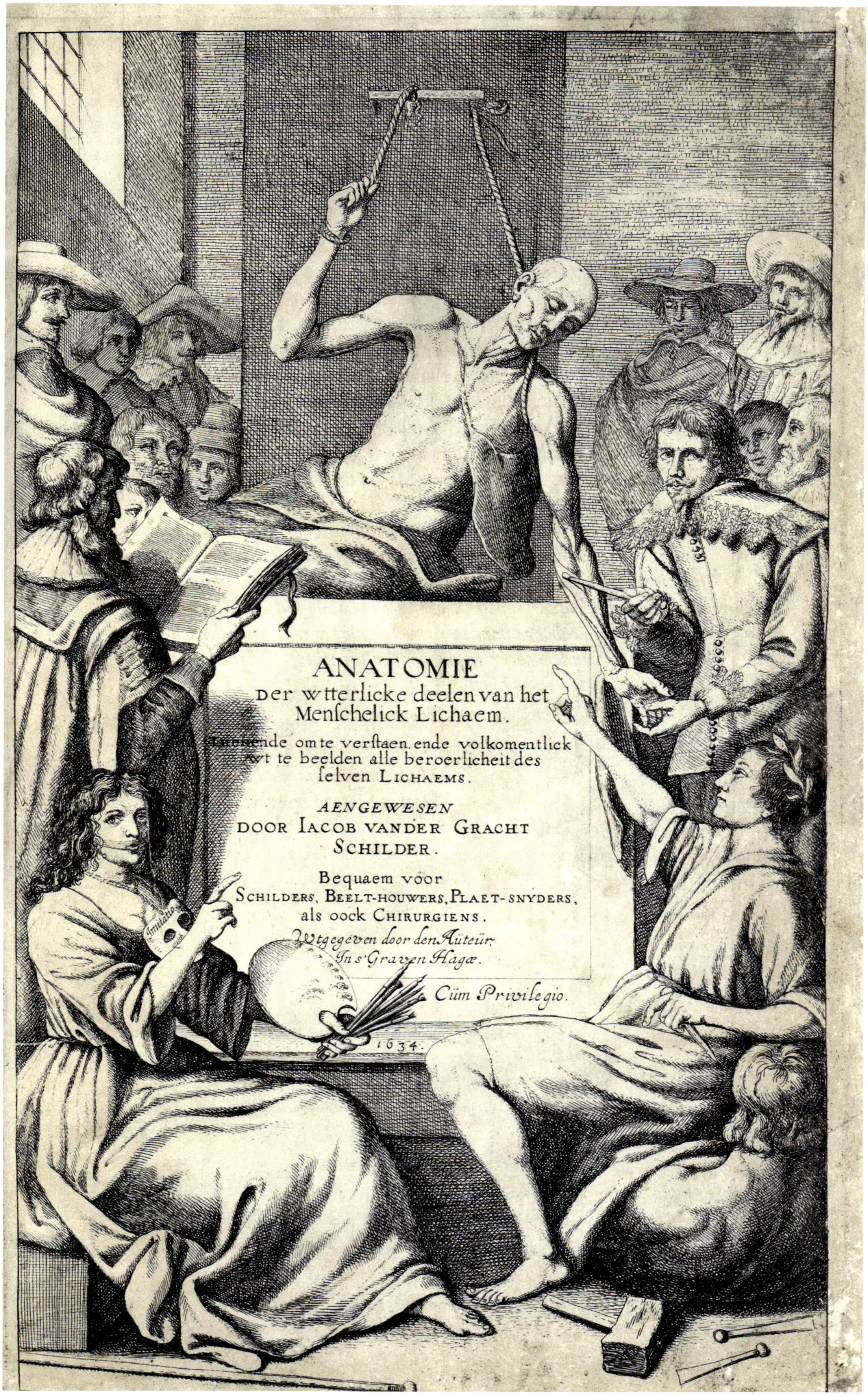
ANATOMIE
Der wtterlicke deelen van het
Menſchelick Lichaem.
om te verſtaen, ende volkomentlick
wt te beelden alle beroerlicheit des
ſelven Lichaems.
AENGEWESEN
DOOR IACOB VANDER GRACHT
SCHILDER.
Bequaem voor
Schilders, Beelt-houwers, Plaet-snyders,
als oock Chirurgiens.
Wtgegeven door den Aüteür,
In s'Graven Hagæ.
Cüm Privilegio.
1634.

20 Carlo Cesi

(Italian, ca. 1622–82)
Foreshortened écorché with a raised leg
Etching
From Carlo Cesi, *Cognitione de muscoli del corpo humano per il disegno* (Rome: Francesco Collignon, 1679), pl. 17
Los Angeles, Getty Research Institute, 87-B13106

BIBLIOGRAPHY
Moehsen 1771, 103–4; Haller 1774–77, I:789; Duval and Cuyer 1898, 162–66; Choulant 1945, 256–57; Bellini 1987, 106–9, nos. .093–.110; Ferraris and Vannugli 1987; Vannugli in Calvesi et al. 1987, nos. 38.D, 44.I; Bridson and White 1990, 204–14; Kornell 1996, 60–63; Vannugli 1997, 263; Röhrl 2000, 128–30, 134; Pierguidi 2017, 10–11; Stetskevich 2019, 154.

The title to Carlo Cesi's anatomy book translates as "The understanding of the muscles of the human body for design," and in the unpaginated preface, the painter and engraver Cesi (Cesio) characterizes such an understanding as "most necessary for those who desire to set out on a sure path to Painting and Sculpture." Cesi, a student of Pietro da Cortona (cat. no. 14), was elected in 1673 as a lecturer on anatomy at the Accademia di San Luca in Rome and, in 1675, as the academy's director (Ferraris and Vannugli 1987, 29). This book and Cesi's drawing manual, *Elementi del disegno* (ca. 1686–96), were likely published with his students in mind.

Despite the title, the bones were not neglected in the book. In addition to the first two plates of skeletons, several illustrations feature a comparison of the bones of a joint in the same pose as the muscles. Composed of sixteen etched illustrations in all, the book concludes with five unlettered muscle figures. Two of these figures in foreshortened repose reflect life-drawing poses and are similar to those found in contemporary drawing manuals (fig. G).

In the preface, Cesi cites Andreas Vesalius as a reference for the anatomical accuracy of the book, although it is noteworthy that Cesi did not copy his illustrations, unlike Giacopo Moro or, in part, Van der Gracht (cat. no. 20; Moro 1679). Cesi also says he consulted Bernardino Genga, surgeon at the hospital of Santo Spirito. Cesi may have been the artist responsible for the frontispiece to Genga's *Anatomia chirurgica* from a few years earlier that features a standing écorché and seated skeleton (Genga 1672). Genga would later collaborate with Charles Errard on an anatomy book for the Académie de France in Rome (cat. no. 32).

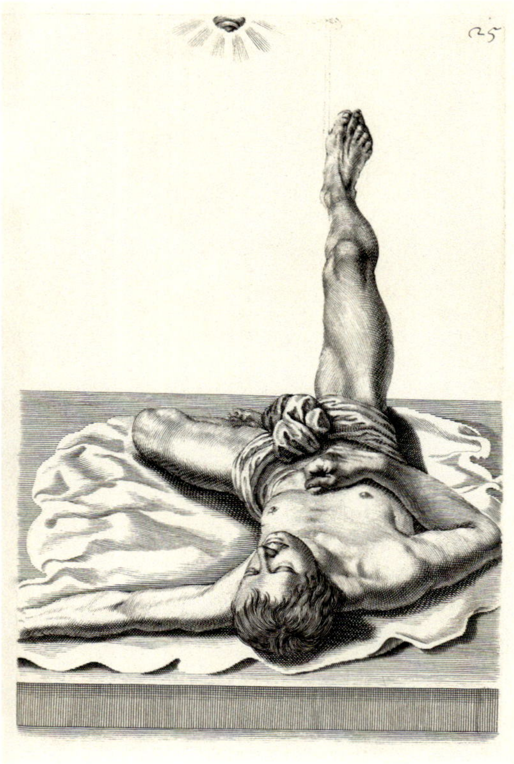

Fig. G
Crispijn van de Passe the Younger (Dutch, ca. 1596–ca. 1670). Foreshortened nude with a raised leg, engraving.
From *La seconda parte, della luce del arte de dissegnare et di depengere* . . . (Amsterdam: Crispijn van de Passe, 1644), 25.
Los Angeles, Getty Research Institute, 86-B25061.

Although Cesi's anatomical errors were criticized (Moehsen 1771; Duval and Cuyer 1898), they did not affect the book's success. In terms of number of editions, it was one of the most successful in the genre. This first edition of 1679, however, exists today in few verified copies, and both Ludwig Choulant and Albrecht von Haller were unaware of it (Choulant 1945; Haller 1774–77). (Other copies are in the Thompson Library, Ohio State University, Columbus [RAR QM51.C4]; the Bibliothèque Nationale de France, département Arsenal, Paris [FOL-S-1492]; and the Istituto Centrale per la Grafica, Rome [FC91340].) A second edition was published in Rome by Arnold van Westerhout (1651–1725) in 1697. Many later editions are derived from the 1706 German translation of Daniel Preissler (1666–1737), who ran an academy of art in Nuremberg (Bridson and White 1990, 201–14; Stetskevich 2019). The Getty Research Institute has one such edition from 1759 (84-B7702) that was once in the extensive library of the London journalist and author George Augustus Sala (1828–95). —MK

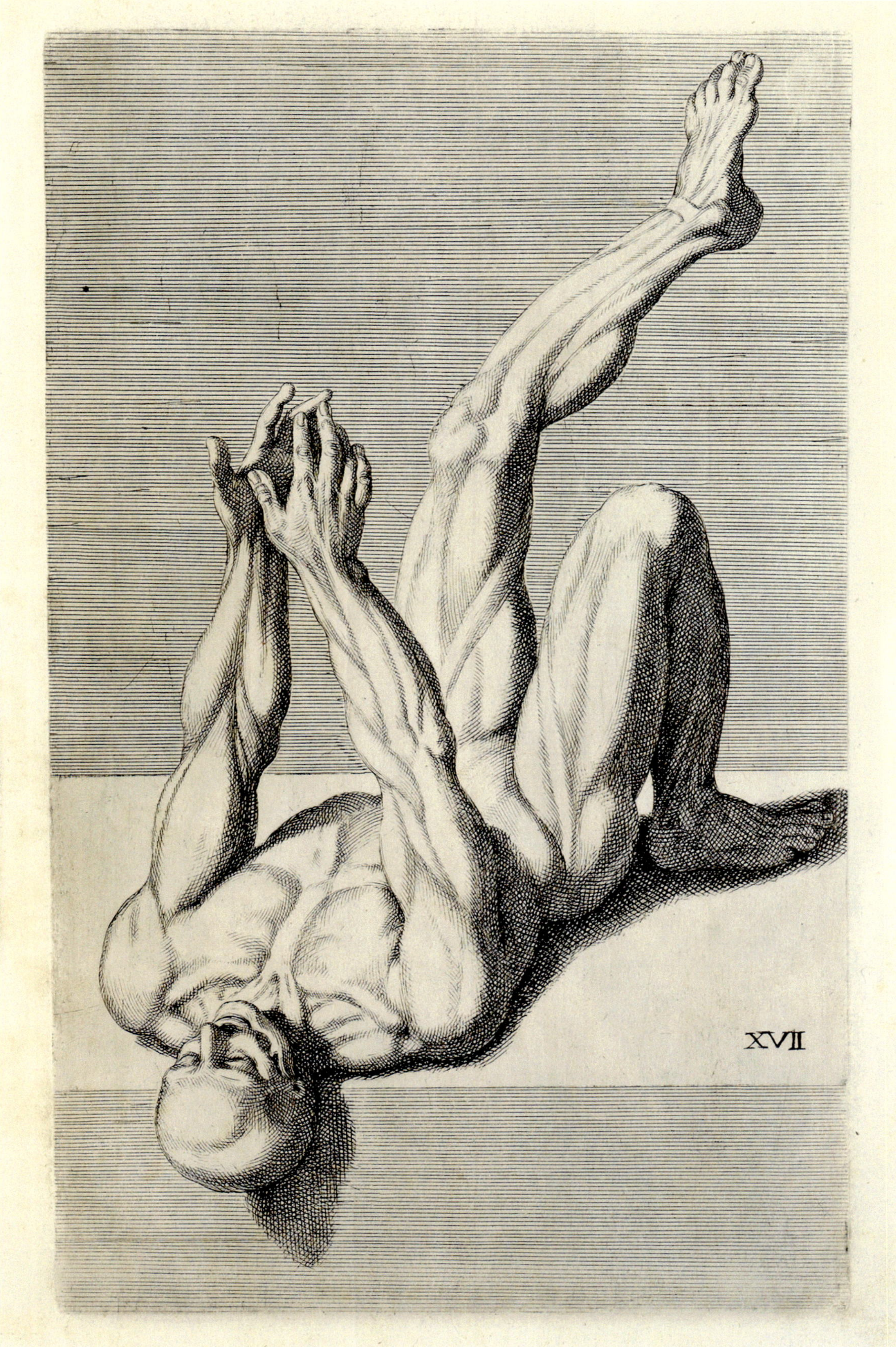
XVII

21 **Gabriel Huquier**

(French, 1695–1772), after Jacques Philippe Bouchardon (French, 1711–53)
Figure of transparent anatomy
1741
Etching
From Edme Bouchardon, *L'anatomie nécessaire pour l'usage du dessein* (Paris: Chéreau & Joubert, [1787–89], pl. 13
Los Angeles, Getty Research Institute, 92-B12692

BIBLIOGRAPHY

Mercure de France October 1741, 2279–80; Duval and Cuyer 1898, 186–94; Ronot 1970; Kornell 1996, 60; Röhrl 2000, 140–41; Comar 2008b, 296–99; Joly 2008, 147–48; Kopp in Desmas et al. 2016, 194, no. 110; Kornell 2016b; Trey 2016; Kopp 2017, 204–5; Chatelain 2018, 534–35.

Edme Bouchardon (1698–1762) was greatly esteemed in his day both as a sculptor and as a draftsman. This book of anatomy for artists, which first appeared in 1741, was one of a series of collaborations with the printmaker and publisher Gabriel Huquier, whom Bouchardon had previously supplied with designs of vases and academy figures. It was the first book in French entirely devoted to anatomy for artists since Roger de Piles and François Tortebat's *Abregé d'anatomie* (1668, with a second French edition in 1733). The address to the reader tells us that Huquier originally planned only to publish prints after a series of drawings by Bouchardon of écorché models. However, these were deemed not to be instructive enough on their own, and so a single general table with the names of the bones and muscles, along with their functions, was drawn up, and six prints of skeletons and figures of transparent anatomy by Bouchardon's younger brother, Jacques Philippe, were added to make a total of fifteen plates, including the title page—all etched by Huquier (for counterproofs in Besançon, France, of two of Philippe's skeleton drawings, see Chatelain 2018, 534–35, nos. R2, R3).

Plate 13 illustrates the outline of the muscles (with an attempt to show their origins and insertions) over the skeleton. The immediate inspiration for this mode of presentation may have been a print by Crisóstomo Martínez first published in 1690 and republished with a commentary by Jacques-Bénigne Winslow in Paris in 1740, the year before Bouchardon's *Anatomie* appeared (Martinéz 1740, pl. 2). This type of transparent anatomy is again seen in Cornelis Ploos van Amstel's anatomy book for artists (1783) (see fig. 1.3). Bouchardon was also deeply committed to the anatomy of the horse and his many drawings of this subject after Carlo Ruini survive (cat. no. 10), a copy of which he owned, and of a related écorché model (Sørenson 2002; Kopp 2017, 202–7; Trey 2016).

The address to the reader is modeled on de Piles's preface to the *Abregé d'anatomie,* editions of which both Huquier and Bouchardon owned and, incidentally, a copy of which is bound with this book. A simplified approach to anatomy is offered, concentrating only on those aspects useful to the artist, and it is suggested that an écorché model be drawn from several angles. Not missing a marketing opportunity, the reader is informed that for this purpose such a model is available for purchase in Huquier's shop on the rue Saint-Jacques in Paris.

In all, six editions of Bouchardon's *Anatomie* are recorded. The edition in the Getty Research Institute dates to 1787–89, during the short period after François Étienne Joubert bought the printing business of Jacques-François Chéreau, when works with both their names appeared on the imprint (Kornell 2016b, 47). —MK

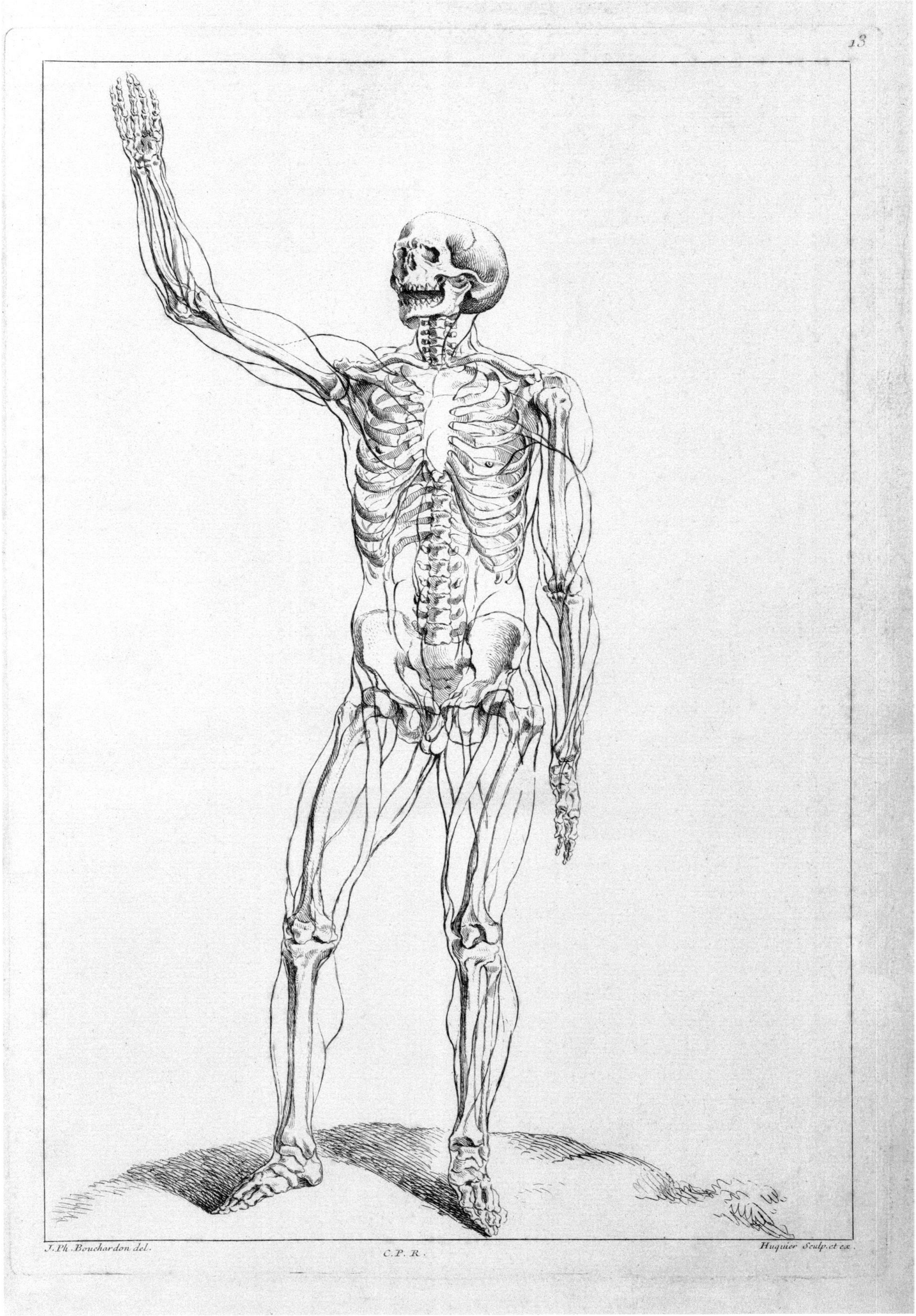
13
J. Ph. Bouchardon del.
C. P. R.
Huquier Sculp. et ex.

22 James Caldwall

(English, 1739–1822), after John Brisbane
(Scottish, ca. 1720–ca. 1776), after Jan Wandelaar
(Dutch, 1690–1759)
Outline plate with lettering (*left*) and walking écorché in profile (*right*)
Engraving
From John Brisbane, *The Anatomy of Painting; or, A Short and Easy Introduction to Anatomy: Being a New Edition, on a Smaller Scale, of Six Tables of Albinus, with Their Linear Figures; Also, a New Translation of Albinus's History of That Work, and of His Index to the Six Tables; To Which Are Added the Anatomy of Celsus, with Notes, and the Physiology of Cicero; With an Introduction, Giving a Short View of Picturesque Anatomy* (London: George Scott, 1769), pl. 6
Los Angeles, Getty Research Institute, 84-B3777

BIBLIOGRAPHY
Bell 1794, vi; Choulant 1945, 292–93; Punt 1983, 128; Russell 1987, nos. 95–96; Darlington 1990, 1:74, 206, 2:491–92; Röhrl 2000, 147–50; Kornell 1996, 57; Cunningham, A. 2016, 272–73; McCormack 2018, 154; Royal Academy of Arts Archive, SA/34/18.

John Brisbane, a Scottish physician who practiced in London, favored the use of anatomical illustration as an introduction to the study of anatomy for physicians, artists, and dilettantes, rather than what he considered the minutiae of dissection: "How much shorter, more easy, and agreeable it is, to be introduced to that science by means of figures, than by tedious systems, and lectures upon dead bodies alone" (Brisbane 1769, vi). He recommended Bernhard Siegfried Albinus's *Tabulae* (1747) above all other illustrations, and Brisbane himself adapted six of its figures with their lettered outline plates for *The Anatomy of Painting* (see fig. 1.2). Brisbane reduced the scale of Albinus's figures, and, in the corresponding outline figures, he enlarged selected areas in order to ensure that their identifying lettering would remain legible. In this way, Albinus's cool, stately figures gain a dynamism as the viewer's eye glides from detail to figure and takes in the changes in scale.

Brisbane's self-portrait in profile, signed and dated 1763 ("J. Brisbane M.D. inv. et del. MDCCLXIII"), appears on the title page in a roundel paired with his coat of arms, suspended by a ribbon below a vignette of Asclepius, Greek god of medicine (fig. H). The date indicates that Brisbane was at work on *The Anatomy of Painting* for several years. The preface is dated 1 January 1769 and soon after, on 13 February 1769, Brisbane presented a copy of his newly published book to the Royal Academy of Arts. His call for a "grand national academy" for the arts and sciences—with which he concludes the "Introduction to the Tables"—had already been in part realized by the founding of the Royal Academy the previous December, which Brisbane notes in a postscript (Brisbane 1769, xxi–xxii).

Although the title promises "a short and easy introduction to anatomy," Brisbane's text is prolix and repetitive. Brisbane earned the acerbic comment of the anatomist and surgeon John Bell for both his bombast and his copies after Albinus (Bell 1794, vii). Brisbane's book includes his translation of Albinus's detailed history of the preparation of his illustrations and of texts by Aulus Cornelius Celsus and Marcus Tullius Cicero; the latter are offered as good examples of a general introduction on a medical subject. These were jettisoned in the second edition of 1831, and Brisbane's preface and introduction significantly cut down, along with some of the book's meandering charm. —MK

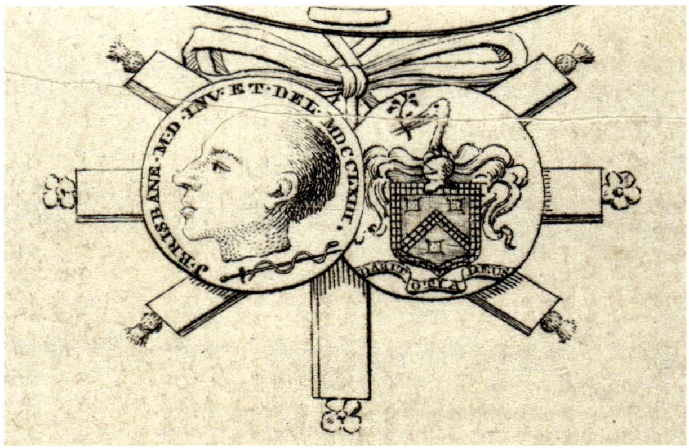

Fig. H
John Brisbane (Scottish, ca. 1720–ca. 1776).
Self-portrait and coat of arms (detail of title page), 1763, etching.
From John Brisbane, *The Anatomy of Painting; or, A Short and Easy Introduction to Anatomy . . .* (London: George Scott, 1769).
Los Angeles, Getty Research Institute, 84-B3777.

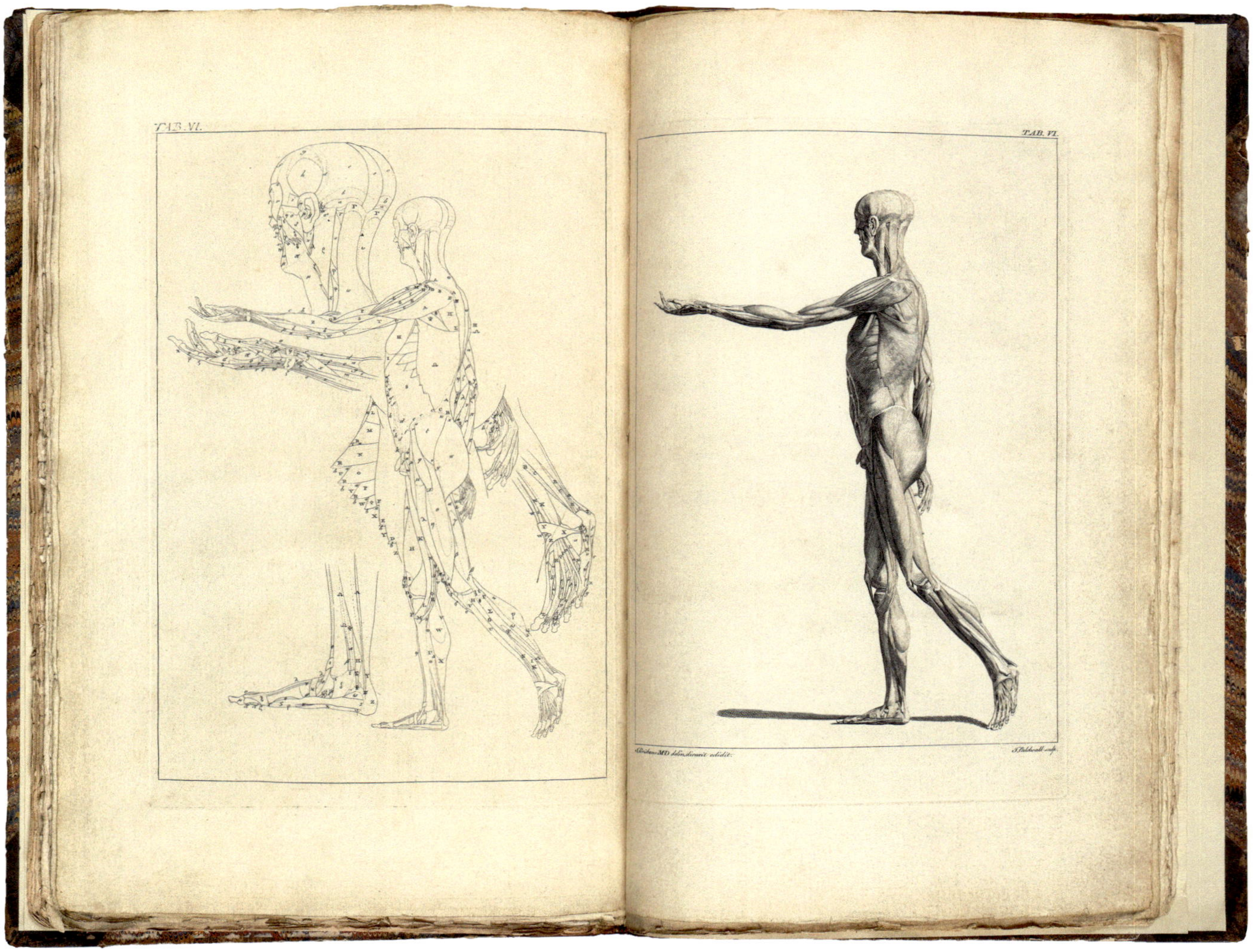
TAB. VI.
TAB. VI.

23 Gilles Demarteau

(Flemish, 1722–76), after Charles Monnet
(French, 1732–after 1808)
Title page
Crayon manner, printed in red ink
From Charles Monnet, *Études d'anatomie à l'usage des peintres* (Paris: Demarteau, 1774)
Los Angeles, Getty Research Institute, 84-B28065

BIBLIOGRAPHY

Mercure de France July 1774, 173–74; Duval and Cuyer 1898, 213–20; Choulant 1945, 352; Roux 1949, 336–37, no. 32; Kornell 1996, 62, 65, fig. 35; Röhrl 2000, 155–59; Joly 2008, 148–51.

Charles Monnet, a history and landscape painter, was of the opinion that only an artist could effectively teach another artist about anatomy, because surgeons were unable to restrict themselves to what was necessary to a painter, nor could they present information in an intelligible order for the artist. Certainly, the viewpoint of the artist is considered in the expressive, lifelike gestures Monnet gives the hands and arms (pls. 22–24, 27–29), poses that could easily be translated to a composition. For the limbs and extremities, Monnet extensively employs a comparison of bone and muscle in the same pose, on the same page—an arrangement that was seen earlier in Carlo Cesi's work (cat. no. 20).

The illustrations in this first edition—issued as seven *cahiers,* or parts, for a total of forty-two plates—are done in crayon manner, printed in red ink. This technique had been perfected by Monnet's illustrator, Gilles Demarteau, who used it to cater to connoisseurs by replicating the character of chalk drawings. On the title page, a winged genius bearing a torch lifts a shroud covering a skeleton, an indication that the secrets of anatomy will be revealed to the artist in Monnet's *Anatomy Studies for the Use of Painters.*

In the Getty Research Institute's copy, the words "du Roi," designating Monnet and Demarteau as painter and printer to the king, have been obscured, reflecting anti-Royalist sentiment. The GRI's copy is annotated in German from the sixth plate onward. —MK

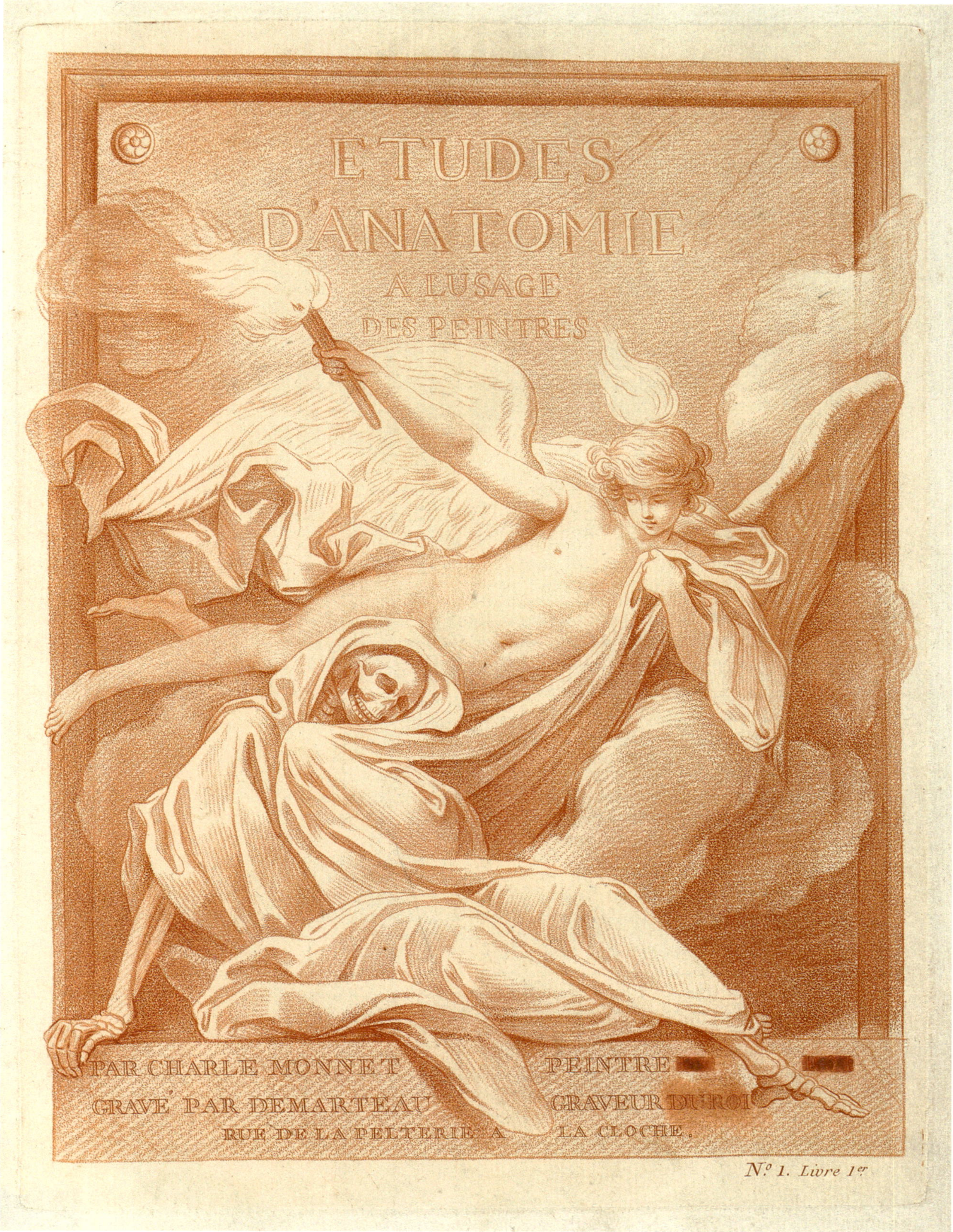
ETUDES
D'ANATOMIE
A L'USAGE
DES PEINTRES
PAR CHARLE MONNET PEINTRE
GRAVÉ PAR DEMARTEAU GRAVEUR DU ROI
RUE DE LA PELTERIE A LA CLOCHE.
N.º 1. Livre 1.er

24 Jacques Lavallée

(French, active 1776–1830), after Jacques Gamelin (French, 1738–1803)
Prone écorché with raised arm
Crayon manner
From Jacques Gamelin, *Nouveau recueil d'ostéologie et de myologie, dessiné d'après nature . . . pour l'utilité des sciences et des arts* (Toulouse: J. F. Desclassan, 1779), pt. 2, unnumbered plate
Los Angeles, Getty Research Institute, 91-B35736

BIBLIOGRAPHY

Barthe 1849, 402–6; Duval and Cuyer 1898, 222–33; Sjöberg 1974, 43, no. 32; Mozziconacci and Michel 1990; Stafford 1991, 70–76; Hildebrand 1992, 10–16; Kornell 1996, 215–17; Röhrl 2000, 160–68; Michel 2003; Joly in Comar 2008b, 192–93, no. 48; Joly 2008, 151–55; Rudy 2013, 51, 208nn62–64.

Jacques Gamelin, an artist from Toulouse, deliberately chose diverse and "sometimes strange" poses for his figures in order to best display the muscles, as he explains in the introduction to the second part of this two-part work, the first section on the bones appearing in 1778 and that on the muscles in the following year ("De myologie," Gamelin 1779). This prone écorché with a raised arm arranged on a tilted platform, the head turned by the hand that grips it, is an example of the unusual poses Gamelin chose for his muscle figures. It is rendered in crayon manner, a technique that simulated drawing. Gamelin offered the option of printing the illustrations of the second volume in red but at a greater cost. A loose sheet in this color of this figure is in the Louvre Museum, Paris (RF 34433.2 recto). According to the original broadsheet prospectus for the second part bound in the Getty Research Institute's copy, the book was aimed specifically at anatomists and draftsmen. Gamelin, however, did not recoup the expenses he incurred and bankrupted himself. Perhaps contributing to the book's lack of financial success was Gamelin's willingness to stray from his subject. To allay boredom and to introduce variety, according to the author, the book is interspersed with illustrations of a range of genres, such as battle scenes, at which he excelled, even though this only added to production costs. The first volume on the bones is punctuated with inventive dance-of-death vignettes. The prone écorché shown here is incongruously followed in this copy by a scene of the return of a prodigal daughter after a painting by Gamelin for a Mr. Daram. More relevant to the book's purpose are the nude academy figures that provide a comparison of surface anatomy with the écorché figures.

The book is warmly dedicated to Gamelin's patron, Nicolas-Joseph Marcassus, baron de Puymarin, who encouraged and supported the painter's artistic studies. This included a period in Italy where Gamelin was elected as a painter of battle scenes in 1771 at the Accademia di San Luca in Rome. Upon his return to Toulouse in the mid-1770s, Gamelin devoted himself to the study of anatomy. According to the preface to the second volume, he was aided in dissections by the son of Bertrand Bécane (1728–ca. 1802); the elder Bécane was a professor of surgery in Toulouse and the censor who gave official approval of Gamelin's book. The deformed skeleton of a five-year-old boy and the diseased femurs (the latter dated 1778) of part one are after specimens in Bertrand Bécane's collection and are mentioned by the surgeon in his book on the effects of cancer (Bécane 1778, 7, 14). An etched portrait of Bécane, signed by Gamelin and dated 1778, is prefixed to the copy of Bécane's 1775 book on the diseases of the bones in the Bibliothèque Universitaire de l'Arsenal in Toulouse (Pf XVIII-398/1) (Bécane 1775). —MK

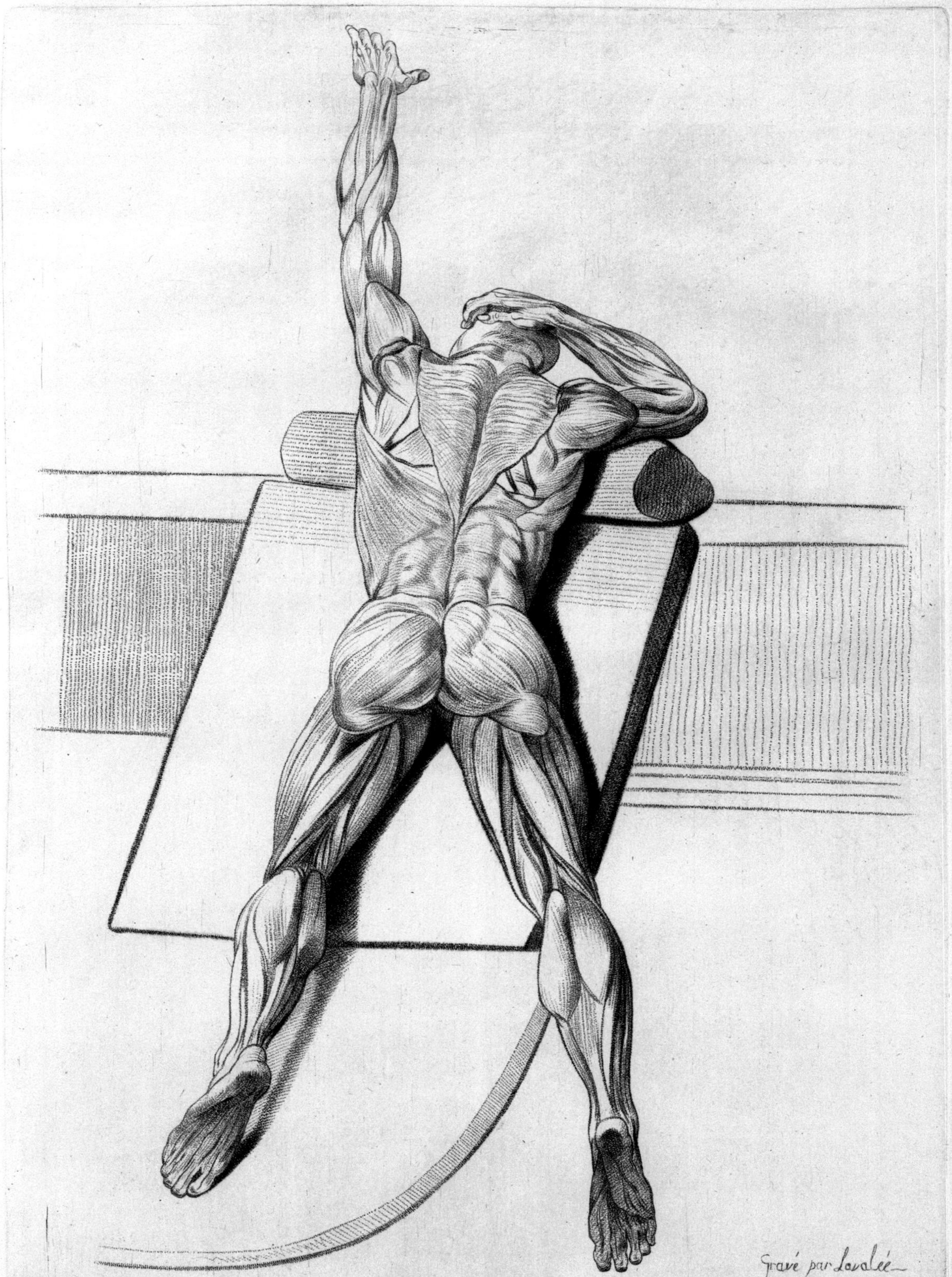
Gravé par Lavalée

25 John Walker

(English, active 1776–1802)

Left: After Anthony Walker (English, 1726–65)
Cadaver posed and flayed by William Hunter for St Martin's Lane Academy
5 April 1783 (print)
Engraving

Right: After Jan Steven van Calcar (North Netherlandish, ca. 1515–ca. 1546)
Copy after the first muscle figure in Andreas Vesalius's *Fabrica*
Engraving with gray wash

From John Walker, *The Artist's Pocket Companion: Being a Compendious Treatise of Anatomy, Adapted to the Arts of Designing, Painting, and Sculpture . . . Most Accurately Outlined from the Best Anatomical Tables and Figures Extant*, 2nd ed. (London: John Walker, 1787), pls. [4], [5]
Los Angeles, Getty Research Institute, 84-B29120

BIBLIOGRAPHY
Hammelmann and Boase 1975, 96–101; Kemp 1975, 11; Brock 1983, 9; Bridson and White 1990, E122, E126; Kornell 1996, 69; Röhrl 2000, 185.

This pocket-size book of anatomy for students of the Royal Academy of Arts, London, by the English printmaker John Walker, is a reduced version of John Tinney's anatomy book for artists, *Compendium anatomicum* (Tinney 1743; Russell 1987, no. 816), a folio that Walker describes as being too large to use (Walker 1787, 6). In its stead, he offers a version described on the title page as "being so convenient, gentlemen . . . need never be at a loss for the form of a muscle, to give a proper appearance to any part of a figure." This second edition is dedicated to Sir Joshua Reynolds, president of the Royal Academy. A previous owner added notes on the anatomical terms for the bones and muscles, and on the dimensions of the bones of the male pelvis. The copy's worn condition suggests that the aim of accessibility was achieved. The book was once in the collection of the art historian Ulrich Middeldorf, director of the Kunsthistorisches Institut in Florence.

Tinney's eight plates were based on Andreas Vesalius's *Fabrica* (1543) and William Cowper's *Myotomia reformata* (1724) (cat. nos. 30, 13). Walker maintained the same number in his version, but a skeleton after Bernhard Siegfried Albinus (cat. no. 15) was substituted for one after Vesalius, and a muscle figure after Cowper was replaced with a print after a drawing made by the engraver Anthony Walker, the author's uncle and Tinney's former apprentice. The latter was drawn around 1750 at St Martin's Lane Academy, London, after a cadaver that had been posed and flayed by William Hunter but before a mold was taken of it (Walker 1787, 7; see Kornell, "'As Large as Nature,'" this volume, pp. 55–56). The cast's later presence in the Royal Academy was recorded in two paintings by Johan Zoffany (1733–1810) (see fig. 5.6). It is unknown when the first edition of the book was published, but it likely appeared between 5 April 1783, the date of the print after Anthony Walker's drawing, and 1787.

Jean François Bosio (1764–1827), a professor of drawing at the École Polytechnique, Paris, likewise "saw how much it was necessary to give students a small portable treatise of elementary rules" and came out with an illustrated, pocket-size drawing manual, the *Traité élémentaire des règles du dessin*, which includes an illustrated discussion of the bones and muscles (Bosio 1800–1801, 5–6; GRI, 86-B1423). —MK

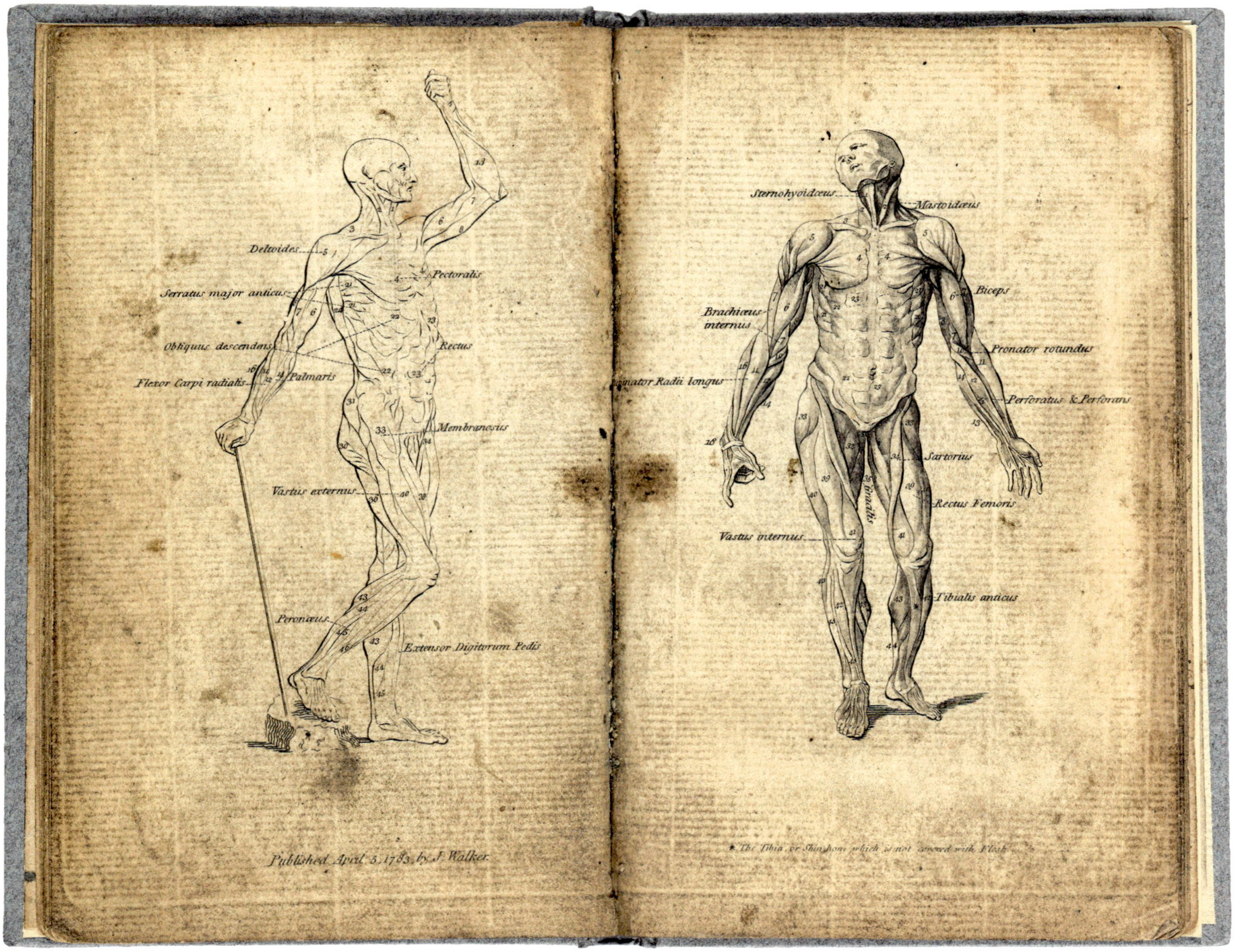
Deltoides
Pectoralis
Serratus major anticus
Obliquus descendens
Rectus
Flexor Carpi radialis
Palmaris
Membranosus
Vastus externus
Peronæus
Extensor Digitorum Pedis
Published April 5. 1783 by J. Walker.
Sternohyoidæus
Mastoidæus
Biceps
Brachiæus internus
Pronator rotundus
pinator Radii longus
Perforatus & Perforans
Sartorius
Gracilis
Rectus Femoris
Vastus internus
Tibialis anticus

26 Carlo Lasinio

(Italian, 1759–1838), after Francesco Carradori (Italian, 1747–1825)
Écorché seen from the back, after a model by Jean-Antoine Houdon (French, 1741–1828)
Etching
From Francesco Carradori, *Istruzione elementare per gli studiosi della scultura di Francesco Carradori professore di dett'Arte nella R. Scuola di Firenze* (Florence: n.p., 1802), pl. 3
Los Angeles, Getty Research Institute, 85-B9009

BIBLIOGRAPHY
Venturoli 1977; Carradori 1979; Carradori 2002; Poulet 2003, 63–71, nos. 1–2; Roani 2016.

Four years after Francesco Carradori was appointed professor of sculpture at the Accademia di Belle Arti in Florence in 1798, he published the *Istruzione elementare per gli studiosi della scultura,* a practical manual for sculpture. Illustrated with seventeen etchings by Carlo Lasinio, who also taught at the academy, it covers topics such as sculpting in stucco, measuring sculpture for reproduction, and making molds for casting. The book also includes a chapter on how to restore ancient sculptures, one of Carradori's specialties.

Carradori places anatomy high on the hierarchy of fundamental principles that a sculptor should master, stating that it is important to know the proportions and all of the parts of the body, as well as their mechanical functions, in order to understand their variations according to their actions and the motions of the muscles (Carradori 1802, I). The text that accompanies the skeleton and the two muscle figures (pls. 1–3), however, does not supply this information and is restricted mainly to giving the names of the bones and muscles, except for a couple of experiential observations, such as the sharp pain that will result from pressure placed on the nerve that passes just behind the elbow (pl. 2, "s"). Carradori considered these first three etchings, together with an illustration of the proportions of a man and child (pl. 4), "more than sufficient, without a much longer anatomical discussion, to demonstrate all that is necessary for a youth to know" (Carradori 1802, I). In a note to the third table, he suggests that those readers seeking more detailed information on anatomy consult "the great work" of Bernhard Siegfried Albinus, referring to the *Tabulae sceleti et musculorum corporis humani* (1747) (cat. no. 15).

All three of Carradori's anatomical illustrations (pls. 1–3), including the skeleton, are not based on Albinus, nor are they related to the illustrations of the anatomist Paolo Mascagni (1755–1815), as it is sometimes stated (Carradori 1979, 23; Roani 2016, 127); rather, they are after an écorché model by the French sculptor Jean-Antoine Houdon (see fig. 5.5). Houdon's écorché was executed in Rome in 1767 as a life-size preparatory study for his statue of Saint John the Baptist. A cast after the écorché soon entered the collection of the Académie de France in Rome, and the subsequent casts Houdon produced for art academies across Europe attest to its great success and popularity (Poulet 2003, no. 1). Lasinio's etchings in Carradori's book are after the original version with a tree trunk. Francesco Bertinatti, in his *Elementi di anatomia fisiologica applicata alle belle arti figurative* (1837–39), included figures inspired by a later version from the 1770s of the écorché with a raised arm (Bertinatti 1837–39, vol. 2, pls. 29, 30; Poulet 2003, no. 2). —MK

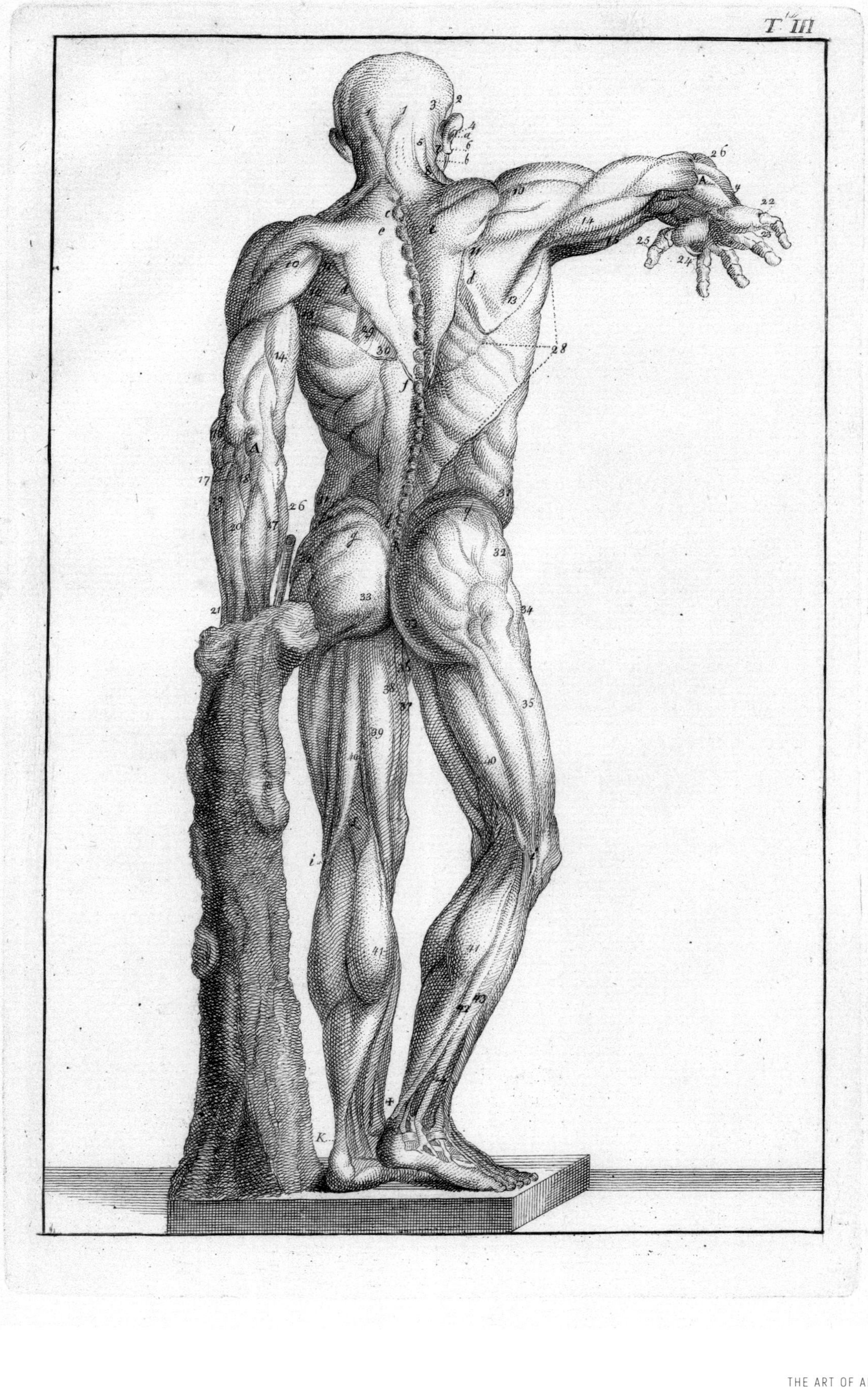
T. III

27 SKULL AND MUSCLES OF THE NECK

Etching and engraving inked *à la poupée* in red and black ink
From Giuseppe del Medico, *Anatomia per uso dei pittori e scultori* (Rome: Vincenzo Poggioli, 1811), pl. 16
Los Angeles, Getty Research Institute, 84-B28069

BIBLIOGRAPHY

Cicognara 1821, 1:59, no. 336; Missirini 1823, 362, 364; Duval and Cuyer 1898, 256–62; Choulant 1945, 331; Kornell 1996, 56, 60, 66; Röhrl 2000, 198–201.

The aptly named Giuseppe del Medico (1772–1825), a professor of surgery, was appointed lecturer on anatomy to the Accademia di San Luca in Rome soon after the election of Antonio Canova (1757–1822) as its president in 1810. The following year, Del Medico published his *Anatomia per uso dei pittori e scultori* and dedicated it to the academy, to which he presented a copy (Missirini 1823, 362). The Italian art historian and bibliophile Leopoldo Cicognara (1806–33) praised its illustrations and considered it to be the best work on the subject that had appeared so far (Cicognara 1821, 1:59, no. 336). The inscription on Del Medico's tomb in the church of San Rocco, Rome, topped with his portrait bust, refers to his authorship of the book. A second, posthumous edition of 1833 was reengraved by Francesco Bosa (died ca. 1870), who in turn dedicated the book to the Accademia di Belle Arti, Venice.

Some copies of the book, like the one in the Getty Research Institute, were printed with black ink for the bones and red ink for the muscles, a distinction also seen in Jean-Galbert Salvage's *Anatomie du gladiateur combattant* of 1812 (cat. no. 34). In the final plate, blue ink is used to distinguish the vena cava from the aorta. The plates are unsigned, and the majority are derived from Bernhard Siegfried Albinus. The sixteenth plate, however, seems to be entirely new. The skull is tilted back in order to more easily display the muscles of the neck and underside of the chin. The connection of the muscles to the skull and to the clavicle and sternum are clearly demonstrated. In the copies with color printing, this effect is all the more visible, and the comparison between the white skull, lit from above, and the red muscles of the neck, is striking. To demonstrate the external parts of the body, Del Medico chose two views of an antique statue, the Borghese Gladiator (pls. 34, 35), considered by Salvage, James Birch Sharpe (cat. nos. 34, 35), and others to be perfect in its depiction of human anatomy. —MK

TAV. 16

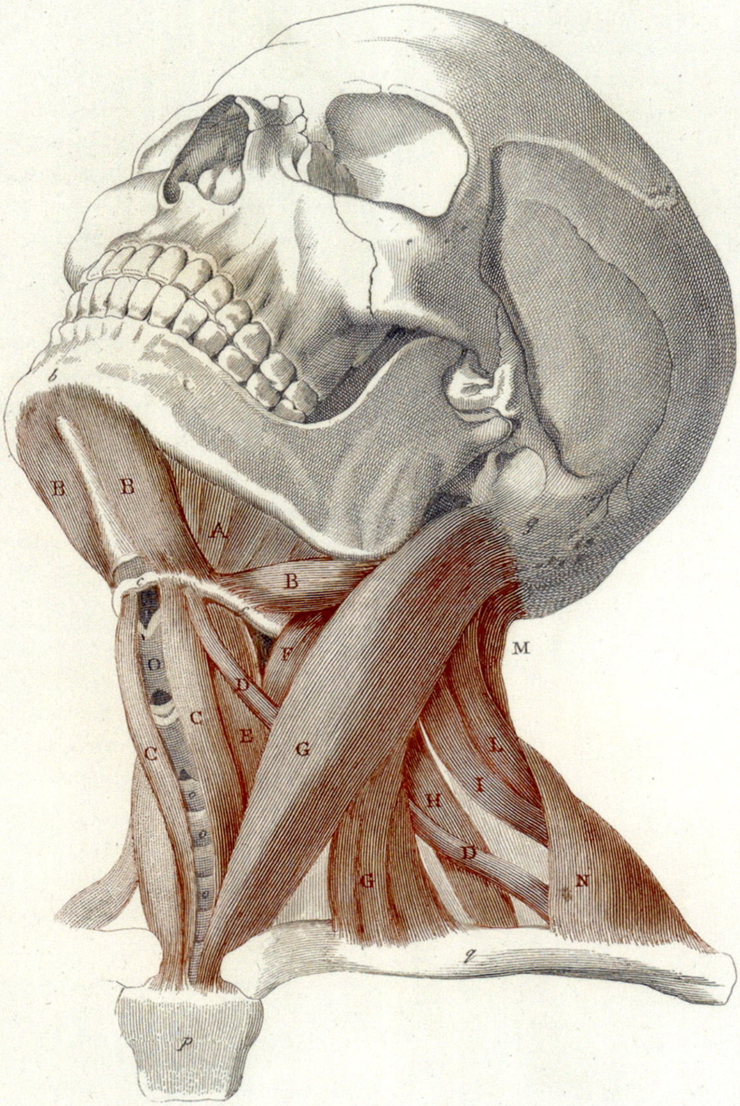

A Muscolo Milo – joideo § 149
B .... Digastrico § 150
C .... Sterno – joideo } V. Tav. 17. fig. 2
D .... Coraco – joideo } V. Tav. 17. fig. 2
E .... Sterno – tiroideo } V. Tav. 17. fig. 2
F .... Tiro – joideo § 154
G .... Sterno – Mastoideo § 155
H .... Elevatore delle pme. coste V. Tav. 17. fig. 2
I .... Elevatore della Scapola V. Tav. 17. fig. 2
L .... Complesso V. Tav. 17. fig. 1

M. Muscolo Splenio V. Tav. 17. fig. 1
N. Porzione del Trapezio V. Tav. 20
O. Cartilagine Tiroide § 148
b... Il Mento § 36
c.... Osso Joide § 148
g... Apofisi Mastoidea § 23
o... Aspera Arteria § 148
p... Sterno § 63
q... Clavicola sinistra § 72

28 MALVINA HOFFMAN

(American, 1885–1966)
Skeleton and outline of a torso in profile; figure in profile bending backward; two heads in different anatomical layers; head and neck seen from behind
1925
Sketchbook, graphite on paper, 28 × 21.5 cm
Los Angeles, Getty Research Institute, 850042
Inscriptions: "Trapezius"; "12th dorsal Vertebra"

BIBLIOGRAPHY

Hoffman 1939; Hoffman 1965.

Malvina Hoffman, a successful American sculptor, studied in Paris, where she was a student of Auguste Rodin (1840–1917). She is best known today for *The Races of Mankind,* her now controversial series of bronze sculptures completed in 1933 for the Field Museum, Chicago. Hoffman was a strong proponent of anatomy: "The understanding of anatomy and bone construction should be part of the mental equipment of every sculptor. It should be studied and memorized, and stored away in the subconscious mind where it may be referred to at any time" (Hoffman 1939, 112). Sketchbooks with anatomical studies from her early student years, 1901–5, and from 1925, when she attended the Art Students League of New York, are in the collection of the Getty Research Institute. While the earlier anatomical drawings are careful and precise, the 1925 drawings, seen here, show greater freedom and dynamism. Concerning her studies of dissection at the College of Physicians and Surgeons at Columbia University, New York, in the early 1910s, Hoffman wrote in her autobiography that they "took all the nerve and willpower I could muster, but they gave me a fundamental understanding of the construction of the human body" (Hoffman 1965, 120).

In the study of the torso at the upper left, Hoffman explores the body's interior structure while retaining a sense of its surface. The drawn outline of the buttocks and legs serves to establish the relative position of the pelvic bones. The sketch just below shows the area studied anatomically above in motion. In the middle of the right of the sheet, the artist has merged interior and surface anatomy by drawing a skull with muscles, a nose, and closed lips over a row of bottom teeth. The sketchbook also contains perspectival studies and drawings of animals.

Hoffman had been encouraged in her anatomical studies by a family friend, Harvey Cushing (1869–1939), the pioneering American neurosurgeon, book collector, and medical historian, and by Rodin, who despite his mastery of the human nude, nevertheless told Hoffman that he regretted not having had experience in dissection himself (Hoffman 1965, 107, 120). —MK

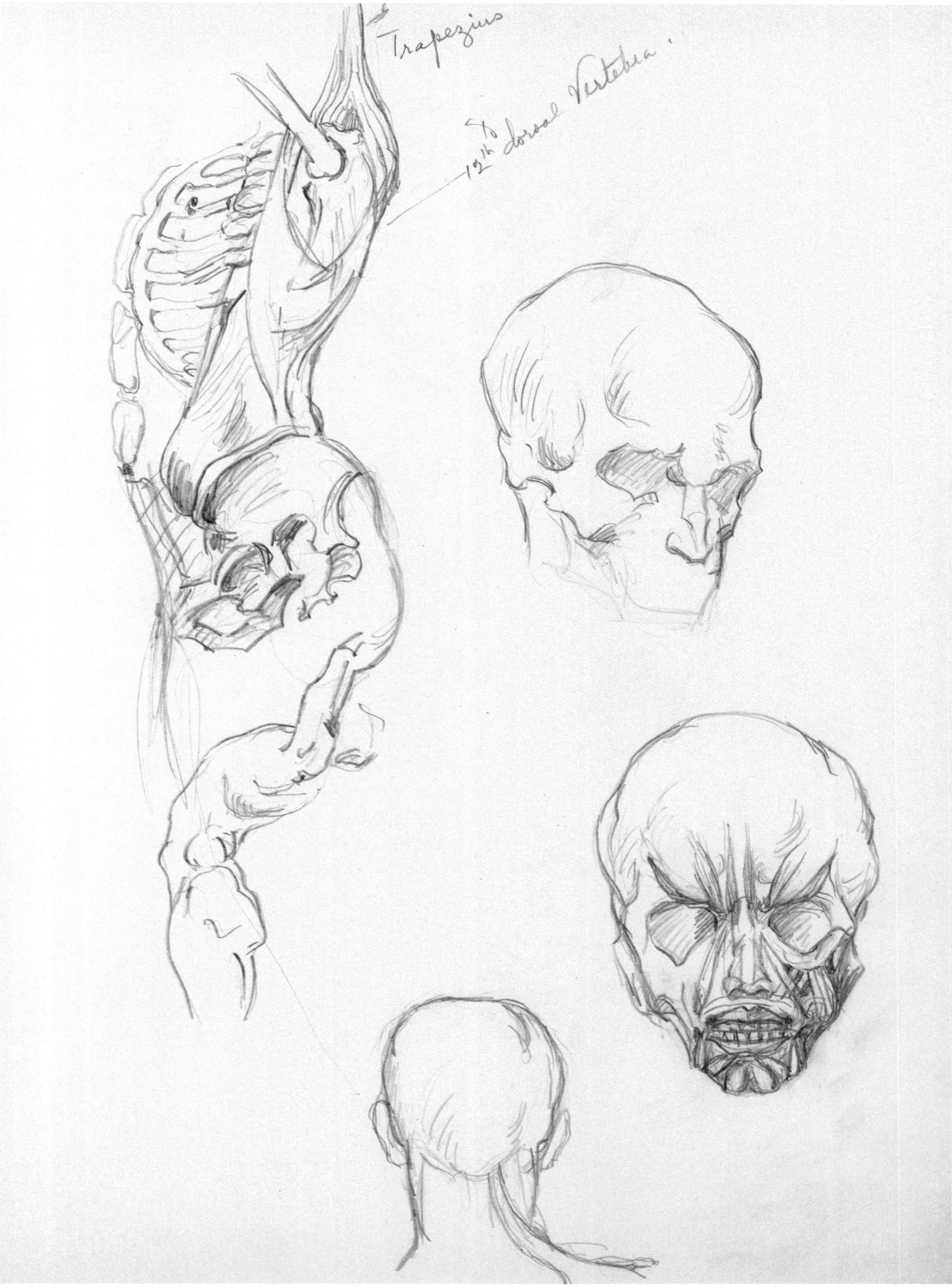
Trapezius
12th dorsal Vertebra.

29 Shelagh Keeley

(Canadian, 1954–)
Vertebrae
Gouache, charcoal, wax with Xerox transfers
From Shelagh Keeley, *Notes on the Body* (New York: Granary Books, 1991), copy 8 of 17
Los Angeles, Getty Research Institute, 93-B17786

BIBLIOGRAPHY
Print Collector's Newsletter 1992, 64; Gale and Dewdney 2010.

At the center of this opening of Shelagh Keeley's artist book, vertebrae are distorted and abstracted so that they appear as flowering stalks. They serve as a visual bridge between the printed Eastern and Western representations of the vertebrae on either side, both reproduced in reverse. On the left, rich black gouache overlaps a Japanese text illustrated with a simplified diagram of the spinal column. Devin Fitzgerald, curator of rare books and history of printing at the UCLA Library Special Collections, notes that the text consists of brief comments on important vertebrae and their uses in Chinese and Japanese medicine. At the upper right are the twelve thoracic vertebrae taken from Bernhard Siegfried Albinus's *Tabulae ossium humanorum* (1753), in which the cool, pristine engraving of the vertebrae by Jan Wandelaar contrasts with the diffuse, shaded outlines of Keeley's pink-tinged drawing.

In another opening, illustrations of the kidney in two canonical texts of Eastern and Western medicine are on facing pages: one is taken from a modern translation of the Huangdi Neijing, an ancient Chinese medical text in which the kidney is associated with the "Gate of Life," and the other is from an edition of Henry Gray's *Anatomy, Descriptive and Surgical,* first published in 1858. Both have been reproduced in reverse, disorienting the viewer and obscuring the text—the latter effect analogous to the mirror text Leonardo da Vinci employed in his notebooks.

A Space for Breathing (1992), another artist book by Keeley in the Getty Research Institute's collection (93-B17816), explores similar imagery of abstracted anatomy mixed with copies from illustrated medical texts and juxtaposed with images of indoor and outdoor spaces.
—MK

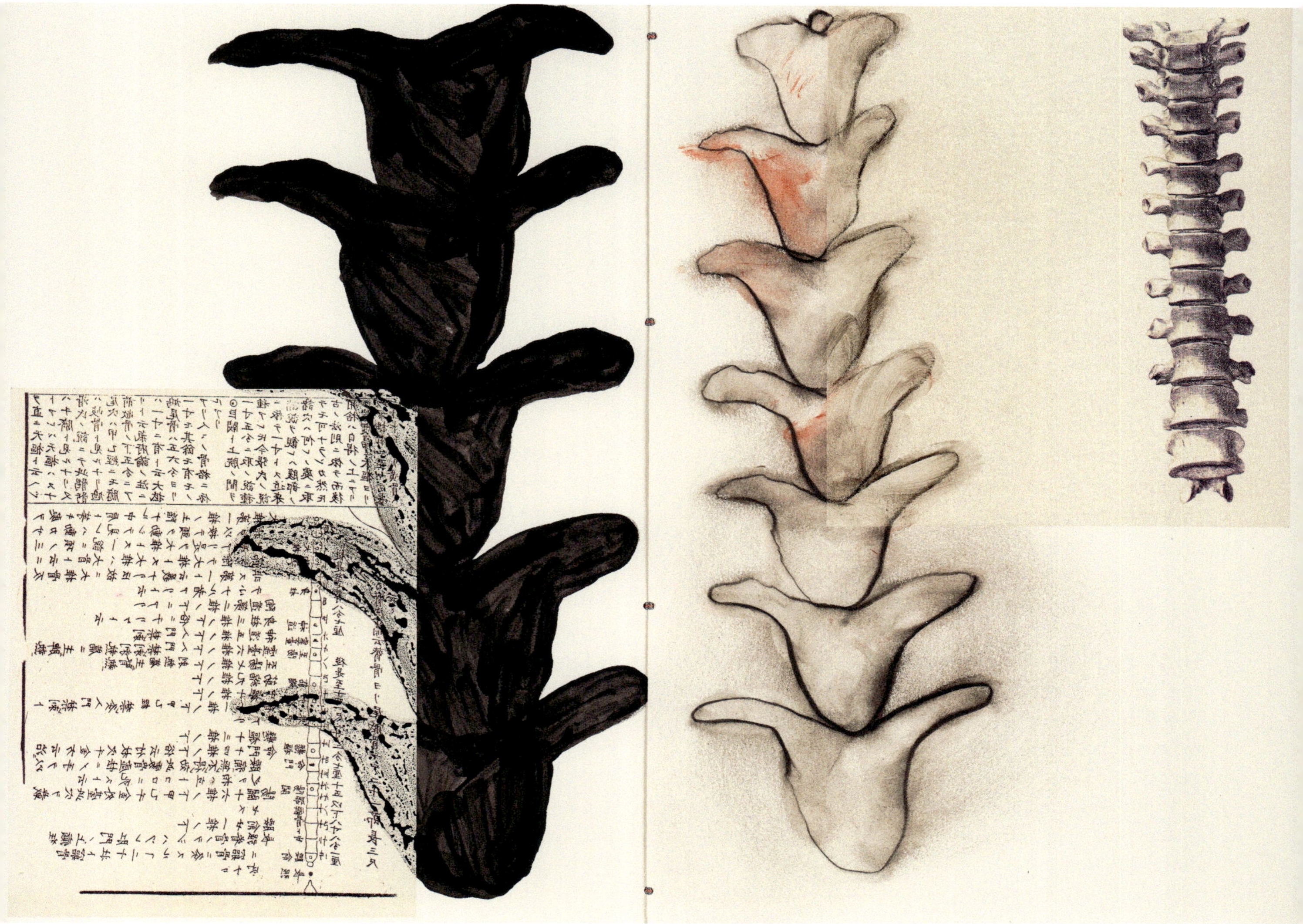

30 After Jan Steven van Calcar

(North Netherlandish, ca. 1515–ca. 1546)
Abdominal dissection; below: gall bladder and bile duct
Woodcut
From Andreas Vesalius, *De humani corporis fabrica libri septem* (Basel: J. Oporinus, 1543), bk. 5, p. 365 [465], figs. 12, 13
Los Angeles, Getty Research Institute, 84-B27611

BIBLIOGRAPHY
Cushing 1962; O'Malley 1964; Kemp 1970; Harcourt 1987; Simons and Kornell 2008; Houtzager 2014; Nutton 2014; Nutton 2016; Kornell 2018; Margócsy, Somos, and Joffe 2018, 1:231.

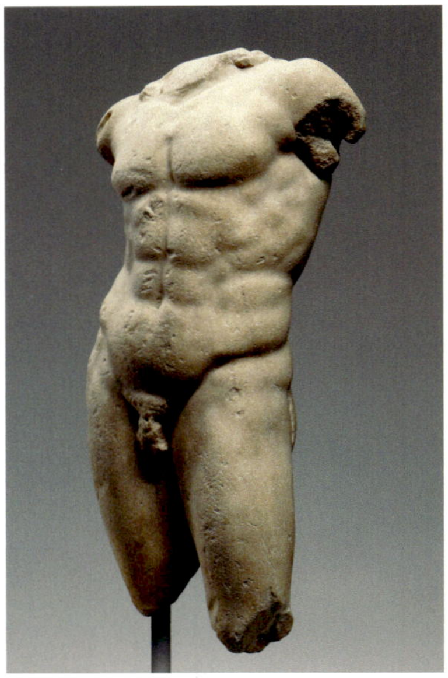

Fig. I
Torso of a male figure, Roman, 1st century BC–1st century AD, marble, 77.5 × 38.1 × 30.5 cm. Los Angeles, J. Paul Getty Museum, 79.AA.146.

Andreas Vesalius (1514–64) was born in Brussels to a medical family and educated in Louvain, Paris, and Padua, where he held appointments in surgery and anatomy while at work on the *Fabrica* from 1538 to 1542. He had previously collaborated with the northern artist Jan Steven van Calcar on the woodcut illustrations for *Tabulae anatomicae sex* (1538), and it is likely that Van Calcar and a team of artists in Titian's circle worked on the woodcut illustrations for the *Fabrica*. Vesalius closely oversaw the production of the woodblocks and paid for them himself. Instead of using one of the printing houses in Venice, he made the remarkable decision to pack up the blocks and send them with the text over the alps to Johannes Oporinus in Basel, with whom he had published before (Clark 1981, 305–7; Nutton 2016).

A landmark work in the histories of anatomy and the printed book, the *Fabrica* is notable for challenging the authority of Galen by favoring evidence of the body and for the number, size, and sophistication of the woodcut illustrations. The innumerable copies of the illustrations that appeared immediately after the *Fabrica*'s publication and in the following centuries, often without Vesalius's original text, is a measure of their success (Cushing 1962; Margócsy 2019).

The fifth book of seven uses antique torsi to showcase a series of abdominal dissections, several in a contrapposto stance similar to one seen in a marble torso in the J. Paul Getty Museum (fig. I). In this illustration, the intestines and the stomach have been moved to the side to better display the mesentery, the portal vein, and the insertion of the bile duct into the intestine. While the anatomy of the abdomen is displayed as if the torso is living flesh, the breaks at the limbs and the neck are smooth and stony and don't show any muscle or bone.

The GRI's copy of the *Fabrica* was once in the library of Sir William Stirling Maxwell (1818–78), a Scottish politician, bibliophile, and historian of Spanish art, whose annotation on Van Calcar as Vesalius's illustrator is found on the flyleaf of this copy. His large bookplate (31 × 27 cm) on the front pastedown takes full advantage of the *Fabrica*'s dimensions. Similarly, large bookplates, each of a different design, adorn Stirling Maxwell's former copy of the *Tabulae anatomicae sex*, now in the University of Glasgow Library (Sp Coll Hunterian Az.1.10), and that of his copy of the second edition of the *Fabrica* (1555), now in Princeton University Library, which also bears an annotation on Van Calcar (Margócsy, Somos, and Joffe 2018, 2:344). In 1874, Stirling Maxwell published a facsimile edition of his copy of the *Tabulae anatomicae sex*, one of only two known complete extant copies.

Two drawings connected with the *Fabrica* are in California: one made for the title page is in the collection of the Huntington Library, San Marino, and the second is a drawing of bones in the Crocker Art Museum, Sacramento. Another California connection is the important biography of Vesalius that the illustrious medical historian Charles D. O'Malley (1907–70) composed while a faculty member of the University of California, Los Angeles (O'Malley 1964). Before the advent of digital imaging, the Vesalian illustrations owed much of their popular modern recognition to their appearance in *Illustrations from the Works of Andreas Vesalius of Brussels* (1950), a book that O'Malley, while still at Stanford University, collaborated on with John B. deC. M. Saunders and that used reproductions of prints made from the original *Fabrica* woodblocks. These had been published in 1934 in *Icones anatomicae* just before the tragic destruction of the blocks during World War II (Goree 2014, 25–41; Vesalius 1934). —MK

DE HVMANI CORPORIS FABRICA LIBER V. 365

DVODECIMA QVINTI LIBRI FIGVRA, QVAE

IAM PRAEPOSITIS INTEGRIS FIGVRIS SECTIO-nis serie succedit. Habet enim dissectũ peritonæum, & omentum quoq; in hac ablatũ est, & hic costas aliquot etiã effregimus, quò iecoris cauũ opportunius delineari posset. Hic namq; conspicitur uniuersum iecoris cauũ, ipsiusq; iecoris forma. Dein uẽtriculi quoq; apparẽt orificia. Intestina autẽ perinde ac uentriculũ in sinistrũ latus depreßimus, ut in cõspectu esset mesenterij pars, ac uenæ portæ in ipsum series: dein bilis meatus in intestinũ insertio, et si quæ sint reliqua quæ characteribus seriatim adnotabimus, mox atq; quid decima tertia ostendat figura, expresserimus.

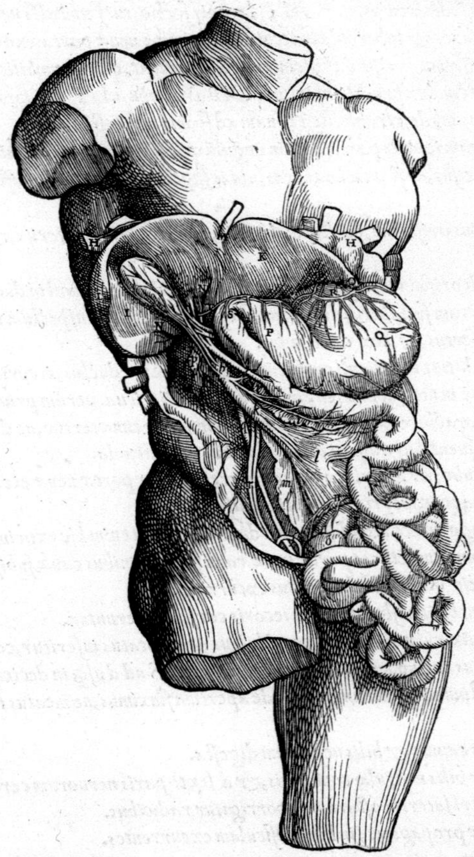

DECIMATERTIA QVINTI LIBRI FIGVRA,

NVDAM BILIS VESICVLAE EIVSDEMQVE meatuum delineationem exprimens.

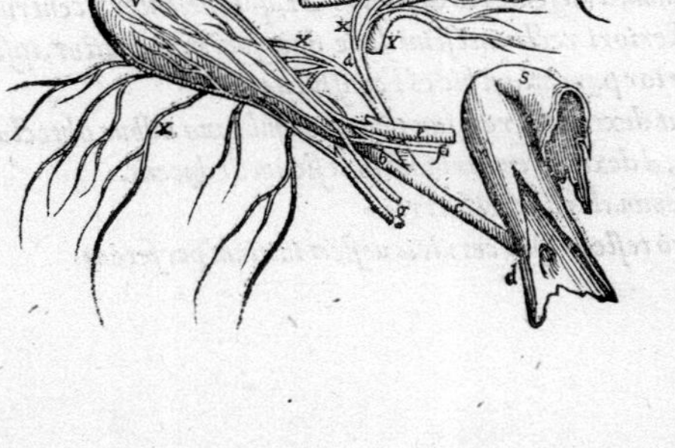

CHA-

31 **Frans Huys**

(Flemish, ca. 1522–before 1562) or Pieter Huys (Flemish, ca. 1520–ca. 1584), after Nicolas Beatrizet (French, 1507 or 1515–ca. 1565), after Gaspar Becerra (Spanish, ca. 1520–ca. 1570)
Anatomized pregnant woman in a Venus pudica pose
Engraving
From Andreas Vesalius, *Vivae imagines partium corporis humani aereis formis expressae* (Antwerp: Christophe Plantin, 1566), bk. 3, p. 99, pl. 5
Los Angeles, Getty Research Institute, 1387–773

BIBLIOGRAPHY
Valverde 1556; Turner 1878, 75–78; Meyer and Wirt 1943, 674; Cushing 1962, 151, VI.D.-43; Kusukawa 2012, 53–54; Bowen and Imhof 2008, 46–47, 67–83; Massey 2008, 84; Houtzager 2014, 201; Skaarup 2015, 242–46; Van Hee 2020a, 165–66, 169.

The *Vivae imagines* is a mélange of text and images that, filtered through copies, derive from Andreas Vesalius's *Fabrica* and *Epitome* of 1543 (cat. no. 30) and the 1559/60 edition of Juan Valverde de Amusco's *Anatomia del corpo humano,* an Italian translation of the first Spanish edition of 1556 with engravings mostly adapted from the *Fabrica* (Cushing 1962, 151, VI.D.-43). Selected and repackaged by the printer and publisher Christophe Plantin (ca. 1520–89), *Vivae imagines* was published in Antwerp in 1566 with a title page that does not credit an author and a title that emphasizes the plates rather than the text: "Lively images of parts of the human body expressed in copper."

Valverde (ca. 1525–ca. 1588) was a Spanish anatomist active in Italy, where he was a student of Realdo Colombo (ca. 1515–59). Vincente Carducho identifies Gaspar Becerra, a Spanish painter working in Rome from the mid-1540s to mid-1550s as Valverde's artist (Carducho 1633, 2r). Several of Valverde's illustrations show the inspiration of antique sculpture (San Juan 2008). Among the figures original to Valverde's book is this anatomized pregnant woman set in the pose of a Venus pudica, an antique pose of modesty paradoxically employed here to showcase the interior of the body. A fetus shown removed from the uterus and seated on the ground behind her is also new, while the remaining illustrations of the uterus, placenta, and fetus were taken from the second edition of the *Fabrica* (1555), in which Vesalius had replaced the canine anatomy of the first edition with human anatomy (Saunders and O'Malley 1950, 174–75).

Engraved copies mainly after the *Fabrica* illustrations and paired with text from Vesalius's short compendium of anatomy, the *Epitome* (1543), first appeared in Thomas Geminus's *Compendiosa totius anatomie delineatio,* published in London in 1545 (Geminus 1959). Valverde followed the example of Geminus by simplifying the landscape setting and grouping together Vesalius's smaller images, this being an easier and cheaper arrangement in the transition from woodblocks to copperplates (Bowen and Imhof 2008, 76; Lo 2018, 230, 232–33). —MK

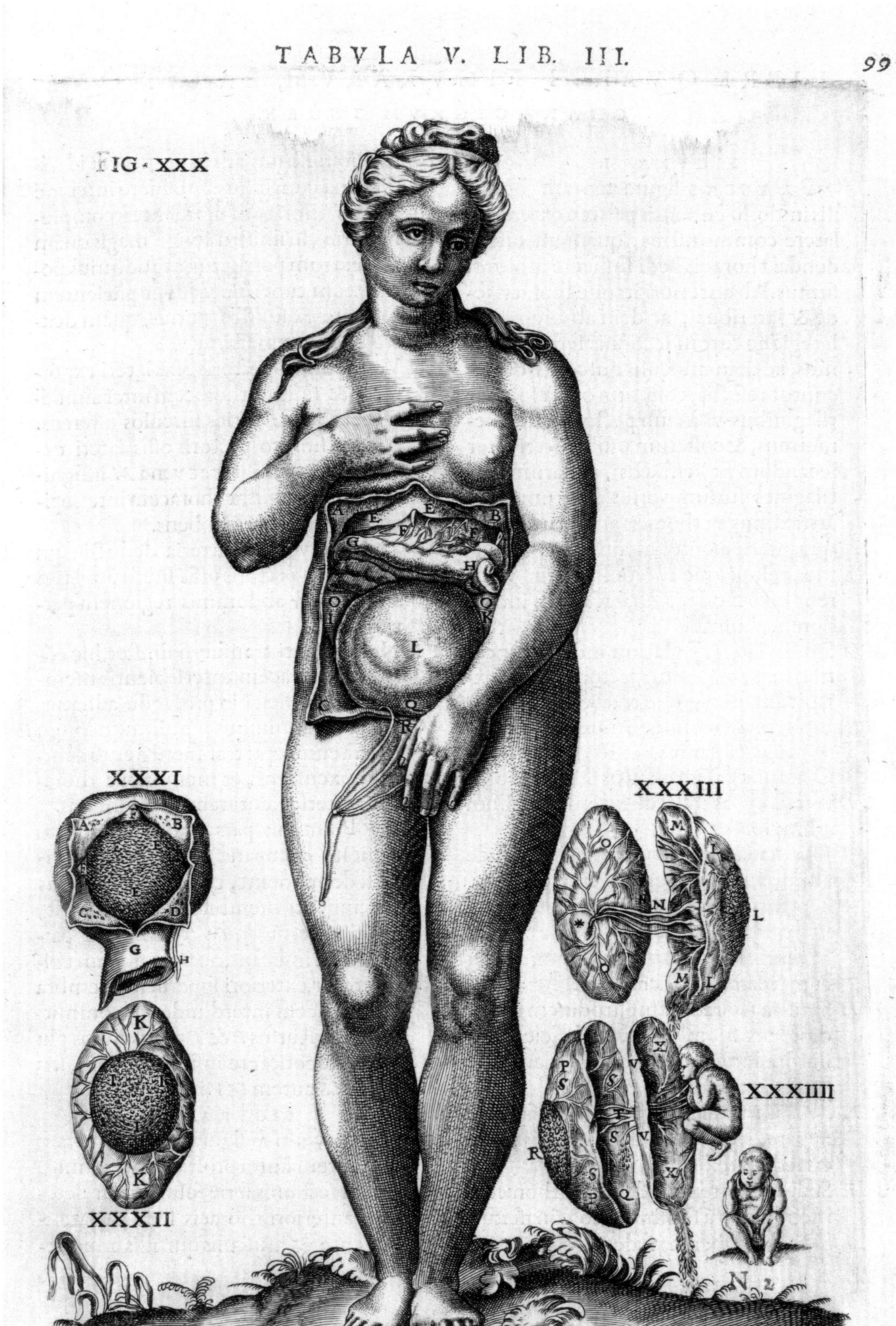
TABVLA V. LIB. III.
99
FIG. XXX
XXXI
XXXII
XXXIII
XXXIIII
N. 2

32 After Charles Errard

(French, ca. 1606/9–89)
Statue of the Gladiator in the Villa Borghese, . . . Considered Anatomically
Engraving
From Bernardino Genga and Giovanni Maria Lancisi, *Anatomia per uso et intelligenza del disegno: Ricercato non solo su gl'ossi, e muscoli del corpo humano; Ma dimostrata ancora su le statue antiche più insigni; Di Roma delineata in più tavole con tutte le figure in varie faccie, e vedute; Per istudio della regia academia di Francia pittura e scultura sotto la direzzione di Carlo Errard già direttore di essa in Roma; Preparata su'i cadaveri dal dottor Bernardino Genga regio anatomico; Con le spiegazione et indice del Signor Canonico Gio. Maria Lancisi già medico segreto della Santa Memoria di Papa Innocentio XI; Opera utilissima à Pittori, e Scultori, et ad'ogni altro studioso delle nobili Arti del Disegno* (Rome: Domenico de Rossi, 1691), pl. 30
Los Angeles, Getty Research Institute, 84-B2835–1

BIBLIOGRAPHY
Choulant 1945, 254–55; Kornell 1996, 60, 65; Kornell in Cazort, Kornell, and Roberts 1996, 216–19, no. 104; Sénéchal in Cuzin, Gaborit, and Pasquier 2000, 391–92, no. 197; Coquery 2002; Brink 2012, 190–94; Coquery 2013, 197–200, D. 361–65, G. 373–424; Aymonino 2015, 48–49.

The *Anatomia per uso et intelligenza del disegno* was produced for the Académie de France in Rome under the direction of Charles Errard, the Académie's founding director (Coquery 2002, 142). Illustrated with engravings after Errard, most of the thirty-nine plates in this copy plus the frontispiece were adapted from sculpture. In the concluding plates, the Farnese Hercules, the central figure of the Laocoön group, the Borghese Gladiator, and adaptations of the Borghese Faun are all depicted in écorché, as if their stony skin had been removed to reveal their underlying muscles. The Laocoön also served as the model for the musculature of the torso in plates 10–14. An anatomical model belonging to the Académie de France appears on plates 21–23 and may record a life-size wax model cast from a flayed cadaver in the hospital of Santo Spirito (Kornell 1996, 60; Coquery 2002, 150). Even the jumble of partially fleshed skeletons in a roundel on the frontispiece has a sculptural source, as it is based on Giacomo del Duca's 1570 bronze relief of the Resurrection that decorates the tomb of Elena Savelli in the church of San Giovanni in Laterano, Rome (Brink 2012).

Almost all the information for the book's production comes from the extended title. The anatomist and surgeon Bernardino Genga (1636–95), longtime professor of anatomy for the academy, is credited with the dissections for the book and Giovanni Maria Lancisi (1654–1720), anatomist and papal physician, with the anatomical explanations and the index. Genga had previously assisted Carlo Cesi on his anatomy book for artists (cat. no. 20), whereas Lancisi's continued interest in art is demonstrated by his key role in the rediscovery and publication in 1714 of Eustachi's plates and by the notable collection of paintings he owned at his death (Lancisi 1714; Campori 1870, 512–15). Some copies of the book, such as the two in the Getty Research Institute, have "Libro Primo" on the title page. Surviving drawings by Errard and related proofs showing skeletons in the poses of antique statues are further indications that another volume had been planned (Coquery 2013, 199–200). The lack of a preface is another sign of the project's unfinished state, likely due to Errard's death. Additionally, by 1690, Genga had been pushed out of his post at the academy by the new director, Matthieu de La Teulière (d. 1702) (Montaiglon 1887, 346).

In 1723, John Senex published an English translation of the *Anatomia* with reengraved plates (Genga and Lancisi 1723), a copy of which is in the GRI (87-B13101). —MK

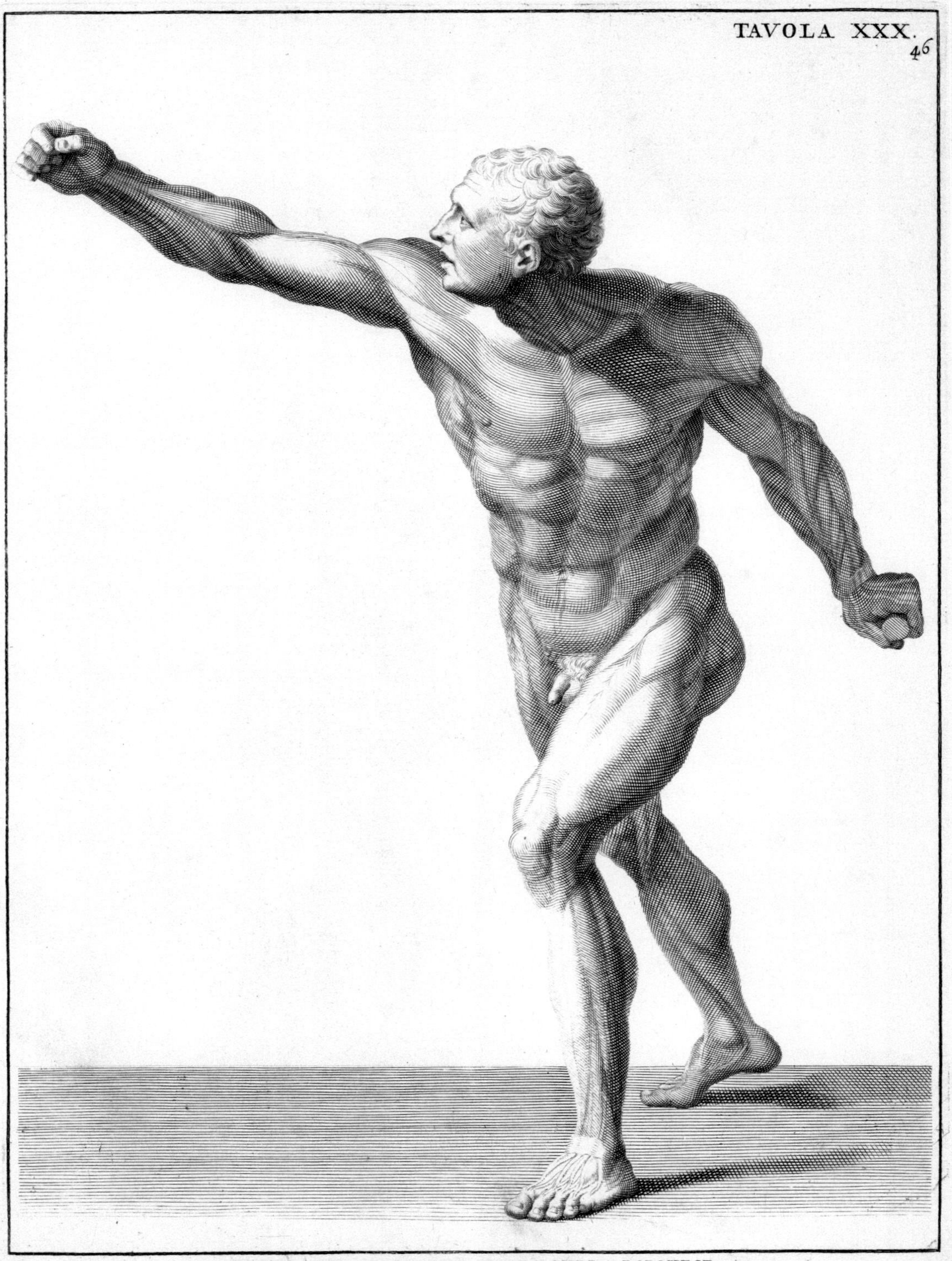

STATVA DEL GLADIATORE NELLA VILLA BORGHESE.
Prima veduta in faccia considerata anatomicamente

33 Gerard Vandergucht

(English, 1696–1776)
The Sceleton of a Woman, in the Same Proportions with the Venus of Medicis
Etching
From William Cheselden, *Osteographia; or, the Anatomy of the Bones* (London: n.p., 1733), pl. 34
Los Angeles, Library Special Collections for Medicine and the Sciences, Louise M. Darling Biomedical Library, UCLA, BIOMED ** WZ 260 C4240 1733

BIBLIOGRAPHY

Douglas 1735; Haller 1774–77, 2:35; Cope 1953, 66–70; Russell 1954; Haskell and Penny 1981, 325–38, no. 88; Schiebinger 1986, 14, 18; Russell 1987, 42–43, no. 172; Schiebinger 1989, 194; Roberts and Tomlinson 1992, 422–30; Kemp 1993, 104–7; Roberts in Cazort, Kornell, and Roberts 1996, 77; Kornell in Cazort, Kornell, and Roberts 1996, 64, 190–93, nos. 75, 76; Kornell 2011; Bertoloni Meli 2017, 53–61; Wagner 2017, 5–6; Flis in Campbell and Flis 2018, 218–19, no. 35.

The female skeleton differs from the male in the shape of the pelvic bones and in the angle of the femurs, yet there were few depictions of it in the early modern period (Schiebinger 1986; Schiebinger 1989). One of these is in the *Osteographia,* a luxurious publication on the bones by William Cheselden (1688–1752), a successful London surgeon. Cheselden may have been aware of another illustration of the female skeleton by his former teacher, the anatomist and surgeon William Cowper (1666/67–1710), which appeared in James Drake's *Anthropologia nova* (Drake 1707, vol. 2, fig. 22).

Cheselden's female skeleton has been given the pose of the antique statue the Venus de' Medici. Here, Cheselden adapts the pose, moving the lower arm to the side to allow an unobscured view of the wider hips of the skeleton. This allows a comparison with the pelvis of the male skeleton, seen in the following illustration, set in the pose of the Apollo Belvedere. Cheselden selected this pose himself since, as he points out in his address to the reader, "The actions of all the sceletons both human and comparative, as well as the attitudes of every bone, were my own choice." Cheselden, whose brevity of description was faulted by contemporaries, does not discuss the differences between the female and male skeletons.

In the *Osteographia,* individual bones are illustrated in life-size. The book was sumptuously printed with a double set of plates, one with lettering and one without (the UCLA copy only has the lettered set). The chapters are decorated with charming initial letters, following the example of Vesalius and Cowper (cat. nos. 30, 13). The head- and tailpieces of vignettes of comparative anatomy, such as the skeleton of a crocodile before a pyramid, are mainly by the Dutch artist Jacobus Schijnvoet (active 1704–33). The final section is devoted to diseased bones (Bertoloni Meli 2017). Cheselden did not recoup the great expense of producing the book, and the two planned volumes of anatomy that were to follow never appeared.

Gerard Vandergucht's father, the Antwerp-born Michael Vandergucht (1660–1725), had worked on many of Cowper's publications, including the *Myotomia reformata* (1724) (cat. no. 13). Father and son both contributed prints to the 1723 English edition of Bernardino Genga and Giovanni Maria Lancisi's anatomy book, which served as an immediate example for the presentation of anatomy in antique poses (cat. no. 32). Cheselden admired Gerard's "open free style" and "the inimitable manner of expressing the different textures of the parts" (Cheselden 1733, address to reader). Gerard worked again with Cheselden on the fifth edition of *The Anatomy of the Human Body,* published in 1740; as a sign of the author's esteem, he is acknowledged on the title page. —MK

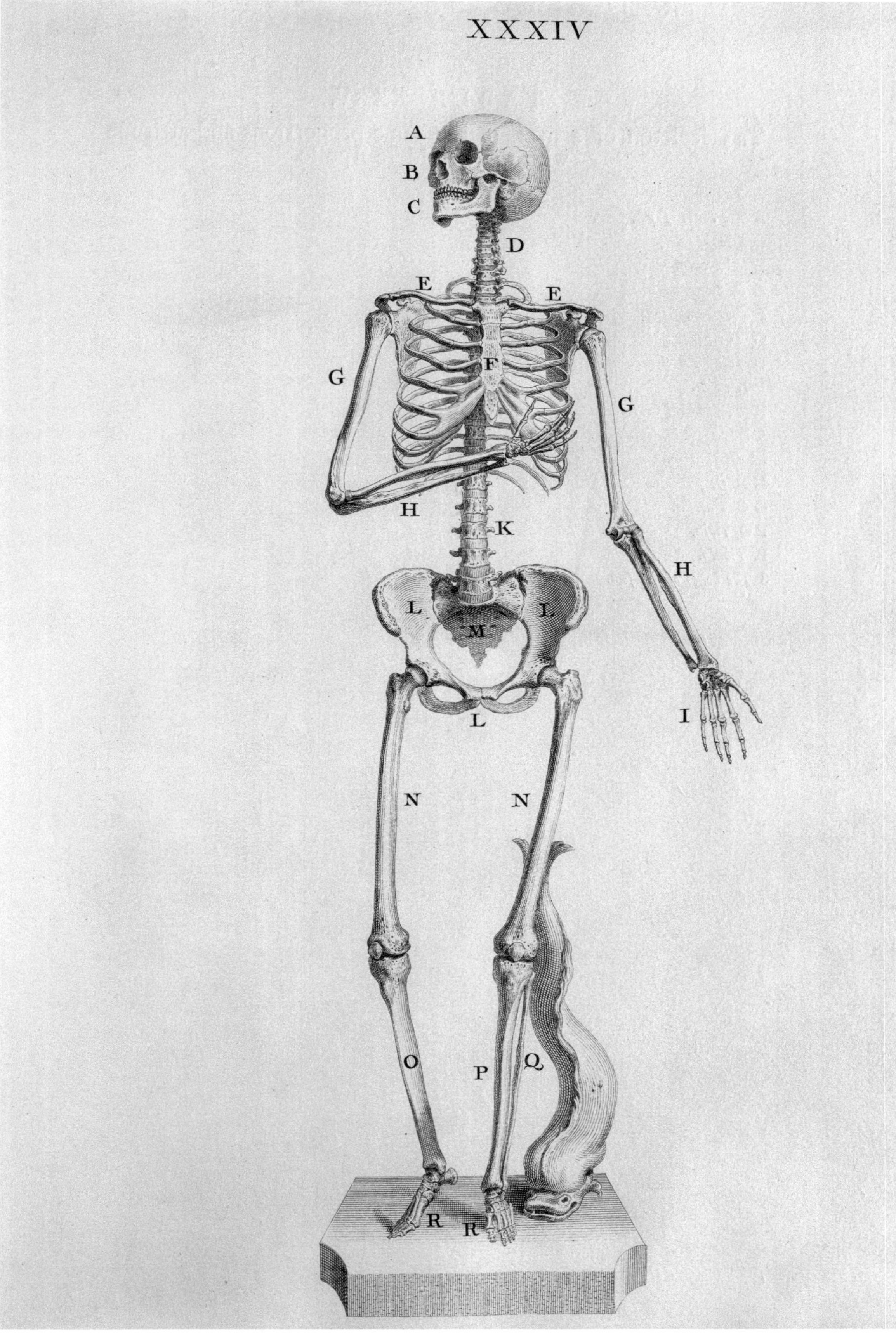
XXXIV
A
B
C
D
E
E
F
G
G
H
K
H
L
L
M
L
I
N
N
O
P
Q
R
R

34 Nikolaj Utkin

(Russian, 1780–1868), after Jean-Galbert Salvage (French, 1770–1813)
The head of the Apollo Belvedere, rendered anatomically
Etching and engraving from two plates printed in black and red ink
From Jean-Galbert Salvage, *Anatomie du gladiateur combattant, applicable aux beaux arts; ou, Traité des os, des muscles, du mécanisme des mouvemens, des proportions et des caractères du corps humain* (Paris: Jean-Galbert Salvage and l'imprimerie de Mame, 1812), pl. 1
Los Angeles, Getty Research Institute, 85-B12146

BIBLIOGRAPHY
Duval and Cuyer 1898, 262–78; Choulant 1945, 332–34; Amerson 1975, 406–9; Haskell and Penny 1981, 221–24, no. 43; Shedd 1991, 98–105; Kornell in Cazort, Kornell, and Roberts 1996, 219–20, no. 105; Sénéchal 1998; Sénéchal in Cuzin, Gaborit, and Pasquier 2000, 392–94, nos. 198–99; Comar and Joly in Comar 2008b, 226–30, nos. 82–86; Lifchez 2009; Fend 2017, 207–15.

A surgeon in the French army, Jean-Galbert Salvage had had an affection for art since childhood (Salvage 1812, i). He provided his own exquisite illustrations to this anatomy book for artists published in 1812, just a year before his death in his early forties. The book is dedicated to Agasias of Ephesos, the sculptor of the Borghese Gladiator, one of the most admired works of art of antiquity since it was unearthed in Italy in the early seventeenth century. Carlo Cesi seems to have taken inspiration from the sculpture for the sixteenth plate in his book of 1679 (cat. no. 20). A flayed gladiator appears in the anatomy book of Bernardino Genga and Giovanni Maria Lancisi (1691) (cat. no. 32). While working at a military hospital in Paris, soldiers who died in duels provided Salvage with subjects in the prime of life, which he dissected in different muscular layers and set in the pose of the gladiator and then cast in plaster. These casts, seen in multiple views, served as the basis for the book's plates. The work was well advanced by 1804, when engravings were submitted to an official committee (Salvage 1812, ij), one of which included Jean-Antoine Houdon (1741–1828), the sculptor of two influential écorché models (see fig. 5.5).

Salvage "anatomized" two plaster casts of the head of the Apollo Belvedere, the subject of plate 1, concluding that the statue's noble and majestic air "depends absolutely" on the bones (Salvage 1812, ijn3; Comar and Joly in Comar 2008b, 228–30, nos. 83, 84). Plate 1 is the result of a forensic deconstruction of the statue, a process akin to Charles Errard's virtual dissection of antique statues on paper in Genga and Lancisi's *Anatomia per uso et intelligenza del disegno* (1691). The skeleton bust of the Apollo Belvedere in profile retains a segment of the statue's distinctive topknot hairstyle. Pinpricked in red is the exterior contour of the head and neck, the eyelids, and the shape of the ear, which floats above the auditory canal. On the bust to the left, the muscles in red are superimposed on the head of Apollo, with areas in black ink showing deeper structures, such as the lower jaw and the hyoid bone and thyroid cartilage under the chin. —MK

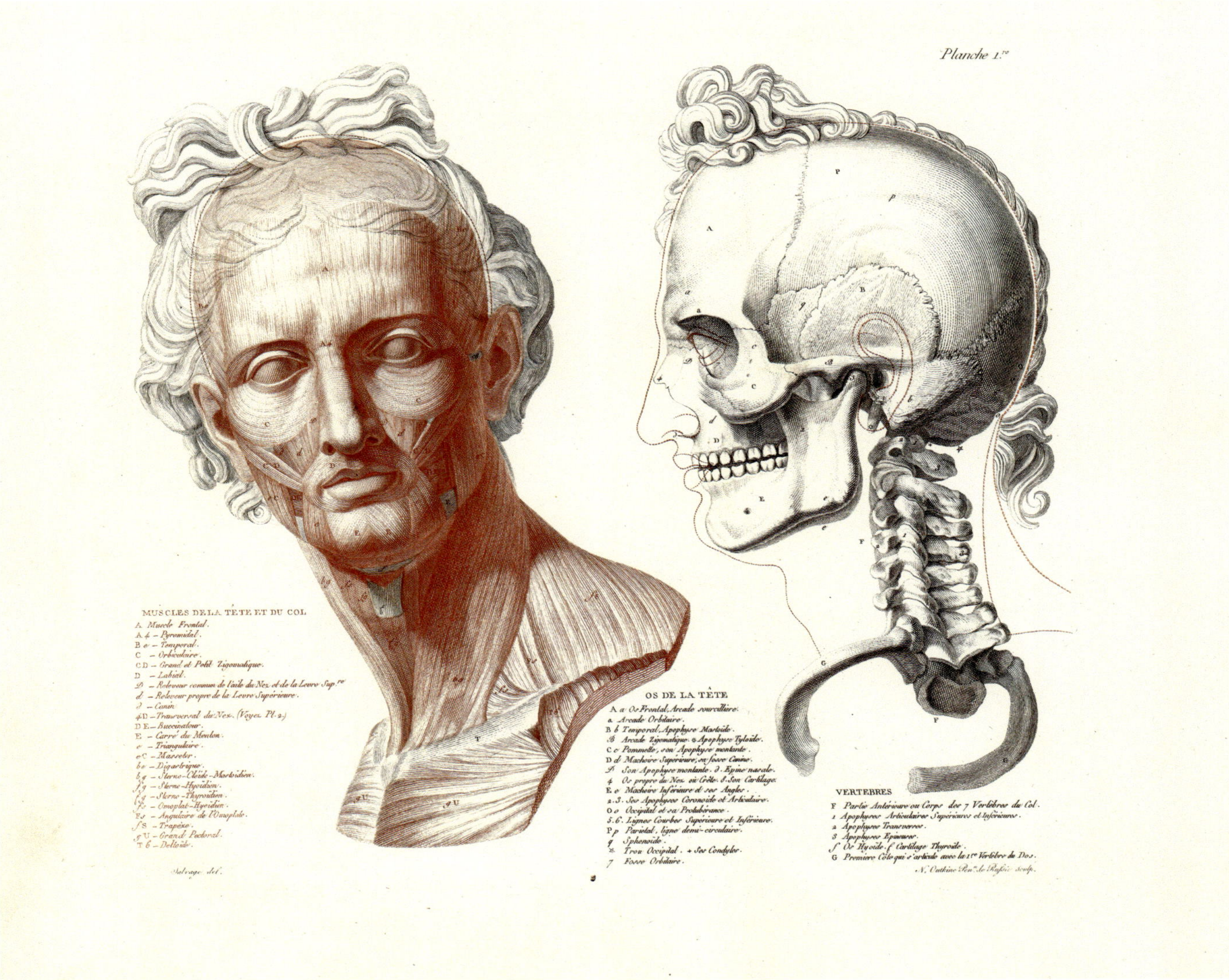
Planche 1.re
MUSCLES DE LA TÊTE ET DU COL
A. Muscle Frontal.
A 4 – Pyramidal.
B e – Temporal.
C – Orbiculaire.
C D – Grand et Petit Zigomatique.
D – Labial.
– Releveur commun de l'aile du Nez et de la Levre Sup.re
d – Releveur propre de la Levre Supérieure.
– Canin
4 D – Transversal du Nez. (Voyez Pl. 2.)
D E – Buccinatour.
E – Carré du Menton.
e – Triangulaire.
e C – Masseter.
b e – Digastrique.
b g – Sterno-Cleïdo-Mastoïdien.
f g – Sterno-Hyoïdien.
– Sterno-Thyroïdien.
– Omoplat-Hyoïdien.
– Angulaire de l'Omoplate.
f S – Trapèze.
g U – Grand Pectoral.
T 6 – Deltoïde.
Salvage del.
OS DE LA TÊTE
A a Os Frontal, Arcade sourcillière.
a Arcade Orbitaire.
B b Temporal, Apophyse Mastoïde.
Arcade Zigomatique. Apophyse Styloïde.
C c Pommette, son Apophyse montante.
D d Machoire Supérieure, sa fosse Canine.
Son Apophyse montante. Epine nasale.
4 Os propre du Nez ou Crête. Son Cartilage.
E e Machoire Inférieure et ses Angles.
2.3. Ses Apophyses Coronoïde et Articulaire.
O o Occipital et sa Protubérance.
5.6. Lignes Courbes Supérieure et Inférieure.
P p Parietal, ligne demi-circulaire.
q Sphenoïde.
Trou Occipital. + Ses Condyles.
7 Fosse Orbitaire.
VERTEBRES
F Partie Antérieure ou Corps des 7 Vertèbres du Col.
1 Apophyses Articulaires Supérieures et Inférieures.
2 Apophyses Transverses.
3 Apophyses Epineuses.
f Os Hyoïde. Cartilage Thyroïde.
G Premiere Côte qui s'articule avec la 1.re Vertèbre du Dos.

35 Joseph Severn

(English, 1793–1879)
Écorché figure in the pose of the Ludovisi Gaul
Etching
From James Birch Sharpe, *Elements of Anatomy; Designed for the Use of Students in the Fine Arts* (London: R. Hunter, 1818), pl. 3
Los Angeles, Getty Research Institute, 84-B9486

BIBLIOGRAPHY

Lavater 1790; Duval and Cuyer 1898, 280–81; Haskell and Penny 1981, 282–84, no. 68; Roberts, A. 1981; Kornell 1996, 66; Röhrl 2000, 192, 245; Brown, Sue 2009, 25–26.

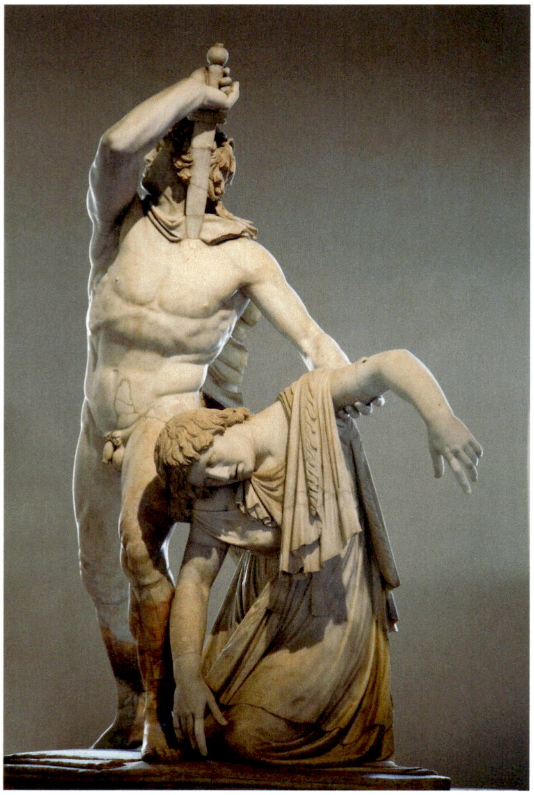

Fig. J
Ludovisi Gaul and His Wife, Roman, 2nd century AD, marble, 2.11 m. Rome, Museo Nazionale Romano, Palazzo Altemps, 8608.

James Birch Sharpe (1789–1863), an English surgeon who cared for inmates at the Hoxton House naval asylum (Roberts, A. 1981), had been a student at the Royal Academy of Arts, London. In *Elements of Anatomy,* Sharpe expresses admiration for anatomy displayed in antique sculpture but also stresses the importance of studying anatomy by dissection, recommending the dissecting theater of Joshua Brookes (1761–1833) for that purpose. The book contains a veiled diatribe against Sir Anthony Carlisle (1768–1840), the then-professor of anatomy at the academy, who felt the study of surface anatomy alone was sufficient for artists. The text in this brief and portable book is a straightforward description of the external muscles and veins, referring to how they appear in living bodies and in antique sculpture. Joseph Severn and Sharpe had attended the Royal Academy of Arts together but had a falling-out, and Severn speaks disparagingly about Sharpe in his memoirs, where we learn that he was his illustrator (Brown, Sue 2009, 25–26). Severn later accompanied the poet John Keats (1795–1821) to Rome, caring for him until his early death three months after their arrival. In later life, Severn served as British consul in Rome.

Three of the five illustrations in the book are after Bernhard Siegfried Albinus. The remaining two are anatomized antique sculptures: the Borghese Gladiator and the Ludovisi Gaul (fig. J). The figure of the Gaul is depicted in isolation, without the body of the woman he originally held up with his left hand. He is also shown without his cape or the trickle of blood from the entry site of the sword. The fig leaf, however, is an addition. Readers sometimes altered early anatomical illustration to censor the genitalia, as has been noted in copies of Andreas Vesalius's *Fabrica* (Margócsy, Somos, and Joffe 2018, 126–30). Fig leaves in ink were added to the illustrations of Bartolomeo Eustachi in the Wellcome Collection's copy of Gaetano Petrioli's *Riflessioni anatomiche* (1740) (see fig. 1.4). Other examples of the judicious placement of leaves over male genitals in anatomical illustration are seen in Delsenbach (1733) (pls. 2, 3) and Squanquerillo (1841) (pl. 21). It is possible, though, that Severn was working from a censored cast. An addition of fig leaves to antique casts was ordered by the Royal Academy of Arts in 1780 (Postle 2004, 60).

The left hand of the figure is Severn's invention, and, as Emily Beeny has remarked (in conversation), it shows the influence of Henry Fuseli (1741–1825), to whom the book is dedicated and whose rendering of hands and feet are particularly praised (Sharpe 1818, 49). Fuseli, who held the posts of keeper and professor of painting at the Royal Academy of Arts, was described as "profoundly conversant in anatomy" by his friend Johann Heinrich Lavater in the dedication to him for his own book of anatomy for artists (Lavater 1790, 4), which was, in turn, based on that of Cornelis Ploos van Amstel (1783) (see fig. 1.3).

Inscribed on the title page of the GRI's copy is "Wm Corden," suggesting ownership by either William Corden the Elder (1759–1867) or the Younger (1819–1900), both of whom were artists to the British royal family. —MK

Plate 3.

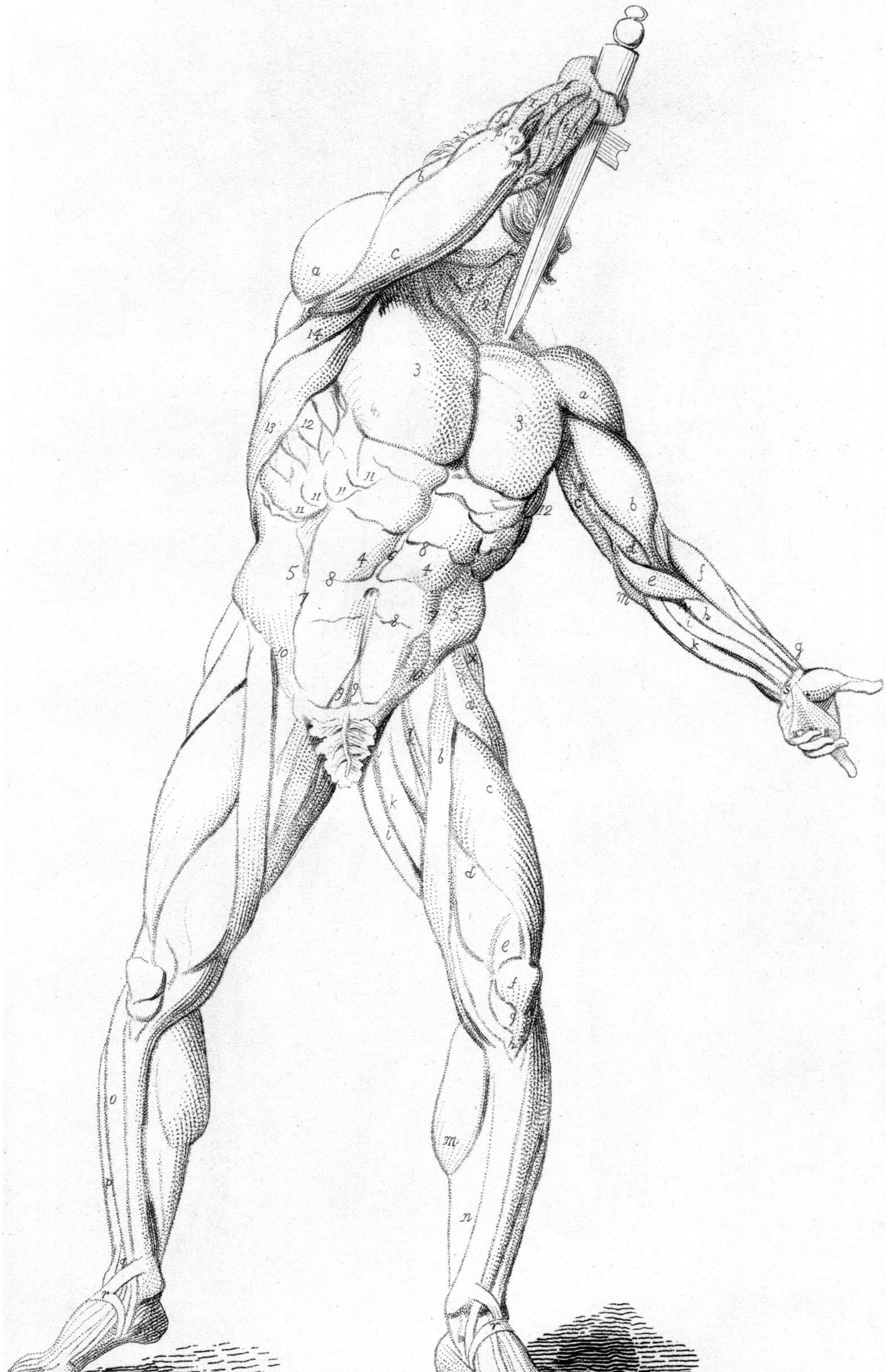

Antonio Cattani

(Italian, active 1777–ca. 1790), after Ercole Lelli (Italian, 1702–66)

36 Écorché figure, seen from the front
1780
Etching and engraving, printed on five sheets, overall size of assembled sheets 187 × 57 cm
Los Angeles, Getty Research Institute, 2014.PR.17**
Inscriptions:
HERCULES LELLI SCULPSIT
ANTONIVS CATTANI PLACENTINVS / INCISIT BONONIAE / MDCCLXXX.
in Bologna nella Stamperia / Cattani e Nerozzi nel Pavaglione

37 Écorché figure, seen from the back
1781
Etching and engraving, printed on five sheets, overall size of assembled sheets 184.6 × 57 cm
Los Angeles, Getty Research Institute, 2014.PR.18**
Inscriptions:
HERCULES LELLI SCULPSIT
ANT. CATTANI PLACENTIAE / INCIDIT BONONIAE. / MDCCLXXXI.
in Bologna nella Stamperia Cattani:Nerozzi / nel Pavaglione

38 Second level of muscles
1781
Etching and engraving, printed on five sheets, overall size of assembled sheets 191.7 × 60 cm
Los Angeles, Getty Research Institute, 2014.PR.19**
Inscriptions:
SECONDO GRADO DE MUSCOLI
Ant. Cattani fece 1781. / In Bologna nella Stamperia Cattani e Nerozzi sotto il portico delle Scuole

BIBLIOGRAPHY

Fantuzzi 1781–94, 5:50; Medici 1856; Cushing 1937; Zeitlin & Ver Brugge 1961, no. 296; Esterquest 1962, 11; Riccòmini 1965, 104–9; Amerson 1975, 376–78; Armaroli 1981; Gilhofer & Ranschburg 1994, no. 45; Cazort in Cazort, Kornell, and Roberts 1996, 224–25, no. 112; Kornell 1996, 58–59; Petherbridge and Jordanova 1997, 90–93, nos. 45, 46; Petherbridge, Ritschard, and Carlino 1998, 42, 221, nos. 139, 140; Christie's, New York 2007, lot 111; Basevi 2014, 496–97; Roncuzzi 2016, 86–88, no. 19.

The Bolognese painter, sculptor, and architect Ercole Lelli established his fame as an anatomical sculptor in 1734 with two écorché caryatids carved of wood with the assistance of Silvestro Giannotti (1680–1750) and installed in the anatomy theater of the Palazzo dell'Archiginnasio, Bologna (fig. K). Known as the "spellati" (the skinned), they are signed "E. LELLI F. 1734" on their bases. They stand on either side of the raised lecturer's chair in the anatomical theater, serving as supports for the canopy above it, and are similar in pose, with the main difference being in the arrangement of their right arms. The models for the statues were two skeletons that Lelli clothed in muscles shaped from various materials impregnated with wax (Medici 1856, 171). The biographer Giovanni Fantuzzi (1718–99) recalls visiting the anatomy school and witnessing "the applause and the admiration with which these statues first came to the public eye" (Fantuzzi 1781–94, 5:50). Soon after this great success, Lelli became master of the mint and then, in 1742, he embarked on a plan, under the patronage of Pope Benedict XIV, to create for the Istituto delle Scienze a series of wax anatomies, now in the Palazzo Poggi, which also had scaffoldings of actual skeletons (Messbarger 2010, 28–51; Dacome 2017, 63–84). He worked on these with the assistance of Giovanni Manzolini (1700–1755) until an acrimonious falling-out over accreditation. The artist and writer Luigi Crespi (1708–79)

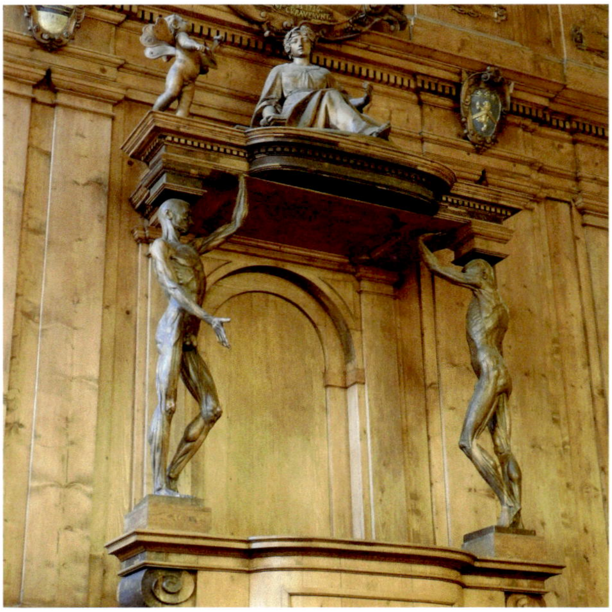

Fig. K
Ercole Lelli (Italian, 1702–66) and Silvestro Giannotti (Italian, 1680–1750).
Écorché caryatids, 1734, wood, 151 cm, without base.
Bologna, Archiginnasio anatomy theater.

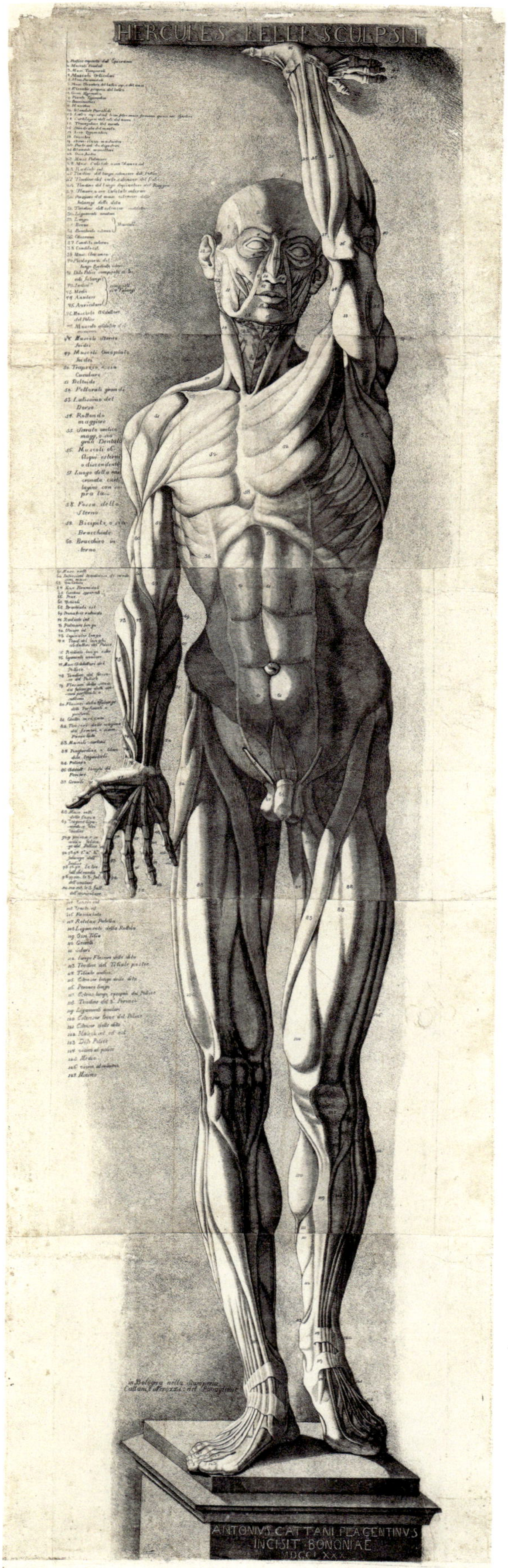
HERCULES LELLI SCULPSIT
ANTONIUS CATTANI PLACENTINUS
INCISIT BONONIAE
MDCCLXXX

Cat. no. 36

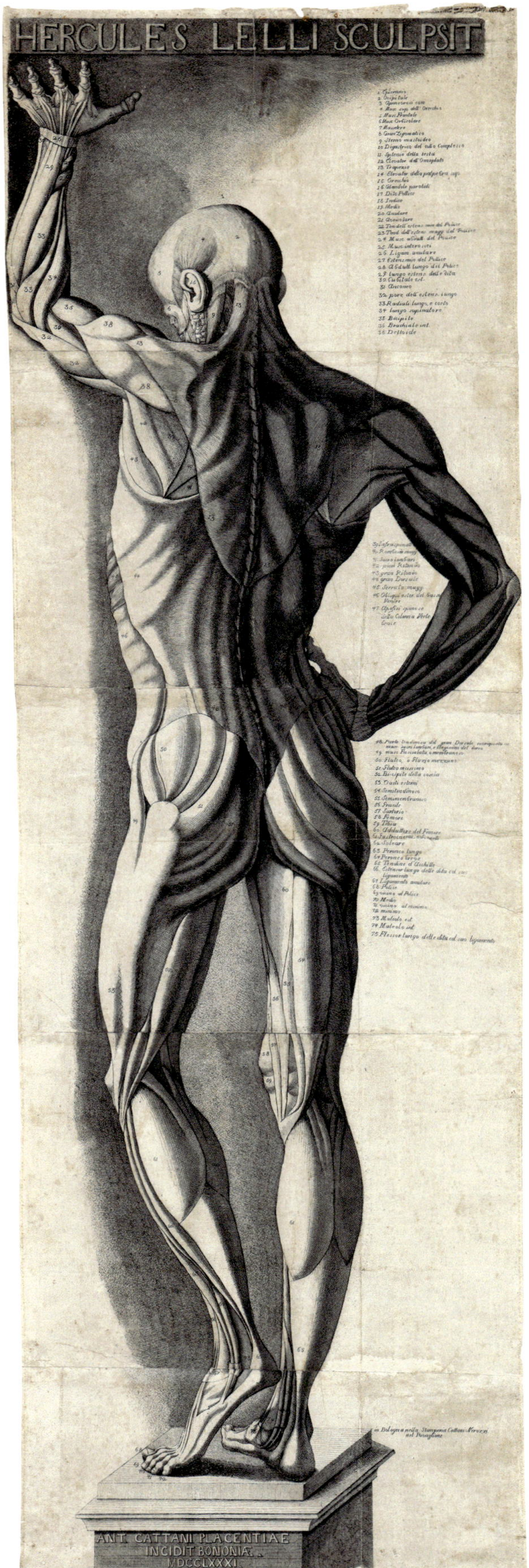
HERCULES LELLI SCULPSIT
ANT. CATTANI PLACENTIAE
INCIDIT BONONIAE
MDCCLXXXI

Cat. no. 37

notably took Manzolini's side, questioning Lelli's contributions (Crespi 1769, 301–3).

Over a generation later, these symbols of Bologna's resurgence as a center for the study of anatomy were the subject of three life-size anatomical figures, each made up of five prints, by Antonio Cattani. Born in Piacenza and active in Bologna, Cattani was a printmaker and publisher who specialized in anatomical prints. Bearing dates of 1780 and 1781, the three figures were issued by Cattani and his partner, Antonio Nerozzi, and were available at their shop in the Pavaglione portico, next to the Archiginnasio, the main building of the Università di Bologna and the site of the anatomy theater. Two of the figures are after the écorché caryatids and are labeled "HERCULES LELLI SCULPSIT" on the canopy above, identifying Lelli as the sculptor (cat. nos. 36, 37). The third figure displays a deeper dissection and is after one of Lelli and Manzolini's wax anatomies (fig. L; cat. no. 38). Cattani has indicated only the top of the wood support upon which the left hand of the figure rests. Slightly larger in scale, it is rendered in a flatter style, without the strong lighting and the background shading of the caryatid figures that heightens their sculptural effect. All three figures are labeled, keyed to an explanation of the parts running down the sides. Cattani's printed figures match the size of the originals, hence the variance in scale. The wood caryatids are small in stature at around five feet in height, and it is possible that Lelli adjusted their dimensions to fit the space below the existing canopy.

Previously undiscussed subscription notices for the prints establish that the figures were the result of Cattani's evolving plans of the late 1770s (appendices A–C). From them, we learn that Cattani was initially aiming at a market of artists and that he originally planned four life-size figures, although only three are known to have been issued. These may now be definitively linked to Cattani's contemporaneously published book, *Osteografia e miografia della testa, mani, e piedi del corpo umano in misura naturale* of 1780 (cat. no. 39).

The first subscription notice, or *manifesto,* dated 19 March 1777 (appendix A) acknowledges a demand by painters for engraved anatomical prints. These, Cattani suggests, could conveniently be kept in artists' portfolios to be consulted when drawing from a nude in the academy so as to quickly resolve an issue of anatomy and, in general, speed up an artist's mastery of the subject. Such portfolios are seen held by artists and resting on benches in Giampietro Zanotti's drawing of the Accademia Clementina life-drawing class in Bologna (see fig. 6.10), an engraving of which appeared on the title page of Zanotti's history of the academy (Zanotti 1739). Cattani promises eighteen prints, issued monthly (starting in May 1777), and priced at ten soldi for subscribers and fourteen soldi for nonsubscribers. These were to be made up of a frontispiece, four heads, two hands, and three feet—all of which would be life-size—and "four nudes," after a small-scale écorché model by Lelli, and four skeletons done on a matching scale. There is no trace of the latter two subjects, but some of the other prints probably form part of Cattani's *Osteografia* (cat. no. 39). Lelli's écorché model is dated to around the time of the wood caryatids for the anatomy theater and was quite popular in the eighteenth century (Amerson 1975, 376–78). Lelli included it in his self-portrait in the Museo di Palazzo Poggi in Bologna, where a cast of the model is also found.

A second previously known subscription notice is found in the text to the first plate of Cattani's *Osteographia* (appendix B; cat. no. 39). Below an etching of a skull, signed and dated 15 November 1778, is the announcement that prints will be issued about every forty days for "a whole course of anatomy of the entire human body," with March 1779 given as a deadline for subscription. The cost of the prints is given as one paolo (papal coinage) for subscribers, a paolo and a half being the cost for nonsubscribers for existing prints, and two paolos for new prints.

New plans are announced in a third subscription notice dated 22 September 1779 (appendix C). Proposed are four life-size figures, made up of five printed sheets each to be glued together. These are described as a follow-up to twenty life-size prints of parts of the body already completed, which may be identified with the collection published by Cattani the following year (cat. no. 39):

> Wishing the undersigned engraver to continue his work of anatomy advanced at this time up to twenty tables containing osteology and myology of the head, hands, and feet of the human body in natural measure with the proper names of all its parts, it is resolved to engrave 4 other anatomical tables in 20 sheets as well as the index of the denomination of all its parts, 10 of which will represent the male human body stripped of its integuments in two large natural figures, one of which will be seen from the front, and the other from the back, demonstrating in the other 10 sheets two human skeletons of equally natural size also seen in the aforementioned views. Each of these 4 tables will be composed of 5 sheets, which glued, one after the other, will form one of the said figures.

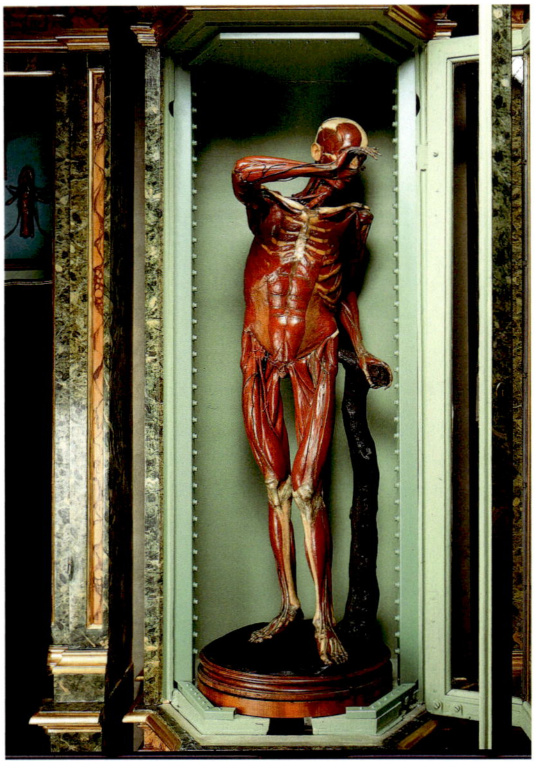

Fig. L
Ercole Lelli (Italian, 1702–66) and Giovanni Manzolini (1700–1755). Second level of muscles over a skeleton, 1742–51, wax, bone, and wood, 175 cm.
Bologna, Museo di Palazzo Poggi.

The two figures of five sheets of a body "stripped of its integuments" in two views correspond to the two prints after Lelli's caryatids (cat. nos. 36, 37). The plan for life-size prints of skeletons seems to have been abandoned, and the third figure showing the "second level of muscles" (cat. no. 38) seen from the front is an indication that Cattani's plan changed again. Since the September 1779 plan is presented as a continuation of Cattani's work, the life-size figures should be viewed as part of Cattani's earlier promise to provide prints of the entire human body (appendix B). Cattani and Nerozzi also published the *Anatomia esterna del corpo umano per uso de'pittori, e scultori* under Lelli's name, providing further evidence of the marketability of Lelli in the late eighteenth century. Undated, it appeared before 1786, when Fantuzzi mentions it (Fantuzzi 1781–94, 5:50). The plates, however, are not original to Lelli but ultimately derive from William Cowper's illustrations in James Drake's *Anthropologia nova* (1707) (Kornell 1996, 58–59).

Cattani had arranged in advance for the distribution of the life-size anatomical prints outside of Bologna. Listed at the end of the subscription notice of 1779 (appendix C) are the outlets where the prints would be available for purchase. In addition to the shop of Cattani and Nerozzi, they could be obtained in Parma, from the printers and booksellers Fratelli Faure, and in Venice, from "Giuseppe Vagner," likely the Swiss engraver and printmaker Joseph Wagner (1706–80). Based in Venice, Wagner died in Munich the same year the figures first appeared.

Despite these distribution plans, few of Cattani's life-size figures survive. Their probable function as wall charts would have contributed to their ephemerality. In addition to the prints at the Getty Research Institute, there are three other sets of all three anatomical figures; they are held by the Harvey Cushing / John Hay Whitney Medical Library at Yale University; the Istituto Centrale per la Grafica, Rome; and the Donazione Putti, Istituto Ortopedico Rizzoli, Bologna. Sets of two figures of different groupings are in the Countway Library at Harvard University (purchased from the Los Angeles bookdealer Jake Zeitlin in 1962; see Zeitlin & Ver Brugge 1961, no. 296; and Esterquest 1962); the Philadelphia Museum of Art, Philadelphia; the Wellcome Collection, London; and the Fondazione Cassa di Risparmio, Bologna. A set of two figures were with the rare-book dealers Gilhofer & Ranschburg in 1994.

Cattani continued to update and correct the prints, resulting in variant states. The figure of a deeper dissection in the Getty Research Institute (cat. no. 38) is an early state. Along with the same figure that was with Gilhofer & Ranschburg in 1994, it has the title *Secondo grado de muscoli,* whereas *Secondo ordine de muscoli* is the title of the same figure in the Wellcome Collection (47217i), the Harvey Cushing / John Hay Whitney Medical Library (print3103), the Istituto Centrale per la Grafica (FC34), and the Donazione Putti, Istituto Ortopedico Rizzoli. The *Secondo ordine* figures also have added explanatory text at the left shoulder ("c Tendine dell muscolo soppra spinoso"), and the letter "c" replaces the number "73" that appears in cat. no. 38; the number "136" indicating the extensor digitorum longus has been canceled out and moved lower down on the foot. Changes were also made to the caryatid figure seen from the back, in which Cattani progressively appended lines of text to the right of the figure (Fondazione Cassa di Risparmio, Bologna, F31083; cat. no. 37; Harvey Cushing / John Hay Whitney Medical Library, print3102).

In the late 1770s, Cattani and Nerozzi's shop had been in the Borgo delle Tovaglie, the address given on prints of the *Osteografia.* The three life-size figures, however, bear the address of Cattani and Nerozzi's new shop, next to the Archiginnasio. Here, medical students passing to and from the anatomy theater could see and purchase the anatomical prints. Cattani and Nerozzi may have relocated with this market in mind. —MK

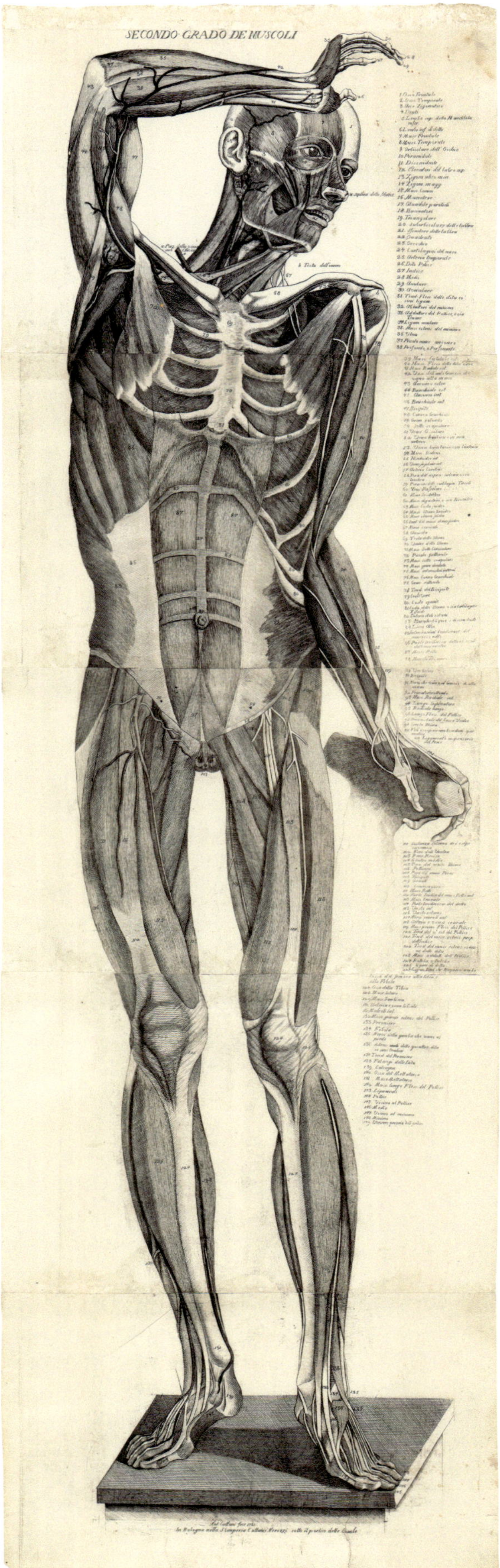

Cat. no. 38

39 Antonio Cattani

(Italian, active 1777–ca. 1790)
Dorsal View of the Muscles of the Hand
Etching and engraving
From Antonio Cattani, *Osteografia e miografia della testa, mani, e piedi del corpo umano in misura naturale con il catalogo de' nomi proprj di tutte le sue parti incise in rame da Antonio Cattani incisore Piacentino* (Bologna: Antonio Cattani and Antonio Nerozzi, 1780), table 14
Los Angeles, Library Special Collections for Medicine and the Sciences, Louise M. Darling Biomedical Library, UCLA, BIOMED ** WZ 260 C2980 1780

BIBLIOGRAPHY

Putscher 1972, 73, fig. 93; Roberts and Tomlinson 1992, 382–83, no. 88; Cazort in Cazort, Kornell, and Roberts 1996, 224–25, no. 112; Simblet 2001, 20–21.

Antonio Cattani's book on the "bones and muscles of the head, the hands, and the feet of the human body in natural dimension," as the title translates, is composed of twenty signed etchings with an image and explanatory letterpress on each sheet. According to a subscription notice with a deadline of March 1779 that appears on table 1, the prints were issued separately at roughly one-month intervals, and the aim was to provide a "course of anatomy of the entire human body" (appendix B). Table 1, a skull resting on a book, is dated 15 November 1778 and is the only print in the collection that bears a date. Work on the prints likely date back to 19 March 1777, when Cattani first announced a project of anatomical prints (appendix A). The prints were described as completed in another subscription notice of 22 September 1779, where the title for the present work is given (appendix C). This same notice gives plans for life-size figures, identifiable with those of Cattani (cat. nos. 36–38), that may be considered pendants to the book.

Inscriptions on the prints indicate that they were available at Cattani and Antonio Nerozzi's Bolognese shop in the Borgo delle Tovaglie (before it moved to the portico of the university) and at the shop of Petronio dalla Volpe (1721–94), who was printer to the Istituto delle Scienze in Bologna. The nonconsecutive numbers at the top left of most of the leaves probably belong to an earlier ordering system. The table numbers were likely added when they were gathered for publication in the book. Table 14 bears the number "1a," suggesting that it was among the first to be issued. Some of the titles and table numbers in the UCLA Library copy have been handwritten. An impression of a skull in profile in the Museum of Fine Arts, Houston (BF.2009.4), that corresponds to table 2, appears with only the number "6a" at the top left.

Cattani uses a variety of shading effects in table 14 to give a strong plastic sense to the dissected hand, the support on which it rests, and the plinth behind it, which is decorated with Greek key and meander borders, then fashionable neoclassical decorative motifs. Tables 14 and 15 give the dorsal and palmar views of the muscles of the hand and are both presented as mounted specimens in a manner similar to the surviving wax reliefs by the wax modelers Anna Morandi (1714–74) and her husband, Giovanni Manzolini (1700–1755), the former assistant of the sculptor and painter Ercole Lelli (Armaroli 1981, 91–92, nos. 85–87; Messbarger 2010, 56–62). The hand itself in table 14 is similar to a plaster cast dating from the late eighteenth century (Comar 2008b, 202–3, nos. 64, 65). —MK

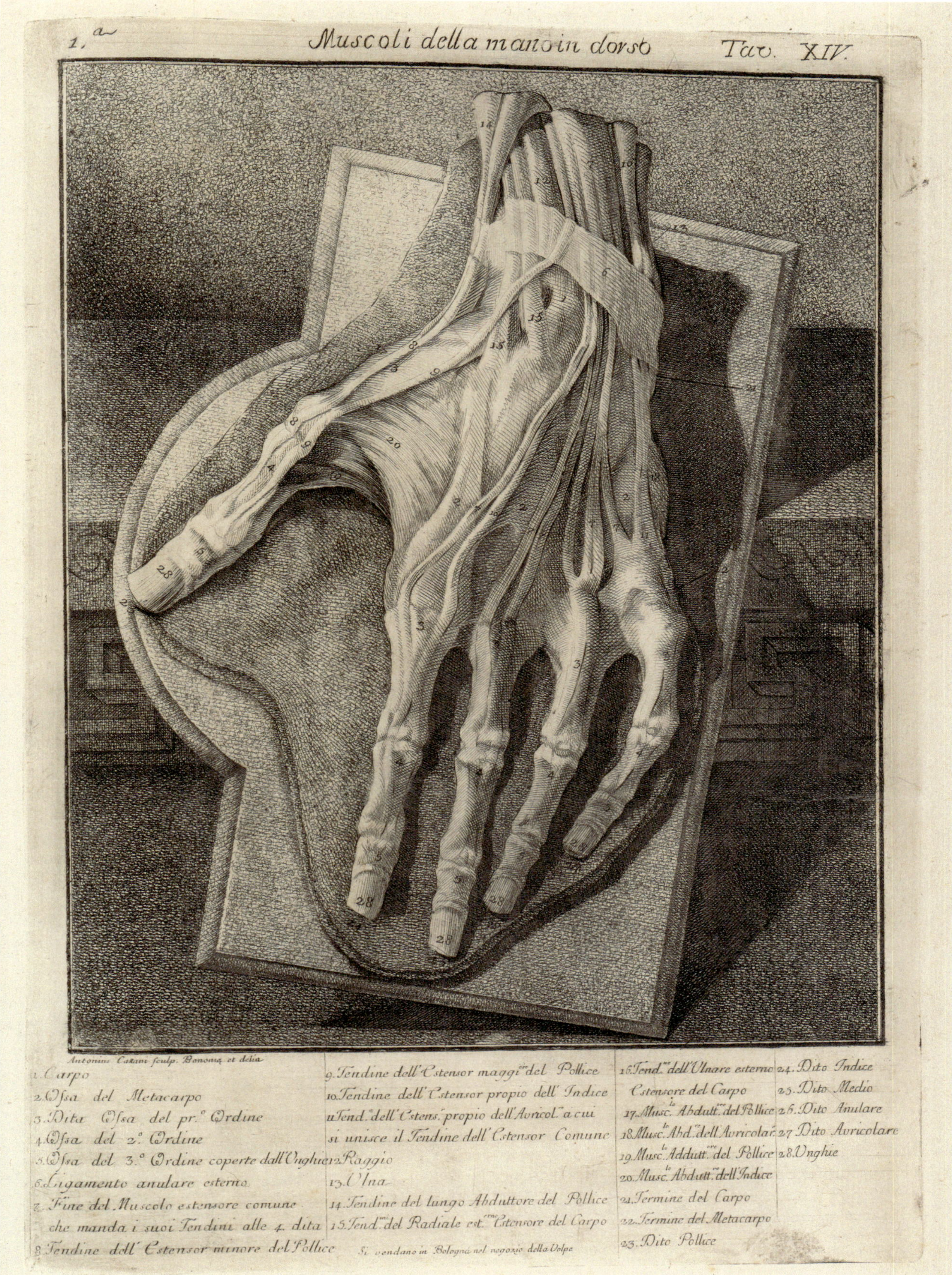
1.a
Muscoli della mano in dorso
Tav. XIV.
Antonius Cattani sculp. Bononiæ et delin.
1. Carpo
2. Osa del Metacarpo
3. Dita Osa del pr.o Ordine
4. Osa del 2.o Ordine
5. Osa del 3.o Ordine coperte dall'Unghie
6. Ligamento anulare esterno
7. Fine del Muscolo estensore comune che manda i suoi Tendini alle 4. dita
8. Tendine dell' Estensor minore del Pollice
9. Tendine dell' Estensor maggi.re del Pollice
10. Tendine dell' Estensor propio dell' Indice
11. Tend.e dell' Estens.e propio dell' Auricol.re a cui si unisce il Tendine dell' Estensor Comune
12. Raggio
13. Ulna
14. Tendine del lungo Abduttore del Pollice
15. Tend.ini del Radiale est.erno Estensore del Carpo
16. Tend.ine dell' Ulnare esterno Estensore del Carpo
17. Musc.lo Abdutt.ore del Pollice
18. Musc.lo Abd.re dell' Auricolare
19. Musc.lo Addutt.ore del Pollice
20. Musc.lo Abdutt.re dell' Indice
21. Termine del Carpo
22. Termine del Metacarpo
23. Dito Pollice
24. Dito Indice
25. Dito Medio
26. Dito Anulare
27 Dito Auricolare
28. Unghie
Si vendano in Bologna nel negozio della Volpe

40 After Gérard de Lairesse

(Flemish, 1640–1711)
Rear view of the muscles of the head and neck
Engraving
From Govard Bidloo, *Anatomia humani corporis* (Amsterdam: The widow of Johannes van Someren, the heirs of Johannes van Dyk, and Henrik Boom and the widow of Theodore Boom, 1685), pl. 16
Los Angeles, Getty Research Institute, 84-B4214

BIBLIOGRAPHY

Dumaître 1982, 88–92; Cazort in Cazort, Kornell, and Roberts 1996, 183–86, nos. 68–71.

A luxury volume, Govard Bidloo's anatomical atlas features 105 full-page, folio-size copperplate engravings designed to rival the model of Andreas Vesalius (cat. no. 30). Following its publication, Bidloo (1649–1713), who was trained as both a surgeon and physician, saw notable advancements in his career, including professorial positions in The Hague and Leiden, and several appointments under William III of England (1650–1702) (Molhuysen et al. 1930, 106).

Bidloo was likely introduced to Gérard de Lairesse through their shared affiliation with the literary society Nil Volentibus Arduum in Amsterdam (De Vries 1998, 123–24). Prior to this commission, De Lairesse does not appear to have undertaken anatomical study, but Bidloo likely viewed De Lairesse's skill with illusionistic techniques and his fame as a history painter as assets, which may account for the rare inclusion of De Lairesse's name on *Anatomia*'s title page. In contrast, De Lairesse makes sparing reference to his collaboration with the anatomist but explains that his measurements for the proportionate human body derived from a skeleton he studied "when [he] drew for Professor Bidloo, all the drawings in his renowned Anatomie-boek, after life [*na het leeven*], following his instructions" (De Lairesse 1712, 1:21). Notably, the title page and preface for *Anatomia* also describe the images as having been done from life, or ad vivum. Modern scholars interpret this Latin phrase as communicating both a method of working in direct consultation with a model and a perceived effect on the viewer (Swan 1995, 354–57; Kusukawa 2012, 174–75; Fend 2019, 304)—in both cases contributing to the understood authority of the image (Parshall 1993, 565–67).

In the plates themselves, pictorial choices reinforce the validity of *Anatomia*'s images as credible tools of study, often presenting the body in great detail and evoking the environment of the dissection hall. For example, in plate 16, swaths of fabric, frayed at the edges, frame our view and only slightly obscure the rope that maneuvers the subject into position (cat. no. 19; cat. no. 8, see fig. A; see fig. 3.2). To the right, an ear escapes the drapery, providing a point of recognition for the viewer. Below, a hint of stubble is one of several individualizing features seen throughout the plates that communicate the variety of subjects consulted by artist and anatomist. The ability of artists to edit their images was a cause for concern in the period and the inclusion of these features and props, such as the pin securing the subject's musculature on the left, were conscious choices designed to reassure viewers of the subject's veracity. —ET

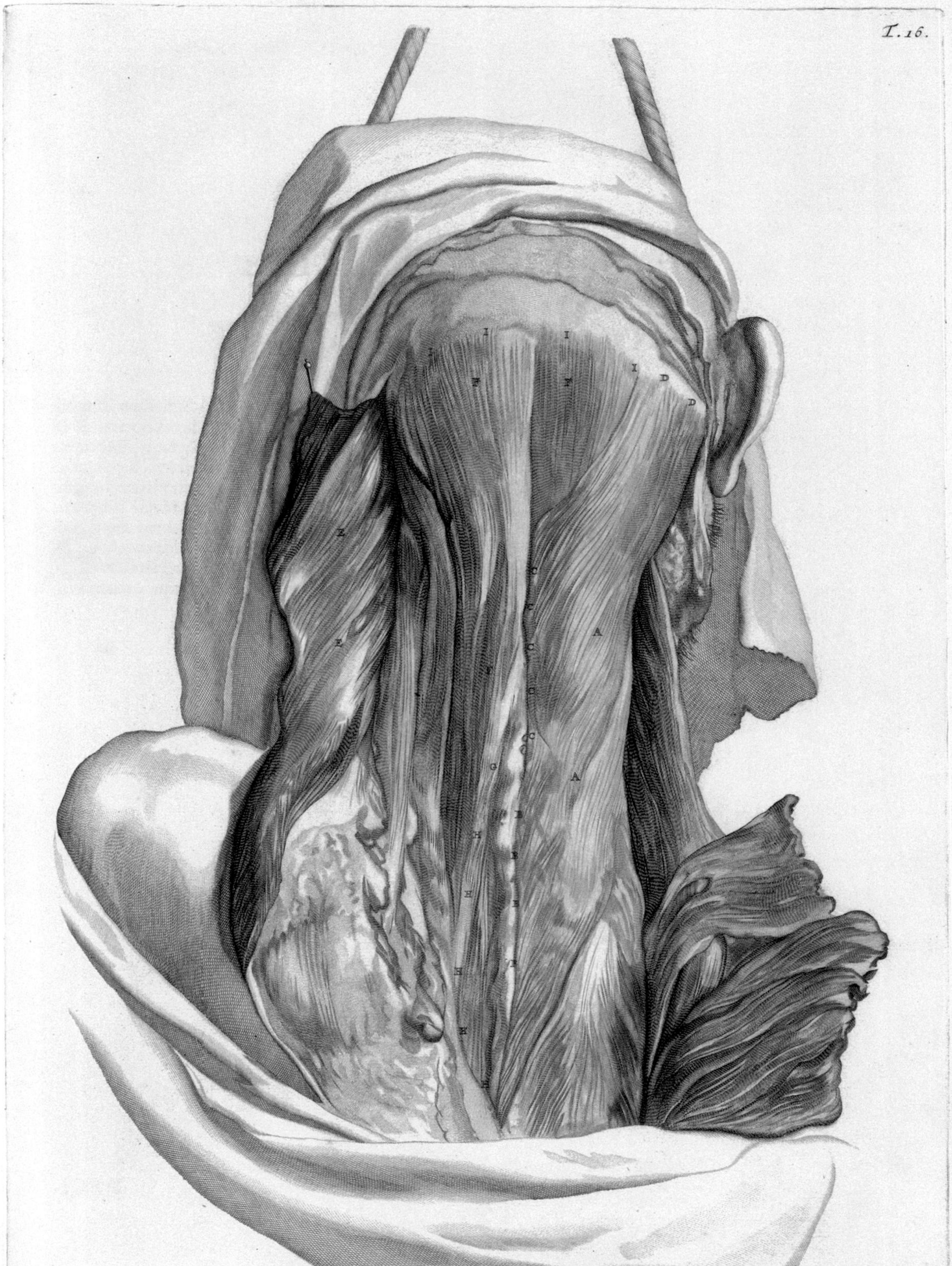
T. 16.

41 Jacques Fabien Gautier Dagoty

(French, 1716–85)
Dissected woman
1750
Color mezzotint, 188.6 × 56.5 cm
From *Anatomie générale des viscères en situation, de grandeur et couleur naturelle, avec l'angeologie, et la neurologie de chaque partie du corps humain* (Paris: n.p., [1752]), pls. 1–3
San Marino, California, The Huntington Library, Los Angeles County Medical Association Collection of Prints and Ephemera, pri LACMA

BIBLIOGRAPHY
Choulant 1945, 270–75; Hébert et. al. 1968, 66–67, nos. 137–139; Roberts and Tomlinson 1992, 523–29; Cazort in Cazort, Kornell, and Roberts 1996, 225–28, nos. 113-116; Rodari 1996, 100–105, 124–32, no. 110; Scott 2003, 169, 175–76, 179–80, 183; Lowengard 2006.

Composed of three joined mezzotint sheets and measuring nearly six feet in length, this rendering of a dissected woman is part écorché, part circulatory diagram. Gazing directly at the viewer, her entreating smile, plump breasts, and carefully preserved pubic triangle invite us to examine her body as though she were alive. Just below the shoulders, however, her skin has been pulled back in a startling display of internal anatomy. Veins and arteries form a dense web of deep reds and blues that creep up her chest, escaping the excision line. These richly saturated colors are offset by the anemic porcelain of her upper body, which appears pallid against an eerie green background. Her abdominal organs seem to hide in the shadows as fuzzy indeterminate shapes behind tangled blood vessels. Both the printing technique and illustrative style seemingly prioritize sensuality and beauty over linear clarity or didactic coherence.

At the figure's feet, a partially legible inscription bears the date of 1750 and credits the dissection to surgeon Antoine Mertrud (d. 1767). The *Anatomie générale* was completed in 1752 after seven years of labor, as Gautier Dagoty explains in a history appended to the work where he also speaks of preparing a second edition. The title adopted here comes from the explanation of the first three plates. The figure appears in a publication of 1754 with a variant title of *Anatomie générale des viscères, et de la néurologie, angéologie et ostéologie du corps humain, en figures de couleurs et grandeurs naturelles* (Paris: Gautier and Delaguette, 1754) (Rodari 1996, 126, no. 110).

Despite early collaborations with the anatomists Jacques-François-Marie Duverney (1661–1748) and Mertrud, the artist Jacques Fabien Gautier Dagoty was known less for his anatomical accuracy than for his pioneering (and controversial) work in color. Having learned trichromatic mezzotint printing from Jacob Christoph Le Blon (1667–1741) as an apprentice in Paris in 1738, Gautier Dagoty later added a fourth, black plate to increase tonality—an adaptation for which he claimed invention of the entire color printing process (Scott 2003, 169). He successfully campaigned to inherit Le Blon's royal privilege on the technology, despite ardent (though fruitless) opposition from the family, and then vaunted this royal seal of approval on all his publications.

Using this monopoly, Gautier Dagoty worked tirelessly to assert himself as a man of science, publishing prolifically on anatomy, color, and natural history, including the self-funded *Observations sur l'histoire naturelle, sur la physique et sur la peinture* (1752–55), the first periodical to feature color plates. In this journal, he printed anatomical images of hermaphrodites, exotic animals, and flowers, and staged his attack on Sir Isaac Newton's theory of color. He also contributed images of the organs of speech and larynx to Antoine Court de Gébelin's *Histoire naturelle de la parole; ou, Précis de l'origine du langage & de la grammaire universelle* (1776).

By the 1770s, Gautier Dagoty had begun signing his works as "G. Dagoty père, Anatomiste pensionné du Roi," using his mother's maiden name (d'Agoty or Dagoty), and taking on the title of anatomist as well as artist and engraver. While these publications were often met with public refutation and critical disdain (Scott 2003, 176; Lowengard 2006), Gautier Dagoty nevertheless succeeded in exploiting the popularity of anatomical imagery for both scientific and lay audiences alike through his haunting color prints. —TG

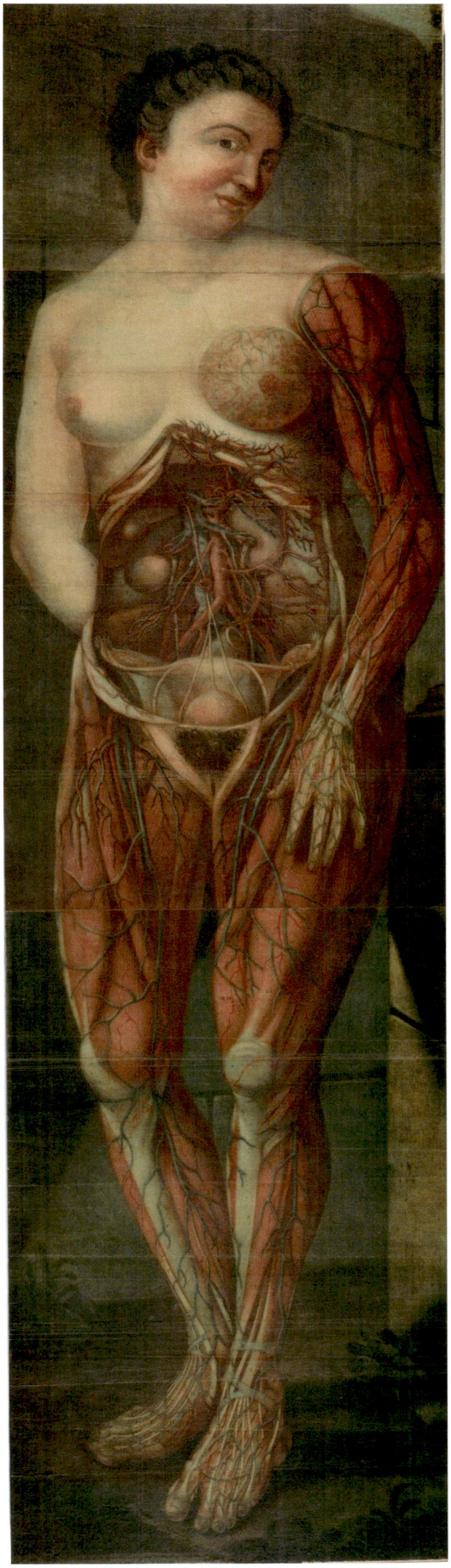

42 Robert Strange

(Scottish, 1721–92), after Jan van Rymsdyk
(Dutch, active 1750–90)
Foetus in utero
Engraving
From William Hunter, *Anatomia uteri humani gravidi tabulis illustrata = The Anatomy of the Human Gravid Uterus Exhibited in Figures* (Birmingham: John Baskerville, 1774), pl. 6
Los Angeles, UCLA Library Special Collections, Charles E. Young Research Library, BELT***RG520.H9

BIBLIOGRAPHY
Jordanova 1985, 386–90; Kemp 1993, 113–21; McGrath 2002; Massey 2005; Massey 2008; Berkowitz 2015; Gamer 2018; Røstvik 2019.

William Hunter (1718–83) served as physician extraordinary to Queen Charlotte, was the first professor of anatomy at the Royal Academy of Arts, and made pioneering discoveries in anatomy and surgery. One of his most enduring contributions is this elephant folio of engravings, which he worked on for over two decades, employing multiple artists, engravers, and printers to produce a sumptuous (and exceedingly expensive) book.

The life-size prints illustrate the morphology of gestation in reverse chronological order, beginning with full-term and ending with conception. These vivid and realistic depictions of pregnancy portray the uterus and fetus, and reduce maternal bodies to dead, fleshy vessels in varying states of dismemberment. Unlike his teacher and colleague William Smellie (1697–1763), whose contemporaneous obstetrical atlas (also illustrated by Jan van Rymsdyk) served as a midwifery manual, Hunter never visualizes nor provides instruction on birth, offering only an anatomical study of assuredly dead subjects. It is in this period that male physicians began to overtake the traditionally female-dominated practice of midwifery, doing so by invoking science and shifting focus onto the fetus. Through the visual vocabulary of anatomy, the so-called man-midwives medicalized (even pathologized) childbirth and thereby assumed authority over a once domestic experience.

Hunter's stunningly realistic illustrations focused on the pregnant cadaver. The fine and scrupulous details demonstrate Hunter's commitment to naturalism, the epistemological and aesthetic imperative of his project. In the preface, he extols the art of engraving as "an universal language" that is better able to communicate ideas than words. This Enlightenment empiricism favored direct observation of nature, free of idealization, allegorical figures, or imaginary landscapes in order to secure the "mark of truth."

In plate 6, Hunter reassures the reader of the image's faithful reproduction of nature, writing, "Every part is represented just as it was found; not so much as one joint of a finger having been moved to shew any part more distinctly, or to give a more picturesque effect." Indeed, the curled fingers of the fetus are tenderly and carefully articulated to serve this stylistic ideal. While the fetus is rendered untouched and in the round, its body still slick with amniotic fluids, the maternal body is gruesomely partitioned—her torso removed and legs sawed off into meaty stumps. Layers of tissue and fat spill out around the fetus like drapery, providing a messy cross-sectional view of musculature and veins. The mother is further excised from view in later plates, which show uteri and fetuses at earlier stages of gestation as material specimens on a table: pregnancy is all but reduced to a womb. —TG

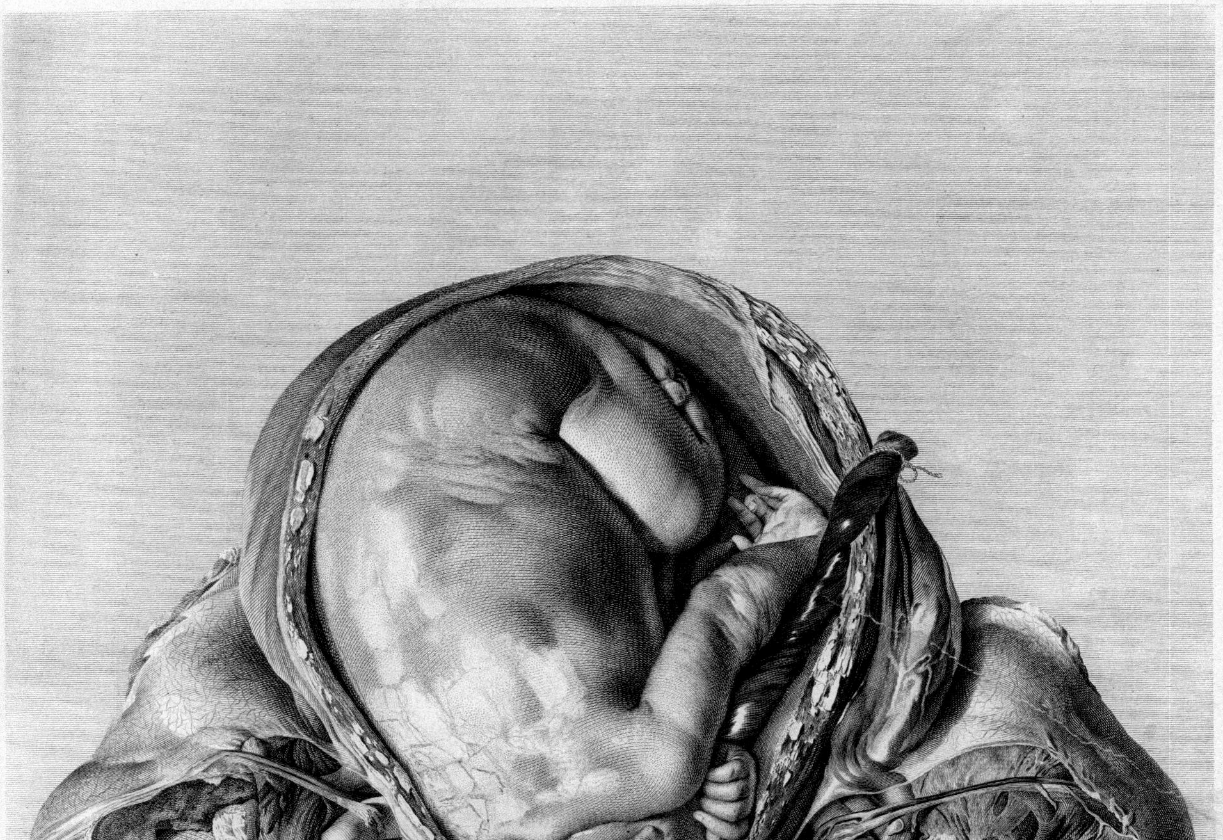

J. V. Rymsdyk delin. R. Strange sculp.

TAB. VI. *Fœtus in utero, prout a naturâ positus, rescissis omnino parte uteri anteriori, ac Placentâ ei adhærente.*

43 Attributed to Jean-Pancrace Chastel

(French, 1726–93)
Écorché bust
After original of ca. 1770
Plaster, 44 × 25 cm
Los Angeles, Private collection

BIBLIOGRAPHY

Rees 1819–20; Cummings 1963; Amerson 1975, 382–85; Amerson 1976, 13–15, no. 1; Bignamini and Postle 1991, 88–89, no. 78; Maral 2003, 64–65, no. 8; Postle 2004, 57; Comar 2008b, 200–201, nos. 61–63; Kornell 2016b, 49.

Either cast from anatomized bodies or modeled after them, écorché models—figures of the body with the skin removed to display the muscles—obviated the need for obtaining a cadaver, avoiding the issue of decay, and had the advantage of offering multiple viewing angles. The head of this animated écorché bust, its eyes with incised pupils, turns to look off to the side, with a mouth open as if in surprise or about to speak. The bust relates to a life-size complete figure depicted in midstep with a raised left arm that had traditionally been attributed to the French sculptor Edme Bouchardon. Examples of the complete figure are found in Paris, Aix-en-Provence, and Lyon. Alexandre Maral, however, reattributes the figure, on the basis of documentary evidence, to the sculptor Jean-Pancrace Chastel, who in 1774 was appointed professor of sculpture in Aix-en-Provence (Maral 2003, 64–65, no. 8). The model, here with a column support, is the focus of an anatomy lesson for artists in a 1795 drawing by François-Marius Granet (1775–1849), an artist also from Aix-en-Provence (fig. M). Seen lying on the ground in the right foreground next to a partial skeleton is another model of an écorché head and neck, all objects of anatomical study for the school.

It is probable that the present bust is the same écorché "head and neck as large as life from the French" recommended to artists in Abraham Rees's *Cyclopaedia* (Rees 1819–20). In the summer of 1815, while attending the school of Benjamin Robert Haydon (1786–1846), the young Charles Landseer (1799–1879) made drawings, now in the Wellcome Collection, London, in different views after an example of the bust (Bignamini and Postle 1991, 88–89, no. 78; see Comar 2008b, no. 62, for a drawing of a variant bust). Another cast of the bust—cataloged by the Science Museum, London, as "St. Bartholomew" (inv. A241443)—has a misaligned repair at the neck.

The pose with a single raised arm was a popular one for écorchés, dating back to early examples such as the small-scale model by Ludovico Cigoli (1559–1613). It allowed a comparison of the changed position of the scapula and may have reflected the practical business of suspending the cadaver for flaying (see fig. 3.2). William Hunter cast a flayed body with a raised arm for the St Martin's Lane Academy around 1750 (see fig. 5.6). In the last quarter of the century, Jean-Antoine Houdon (1741–1828) produced a life-size variant of his first écorché with a raised arm (Poulet 2003, no. 2). Portable small-scale versions were produced of both. —MK

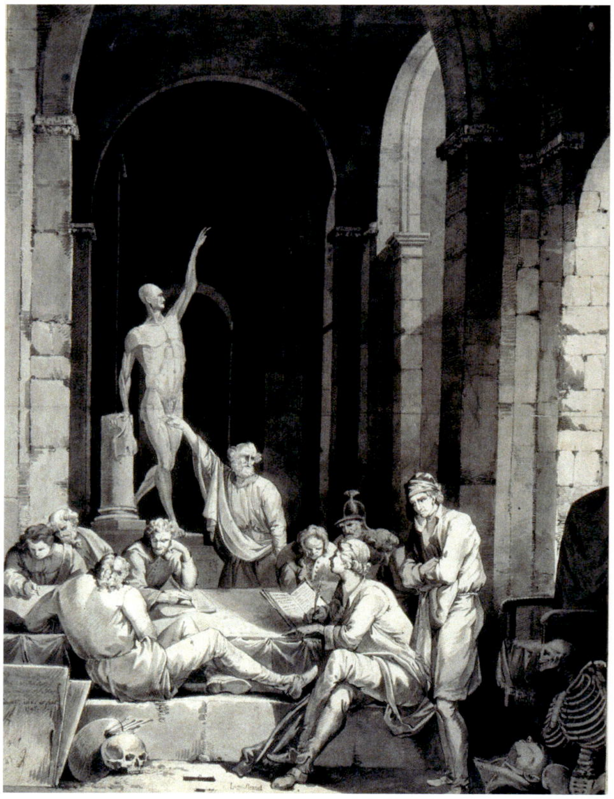

Fig. M
François-Marius Granet (French, 1775–1849).
The Anatomy Lesson, 1795, ink and gray wash.
Aix-en-Provence, Musée Granet, D-849.1.3.P.2.

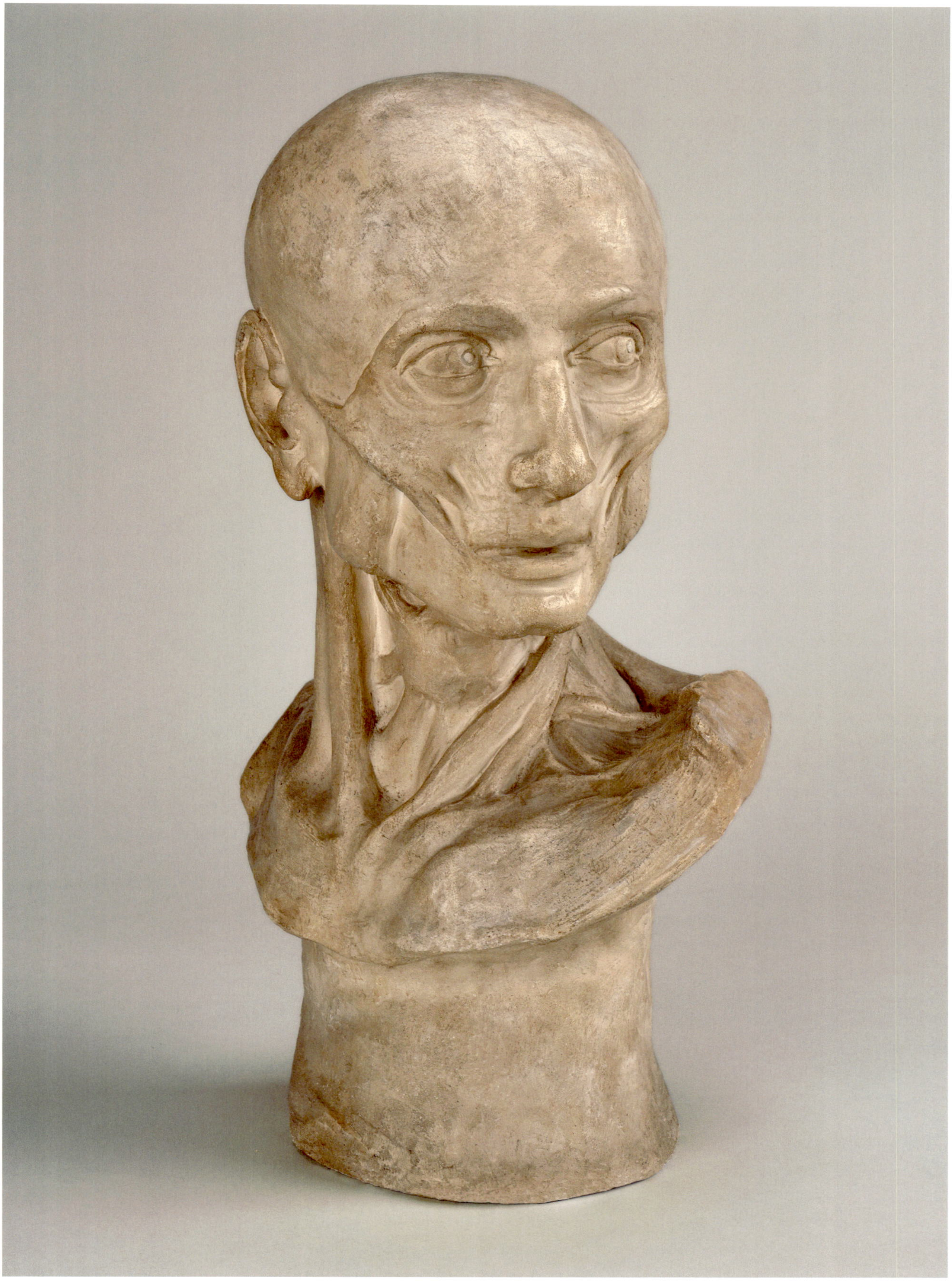

44 Johann Friedrich Bause

(German, 1738–1814), after Adam Friedrich Oeser (Austrian, 1717–99)
Human infant skull
1791
Color etching and stipple, inked *à la poupée* with hand coloring, plate mark: 12.4 × 18.6 cm; sheet: 18.5 × 26.2 cm
Los Angeles, Getty Research Institute, P990016** (2)

BIBLIOGRAPHY
Martens 1803; Keil 1849, 18, no. 22; Dürr 1879, 228, no. 74; Teeuwisse 2011; Ittmann 2017, 18–19, 364, no. 6.

Georg Keil, in his catalog of Johann Friedrich Bause's prints, describes the subject as "the skull of a seven-month-old child" (Keil 1849, 18, no. 22), but, according to Prof. Dr. Jörg-Elard Otten (pers. comm. 2019), it appears closer in development to that of a newborn infant depicted near life-size. In this delicate rendering, the depressions between the still-unfused sections of the skull are visible, and red ink is used for the branches of blood vessels and the blush of red on top of the skull. An unsigned preparatory drawing in reverse attributed to Adam Friedrich Oeser was on the art market in 2011 (Teeuwisse 2011).

In the collection of the Getty Research Institute, there is another impression in sanguine ink alone (P990016** [1]). Keil reports examples in red, brown, and purple and yellow "in which the injected vessels of the skin become visible." An infant skull had earlier been the subject of full-color mezzotint illustrations by the Dutch etcher Jan L'Admiral (1699–1773), after a preparation by Frederik Ruysch (1638–1731) in 1738 (Rodari 1996, 86–89, nos. 72–81).

Oeser, a painter and sculptor and the first director of the Leipzig Akademie, was known for a style of sentimental classicism (Hüttel 2008; Vogel 2017). He provided book illustrations for mainly literary works and did not specialize in scientific drawings. It would have been late in his career to attempt it, as he was seventy-four at the time the print was produced. Whatever its original purpose, the infant skull was later employed as an anatomical illustration by the young physician Franz Heinrich Martens (1778–1805), who had studied in Leipzig, where he may have met Bause, whom he describes as a friend. A copy of the skull, drawn and engraved by Martens himself, appears on the title page of his epistolary explanation of the phrenological system of Franz Josef Gall (1758–1828), one of many publications in this period promoting Gall's ideas (fig. N). "I thought the skull was fitting and instructive, and that was why I copied it" (Martens 1803, 94). Regarding it, Martens remarks that children are unceasingly observing a world that is new to them and that the bulging forehead in a child's skull indicates a strong "spirit of observation." While this capacity for observation diminishes with age, Martens suggests that the shape of the heads of some physicians and natural historians such as Gall, Johann Friedrich Blumenbach (1752–1840), and Gerard van Swieten (1700–1772) indicate that they maintained their keen powers of observation. —MK

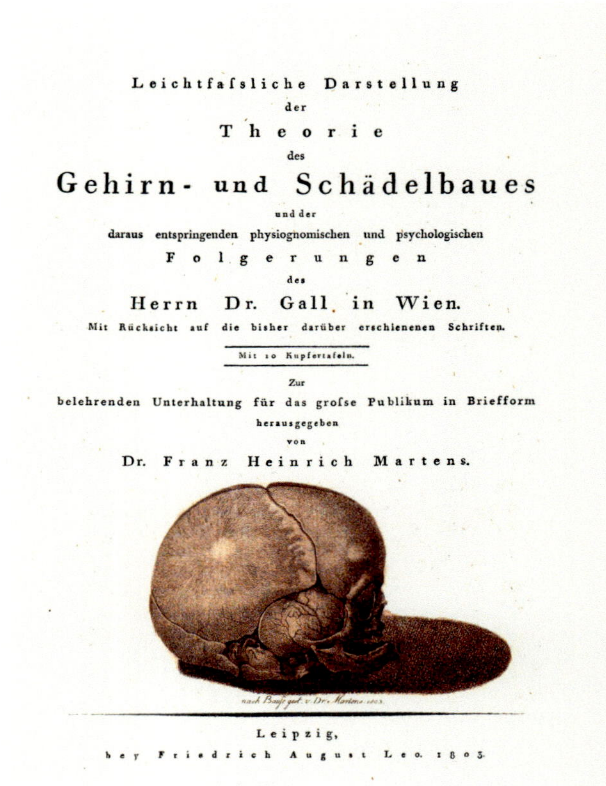
Leichtfaſsliche Darstellung
der
Theorie
des
Gehirn- und Schädelbaues
und der
daraus entspringenden physiognomischen und psychologischen
Folgerungen
des
Herrn Dr. Gall in Wien.
Mit Rücksicht auf die bisher darüber erschienenen Schriften.
Mit 10 Kupfertafeln.
Zur
belehrenden Unterhaltung für das groſse Publikum in Briefform
herausgegeben
von
Dr. Franz Heinrich Martens.
Leipzig,
bey Friedrich August Leo. 1803.

Fig. N
Title page from Franz Heinrich Martens, *Leichtfassliche Darstellung der Theorie des Gehirn- und Schädelbaues und der daraus entspringenden physiognomischen und psychologischen Folgerungen des Herrn Dr. Gall in Wien* (Leipzig: Friedrich August Leo, 1803). Chicago, Galter Health Sciences Library, Northwestern University, 611.81 M36 1803.

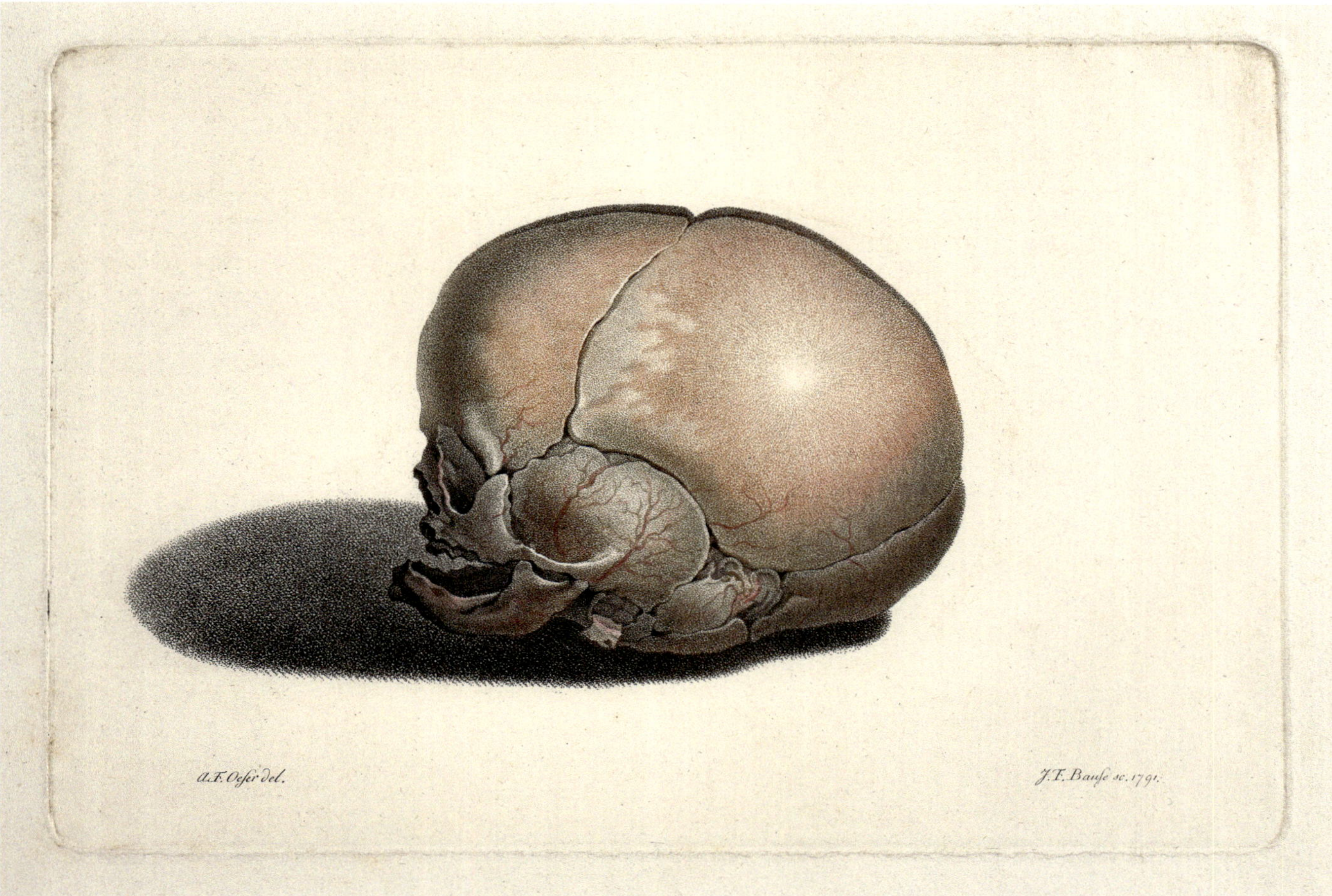
A. F. Oeser del.
J. F. Bause sc. 1791.

45 Antonio Serantoni

(Italian, 1780–1837)
Second level of the human body, seen from behind, showing the muscles, arteries, veins, and parotid gland
Hand-colored engraving
From Paolo Mascagni, *Anatomia universale del professore Paolo Mascagni: Rappresentata con tavole in rame ridotte a minori forme di quelle della grande edizione Pisana per Antonio Serantoni disegnatore, incisore e modellatore in cera* (Florence: V. Batelli e figli, 1833), pl. 7, facing p. 161
Los Angeles, Getty Research Institute, 84-B28034

BIBLIOGRAPHY
Vannoni 1838, 13n2; Choulant 1945, 319–20; Massart, Passeri, and Ciardi 1976; Roberts and Tomlinson 1992, 384–92; Cazort in Cazort, Kornell, and Roberts 1996, 230–33, nos. 119–21; Comparini 1996, 52; Kemp 2000, 49, 52, no. 231; Kemp 2010, 193, 202–3; Maerker 2011, 213.

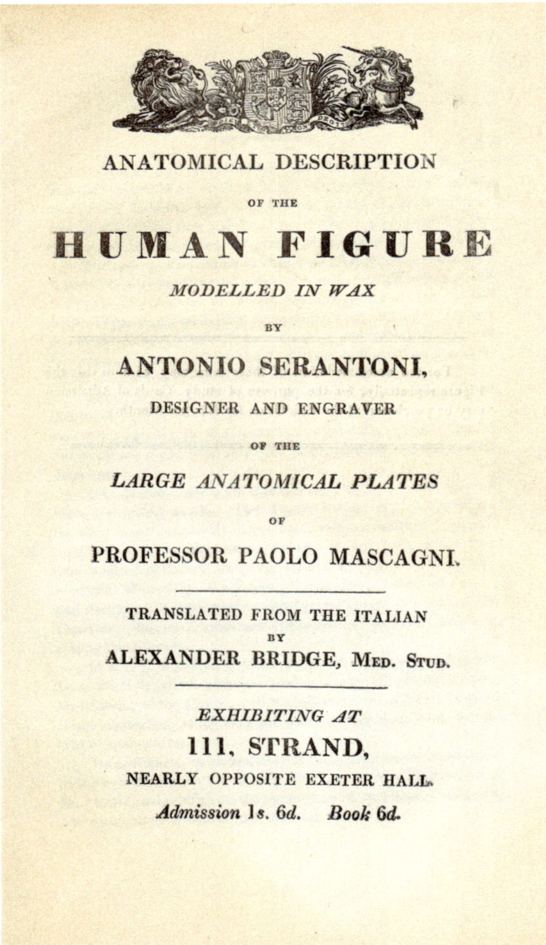
ANATOMICAL DESCRIPTION
OF THE
HUMAN FIGURE
MODELLED IN WAX
BY
ANTONIO SERANTONI,
DESIGNER AND ENGRAVER
OF THE
LARGE ANATOMICAL PLATES
OF
PROFESSOR PAOLO MASCAGNI.

TRANSLATED FROM THE ITALIAN
BY
ALEXANDER BRIDGE, MED. STUD.

EXHIBITING AT
111, STRAND,
NEARLY OPPOSITE EXETER HALL.
Admission 1*s*. 6*d*. *Book* 6*d*.

Fig. O
Anatomical Description of the Human Figure Modelled in Wax by Antonio Serantoni, Designer and Engraver of the Large Anatomical Plates of Professor Paolo Mascagni, trans. Alexander Bridge (London: A. Handcock, n.d.). Los Angeles, Getty Research Institute, 86-B21706.

Paolo Mascagni (1755–1815), a Tuscan polymath and professor of anatomy, published a groundbreaking work in 1787 on the lymphatic system, with illustrations based on mercury-injected preparations (Mascagni 1787). From 1807, he taught anatomy at the Accademia di Belle Arti of Florence. Working mainly with Antonio Serantoni, an engraver and a maker of anatomical wax figures, Mascagni prepared copperplates for colored life-size illustrations for two posthumously published works: an anatomy book for artists published in 1816 (Mascagni 1816) and a general work on anatomy that eventually appeared in two rival editions in Pisa and Paris (Mascagni 1823–31; Antommarchi 1823–26; see Roberts and Tomlinson 1992, 384–92).

In 1833, Serantoni released a reduced version of the Pisa edition, one that was more affordable and easier to handle. In this opening, the body is shown in the second of three anatomical levels into which the figures are divided, with resected muscles attached and floating away from the figure. As in the original, the figure is shown free of shadow and without a landscape background. Serantoni's illustrations are not, however, exact copies. Wishing to compensate for loss in detail resulting from the reduction of the plates, he provided a duplicate set of figures that had the nervous system separated from the venous and arterial systems.

An account of Serantoni's wax anatomies is given in Pietro Vannoni's biography of the artist, published a year after the artist's death in 1837. Serantoni's waxes were exhibited in London in the 1830s, and there is an undated pamphlet (fig. O) that was printed to accompany the exhibition of a wax anatomical Venus with removable parts, which was inspired by Susini's (for a copy, see fig. 8.3). Serantoni is described on the title page as "designer and engraver of the large anatomical plates of Professor Paolo Mascagni," linking the artist to the famous anatomy atlas. —MK

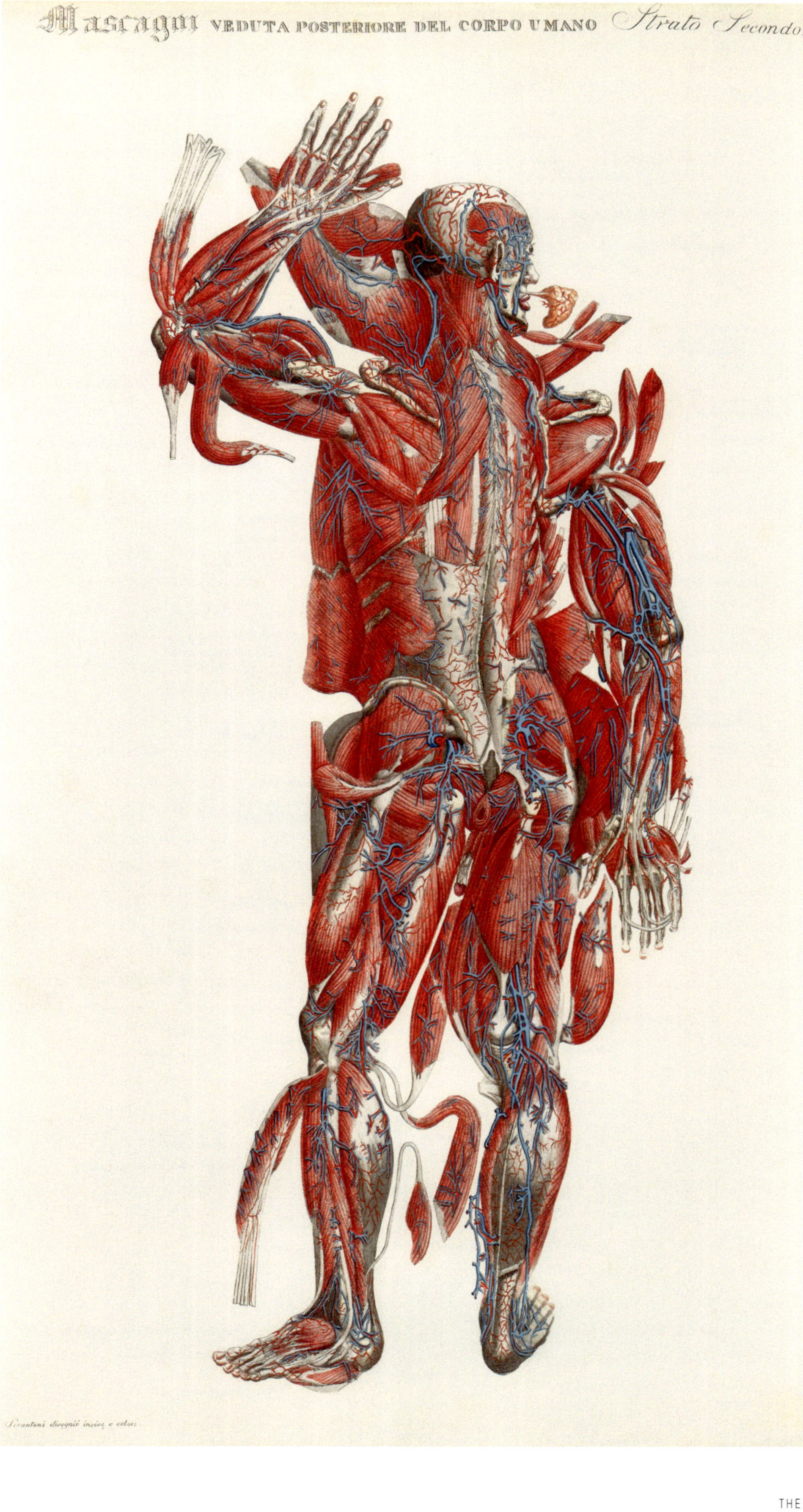
Mascagni VEDUTA POSTERIORE DEL CORPO UMANO Strato Secondo.
Serantoni disegnò incise e colorì

46 Robert Rauschenberg

(American, 1925–2008)
Booster
1967
Five-color lithograph and screenprint on paper,
182.9 × 90.2 cm
Edition 38 of 38
From the series *Booster and 7 Studies*. Published by Gemini G.E.L.
Los Angeles, Los Angeles County Museum of Art, Gift of the Times Mirror Company, M.73.64.1

BIBLIOGRAPHY
Lippard 1967; Foster 1970, no. 47; Fine 1984, no. 4; Gilmour 1986, 48–49; Fine 1997; Mattison 2003, 134–39; Kotz 2004, 153–55; Weems 2008; Brown, Sienna 2010, 90–91; Felsen 2014; Liebert 2016, 361; Roberts, J. 2016, 243–44.

In February 1967, Robert Rauschenberg traveled to Los Angeles to collaborate on his first print edition at Gemini G.E.L., the artist's workshop and publisher established a year earlier by Sidney B. Felsen (b. 1924) and Stanley Grinstein (1927–2014) with master printer Kenneth Tyler (b. 1931). Upon arriving, Rauschenberg determined to make a "self-portrait of inner man" and sought a radiologist who could capture an X-ray of his entire body in a single film. The only machine in the United States with this capacity was at the Eastman Kodak Company in Rochester, New York; the artist therefore resorted to having his body imaged in six one-foot segments (Felsen 2014). Piecing together these radiographs, Rauschenberg created the full-scale anatomical self-portrait that dominates the monumental Gemini print *Booster* and appears again in *Autobiography*, an offset lithograph of 1968.

The artist's standing body, naked but for a pair of boots, is seen frontally, fragmented into six panels that display his head, chest, abdomen and pelvis, upper legs, knees, and lower legs and feet. The X-rays reveal the skeletal structure and shadowy soft tissue, though some elements and details are obfuscated, likely lost in the process of transfer onto the printing stone. Offering a new visual vocabulary for figural representation, X-radiography becomes both a tool and a subject in *Booster*. The work coincided with the beginnings of the organization E.A.T. (Experiments in Art and Technology), which Rauschenberg cofounded in 1966 to foster collaborations between artists and engineers.

In *Booster*, Rauschenberg deployed his distinctive formal language and collage aesthetic to stack the X-ray plates as well as to juxtapose and layer a variety of found images around the figure. To transfer the photographic images onto the lithographic stone, the artist characteristically applied mechanical and manual processes, the latter conveying his gestural marks. Among the familiar Rauschenberg motifs are a wooden chair, an athlete in motion, and arrows indicating the rotational direction of two power drills. *Booster* also discloses an enthusiasm for space exploration, both in the faint image of the NASA launch pad at Cape Canaveral at upper right and in the very title of the work (Mattison 2003, 136–38). An astronomical chart for the year 1967, with its gridded network of lines, is screenprinted in red ink over the bottom two-thirds of the lithographic composition.

Rauschenberg first turned to printmaking in the early 1960s, producing groundbreaking work with Tatyana Grosman (1904–82) at Universal Limited Art Editions. With his seminal Gemini series *Booster and 7 Studies*, Rauschenberg continued to press the technical boundaries of printmaking, challenging conventions of scale, process, and materials. At the time of its publication, *Booster* was the largest hand-pulled lithograph ever made and heralded a shift in twentieth-century printmaking toward works that could rival the visual impact of painting. Measuring some five feet, ten inches, Rauschenberg's composite skeleton exceeded the scale of available lithographic stones and required printing from two separate stones in two successive pulls onto a single sheet of paper. Much like Antonio Cattani, who printed each of his life-size ecorché figures from five plates on five sheets of paper (cat. nos. 36–38), Rauschenberg surmounted the limitations of available technologies and materials to achieve a work of remarkable ambition and complexity. —NT

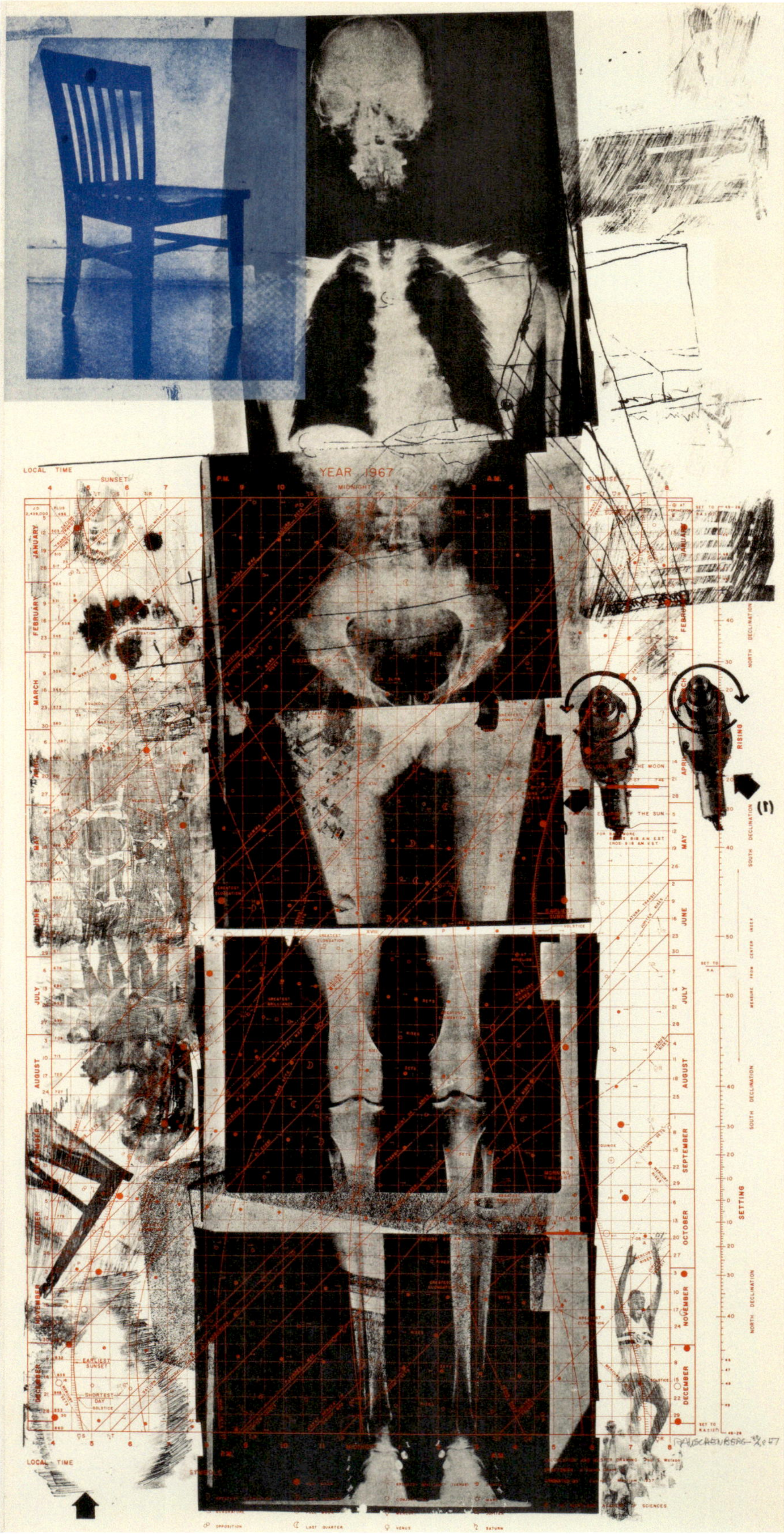
LOCAL TIME
SUNSET
YEAR 1967
MIDNIGHT
JANUARY
FEBRUARY
MARCH
JUNE
JULY
AUGUST
SEPTEMBER
OCTOBER
NOVEMBER
DECEMBER
MAY
RISING
SETTING
NORTH DECLINATION
SOUTH DECLINATION
EARLIEST SUNSET
SHORTEST DAY
LOCAL TIME
OPPOSITION
LAST QUARTER
VENUS
SATURN

47 Tavares Strachan

(Bahamian, b. 1979)
Robert
2018
Neon, Pyrex, transformers, medium-density fiberboard box, dimensions variable
New York, New York, Collection of the artist

BIBLIOGRAPHY
Crutchfield and Hobbs 2013; Griffin 2019, 1:52–53, 330–31, 547.

Pulsating neon tubes illuminate the body's circulatory system in this work by Tavares Strachan (Getty Research Institute artist in residence, 2019–20). This life-size anatomical portrait of the first African American astronaut, Robert Henry Lawrence Jr. (1935–67), who tragically died during a training event in 1967, is seen here as if floating weightless in space. The work was originally exhibited with neon script describing Lawrence's forgotten history, as well as the racist remarks that sought to dehumanize his memory—one that Strachan has sought to restore. In a 2018 collaboration with the Los Angeles County Museum of Art and SpaceX, Strachan launched a gilded canopic jar decorated with Lawrence's features into orbit, finally putting the astronaut into space. *Robert* is part of a decade-long multimedia project examining the nature of invisibility, in which Strachan seeks to increase visibility of marginalized people and histories. An earlier neon anatomy by Strachan, *What Will Be Remembered in the Face of All That Is Forgotten* (2014–15), commemorates Rosalind Franklin (1920–58), a British chemist whose research was critical to James Watson and Francis Crick's discovery of the molecular structure of DNA but who was excluded when they were awarded the Nobel Prize in Physiology or Medicine in 1962. Strachan has also drawn life-size anatomical portraits of Franklin and of another unrecognized pioneer, Matthew Henson (1866–1955), the African American explorer who traveled with Robert Peary to the North Pole in 1909. In addition, Strachan has rendered the cardiovascular system of the Panchen Lama in delicate blown glass. Its submersion in an oil-filled Pyrex vitrine causes it to oscillate from invisible to visible.

Strachan's ongoing efforts to transcend invisibility rehearse the fundamental task of anatomical illustration: to make the obscured internal body visible to observation. Yet his revelatory anatomies do not serve a biological inquiry, nor do they represent an exact record of an individual. Rather, Strachan's anatomies illuminate identity, assert subjectivity, and insist upon a collective universal humanity. Indeed, for many contemporary artists, the invisible object is not the organic tissues or physiological systems under the skin but instead the cultural and social forces pressing upon the human body. Mobilizing the visual vocabularies of science, from anatomical illustrations to diagrams and mathematical proofs, Strachan analyzes and interrogates systems of cultural erasure and absence, giving form to the unseen and unrecognized. —TG

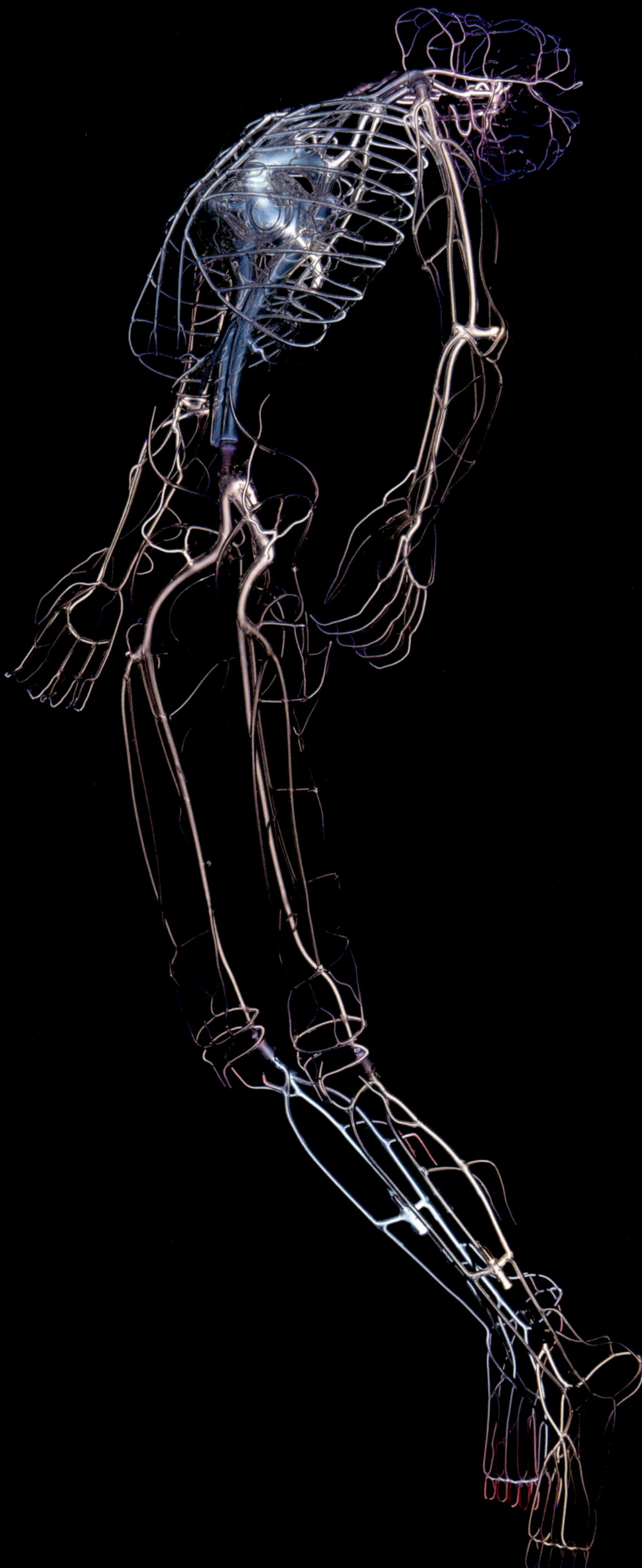

48 Thomas Eakins

(American, 1844–1916)
Wrestlers
Circa 1899
Oil on canvas, 40.8 × 50.96 cm
Los Angeles, Los Angeles County Museum of Art, Mr. and Mrs. William Preston Harrison Collection, 48.32.1

BIBLIOGRAPHY
Brownell 1879; Goodrich 1933; Sewell 1982; Johns 1983; Doyle 1999; Athens 2006; Werbel 2007.

Thomas Eakins's preparatory sketch of about 1899 for *Wrestlers* (LACMA, M.2007.1) presents the dynamic climax of a wrestling match, with one figure holding his opponent in a half-nelson and crotch grip hold. The composition was carefully posed in a studio setting, as attested by a photograph of the modeling session that Eakins shot in preparation for the painting (fig. P). Such a record allowed him to carefully examine the structures and angles of these tangled bodies while he completed the final canvas. Eakins's sporting images were opportunities to showcase the male nude and his extensive knowledge of anatomy. From rowers to boxers, Eakins captured these athletes in physical exertion, reveling in the contours of musculature and articulations of surface anatomy.

Eakins is often remembered for his medical portraits, with *The Gross Clinic* (1875) and *The Agnew Clinic* (1889) considered among his masterpieces. These monuments to surgical expertise and innovation are, moreover, testament to Eakins's close ties to the medical community of Philadelphia at the end of the nineteenth century. Indeed, throughout his career, Eakins was deeply committed to anatomy, incorporating its instruction into his pedagogic practice and founding his meticulous realist aesthetic in its understanding.

His own education in anatomy began at Jefferson Medical College during his years as a student at the Pennsylvania Academy of the Fine Arts. In Paris, he studied painting under Jean-Léon Gérôme (1824–1904) and continued his anatomical studies in the city's hospitals. Back in Philadelphia, Eakins taught anatomy and dissection to art students at the academy, serving as the prosector in anatomy before being promoted to professor of drawing and painting and, ultimately, to director. Forgoing the traditional practice of copying antique casts, Eakins championed an extensive program of anatomical training that included dissecting cadavers and drawing and painting from life. He justified this unorthodox curriculum by stating, "No one dissects to quicken his eye for, or his delight in, beauty. He dissects simply to increase his knowledge of how beautiful objects are put together to the end that he may be able to imitate them" (Brownell 1879, 745). His use of nude models for students of both sexes and encouragement of both male and female students to model for each other provoked his ultimate dismissal in 1886. Despite such condemnations, Eakins refused to alter his practice or pedagogy and remained steadfast in his defense of scientific precision when depicting the body.

Eakins was an early adopter of photography, which he used as a tool to ensure verisimilitude as well as the study of anatomy in the living figure. He employed photography for motion studies and worked alongside Eadweard Muybridge at the University of Pennsylvania, creating his own camera that could capture multiple images on one negative. These photographic aids contributed to his facility with anatomical representation and further served to advance the realism of his painted figures. —TG

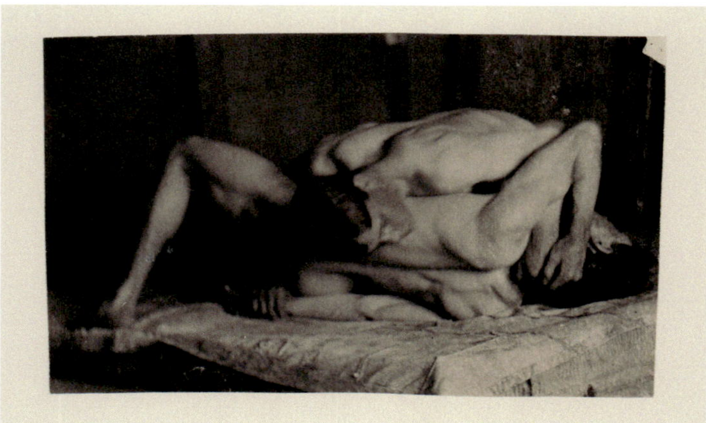

Fig. P
Thomas Eakins (American, 1844–1916).
Wrestlers in Eakins's studio, ca. 1899, platinum print on paper, 9 × 15.2 cm.
Washington, D.C., Hirshhorn Museum and Sculpture Garden, Smithsonian Institution, Gift of Joseph H. Hirshhorn, 1966, 83.96.

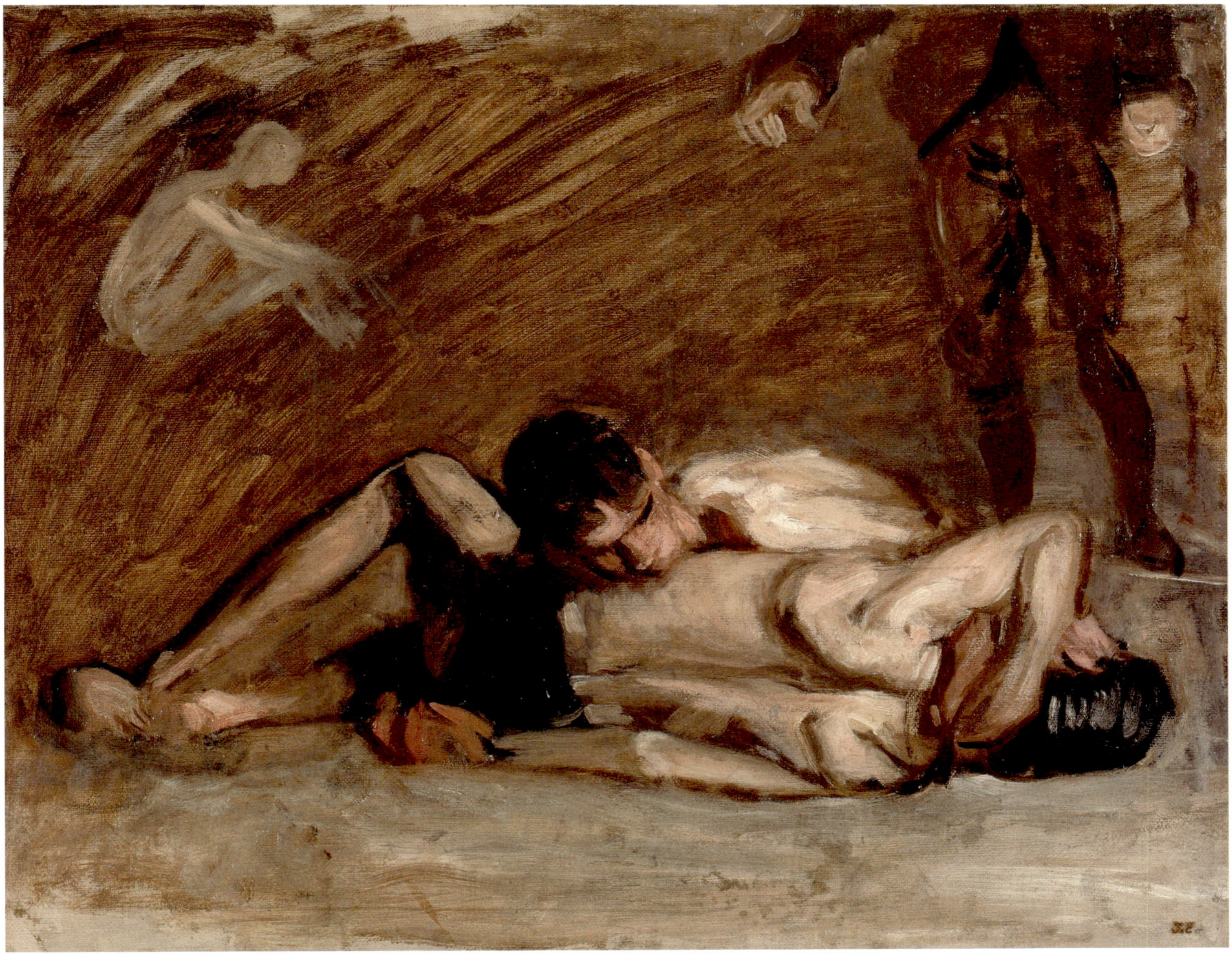

49 Cynthia Maughan

(American, 1949–2019)
Arteries and Veins, in *Short Video Works, 1974–1975*
Circa 1974–75
Videocassette, 1:38 min., black and white with sound
Los Angeles, Getty Research Institute, 2006.M.7 (A.Maug-2), no. 3

BIBLIOGRAPHY
Phillips 2008, 14, 162–65.

Arteries and Veins is from a series of videos the Los Angeles-based artist Cynthia Maughan made in the 1970s. In it, she physically documents the visibility of the veins through the skin of her chest and neck by tracing over them with a marker. While doing so she gives contradictory explanations of the circulation of the blood and implies that deoxygenated blood traveling to the heart is blue and that it turns red instantaneously upon exposure to air. (In fact, venous blood is dark red; veins appear blue through the skin due to the variation in the absorption of wavelengths of light.) As if to test the theory, she raises a razor to her neck just before the video abruptly ends. In a related video, *Suicide* (1973–74), also in the collection of the Getty Research Institute (2006.M.7 [A.Maug-2], no. 2), a razor blade is tossed into a bathtub and dark liquid spurts into the bathwater, eventually turning it completely opaque as the breathing heard on the soundtrack slows (Phillips 2008, 10, 165). —MK

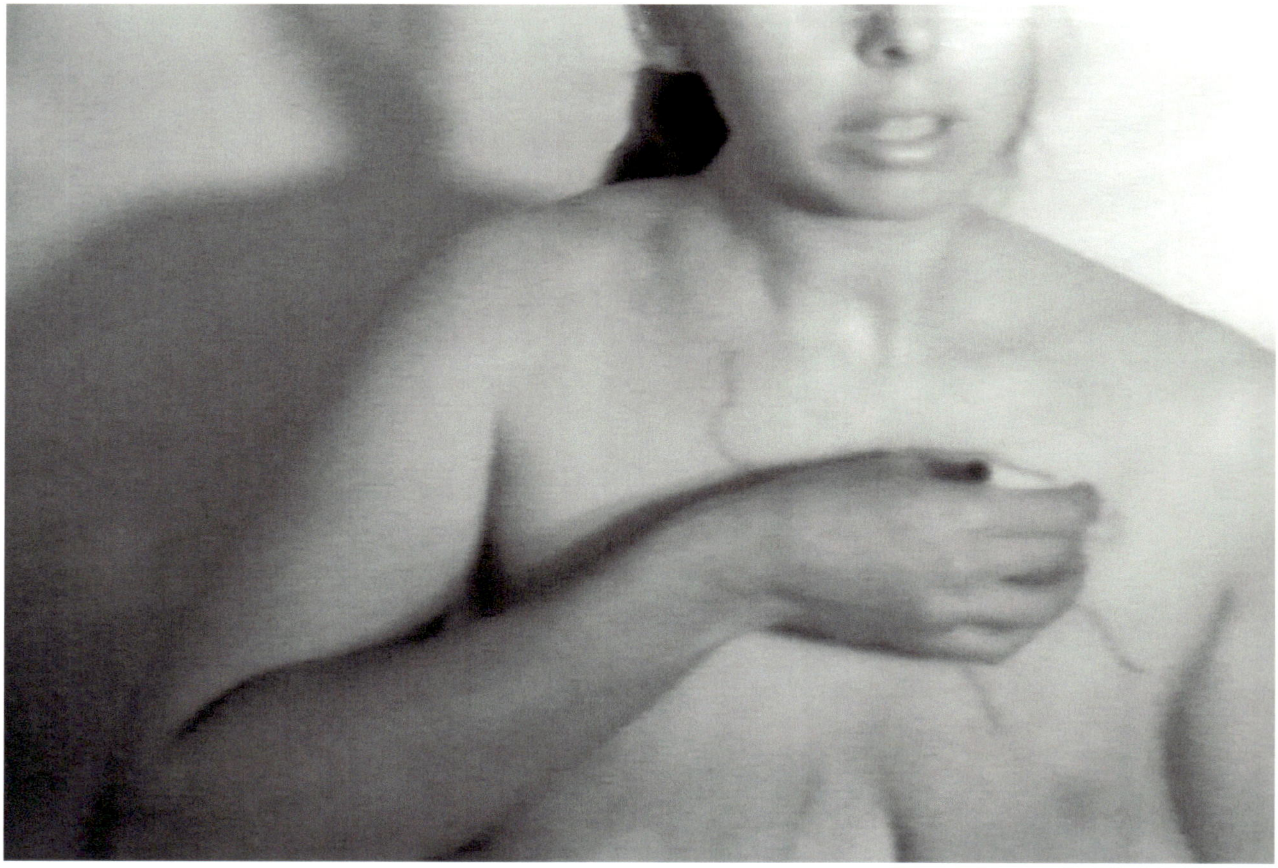

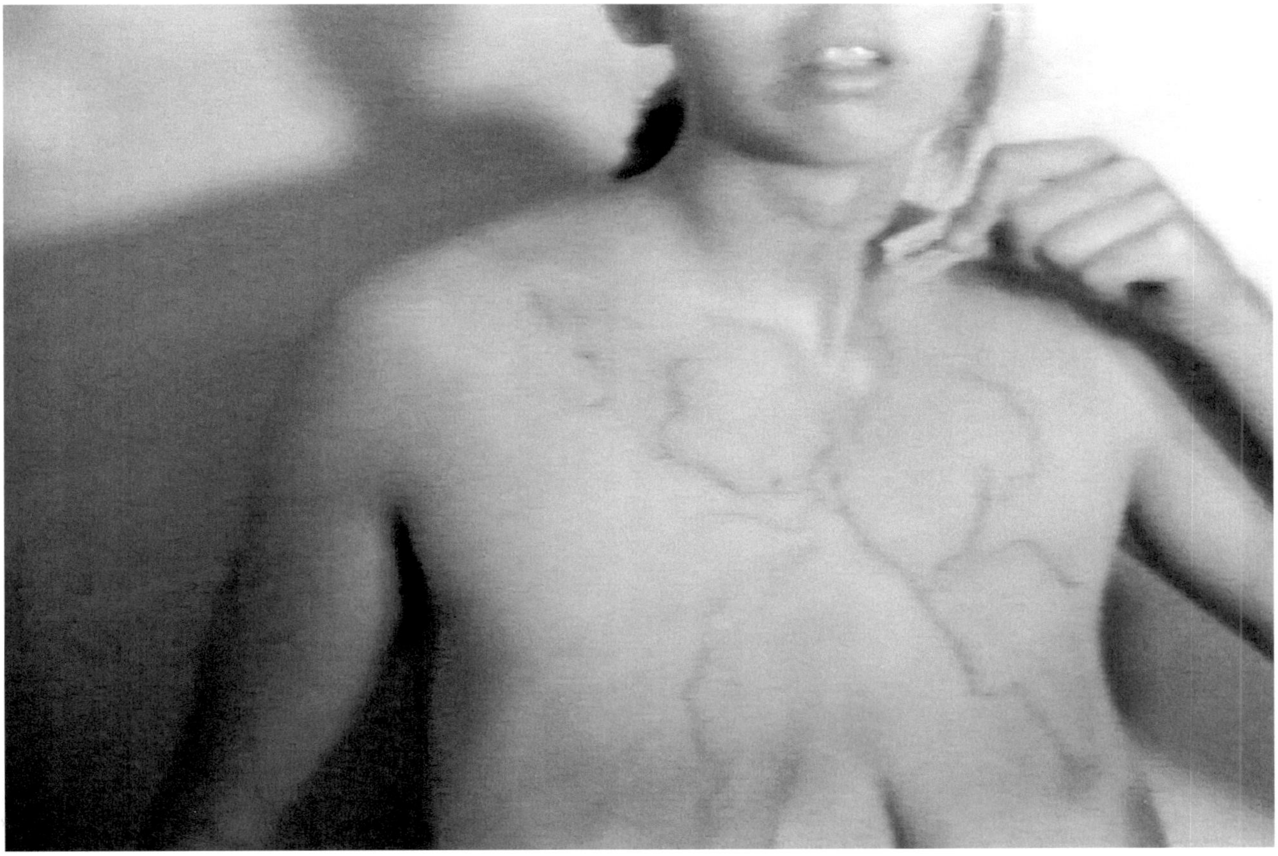

50 ROBERT MAPPLETHORPE

(American, 1946–89)
Ken Moody
Negative, 1985
Gelatin silver print, 2011
Image: 48.4 × 38.2 cm; sheet: 50.4 × 40.5 cm
Los Angeles, J. Paul Getty Museum, 2012.52.10, and Los Angeles County Museum of Art, M.2011.30.554

BIBLIOGRAPHY

Morrisroe 1995, 290; Martineau and Salvesen 2016.

Striated light falls across the nude figure of Ken Moody, modulating a bodily surface in which each dip and hollow is captured and highlighting the course of the veins down the legs and across the feet. The raised-heel pose of the left foot brings out the structures of muscle, tendon, and bone beneath the skin of the lower leg. This same pose with the head cropped out and the play of light are repeated in Mapplethorpe's *Ken and Tyler* from 1985 (Solomon R. Guggenheim Museum, 96.4373). The doubling of pose and contrast in skin color between the two figures in *Ken and Tyler* was one explored in *Ken Moody and Robert Sherman* of the previous year (J. Paul Getty Museum, 2011.7.23) (Martineau and Salvesen 2016, 308, no. 135). Moody was a physical trainer and a frequent model for Mapplethorpe; the definition of Moody's extraordinary physique was enhanced by the condition of alopecia, which rendered his skin hairless. —MK

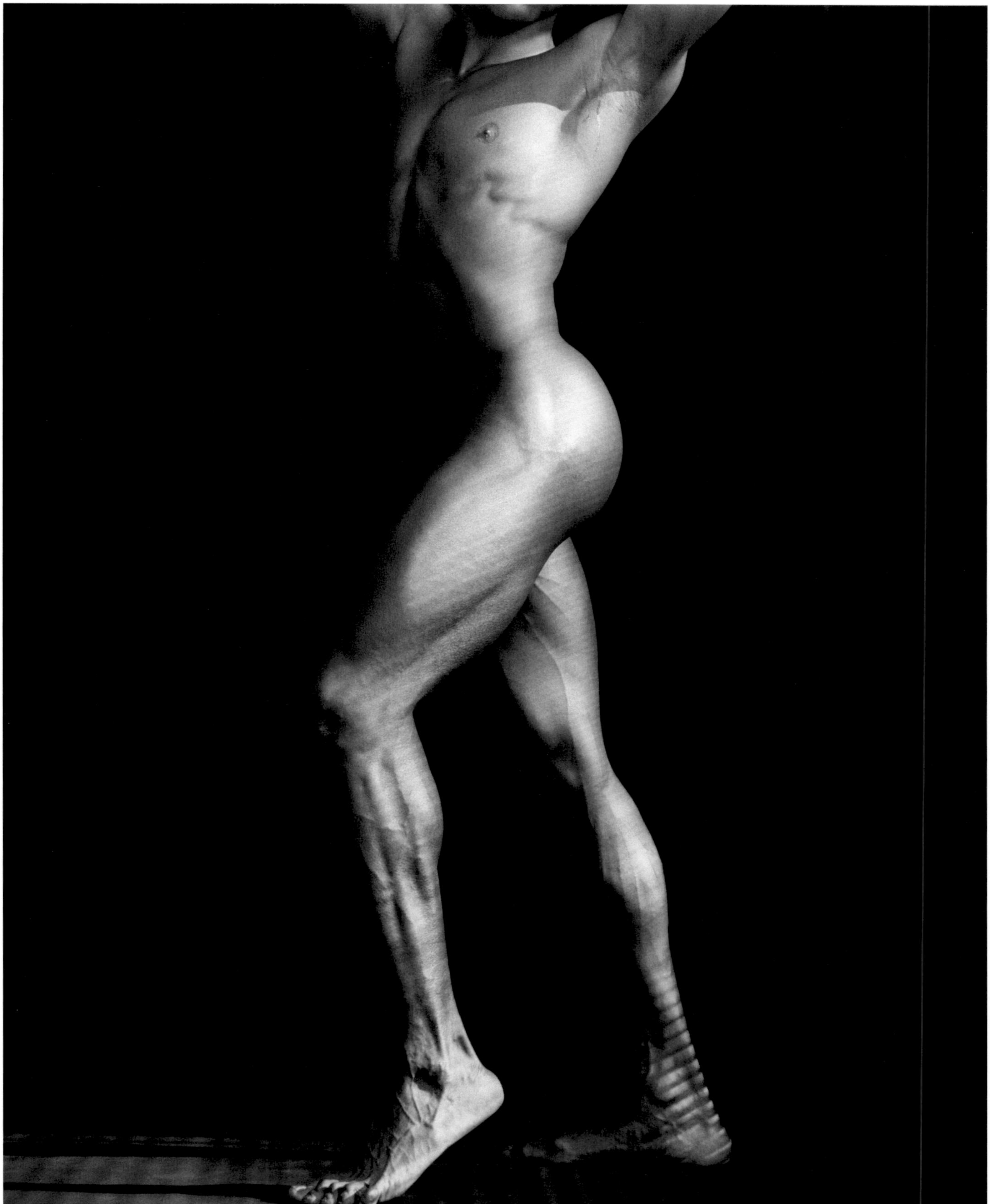

51 Lucas Kilian

(German, 1579–1637), after Johann Remmelin (German, 1583–1632)
Third Vision from Mirrors of the Microcosm
Engraving
From Johann Remmelin, *Catoptrum microcosmicum: Suis aere incisis visionibus splendens, cum historia, & pinace, de novo prodit* (Augsburg: David Franck, 1619)
Los Angeles, Getty Research Institute, 2017-B292

BIBLIOGRAPHY
Roberts and Tomlinson 1992, 52–53, 64; Cazort in Cazort, Kornell, and Roberts 1996, 171–72, nos. 56–58; Carlino 1999b, 71–73; Buckley 2013; Schmidt 2011, 82–91; Massey 2013.

Prints of Johann Remmelin's *Catoptrum microcosmicum* first appeared in 1613, but this edition of 1619 is the first authorized version. Printed with an explanatory Latin index, this edition includes three illustrated flap anatomy sheets that feature male and female figures surrounded by an array of allegorical, alchemical, and religious symbols, as well as subsidiary anatomical details. The print *Visio tertia* depicts a smiling nude woman, who rests her right foot upon an upturned skull. A serpent slithers through the skull's base, a Christian reference to the fall of man; the skull is composed of stacked flaps that conceal layers of cranial anatomy. The woman's pubis is modestly covered by a thick plume of smoke that emerges from the phoenix below in an alchemical reference to Jesus's resurrection.

The bodies throughout are superimposed with paper flaps, which can be lifted to reveal internal organs and physiological systems. This tactile technology invites the viewer to engage in an exploratory survey of human anatomy—a virtual and interactive dissection to demonstrate spatial relationships within the body. Many of the schematic structures are after Andreas Vesalius, who had also used this method in his *Epitome* (1543), enabling readers to construct a three-dimensional representation of the anatomical body by cutting and pasting illustrated parts.

Kilian's fine copper engravings of Remmelin's anatomy represent an advance in the genre: they are embellished with a far more ambitious program of anatomical details, and their construction, which includes attached and free-floating flaps, is complex. While ostensibly promoting a medical understanding of the body, the anatomy of these prints was already out of date by the time of their publication. Moreover, they are replete with esoteric, moral, and theological imperatives. The anatomical scheme is embedded with Christological, alchemical, cabalistic, and other iconographies, with inscriptions in Hebrew, Greek, and Latin, as well as symbolic imagery and emblems. Imbued with soteriological and eschatological references, this work provokes self-reflection on the transience of life and compels the viewer to contemplate the glories of God through the study of the body—a microcosm of the universe.

The density of meaning in these multilayered illustrations suggests that they may have been considered more curiosities than medical teaching devices. Printed in several editions, languages, and pirated forms throughout the eighteenth century, they enjoyed great commercial success and were a precursor to the "golden age" of flap anatomies in the nineteenth century, particularly in obstetrical atlases. —TG

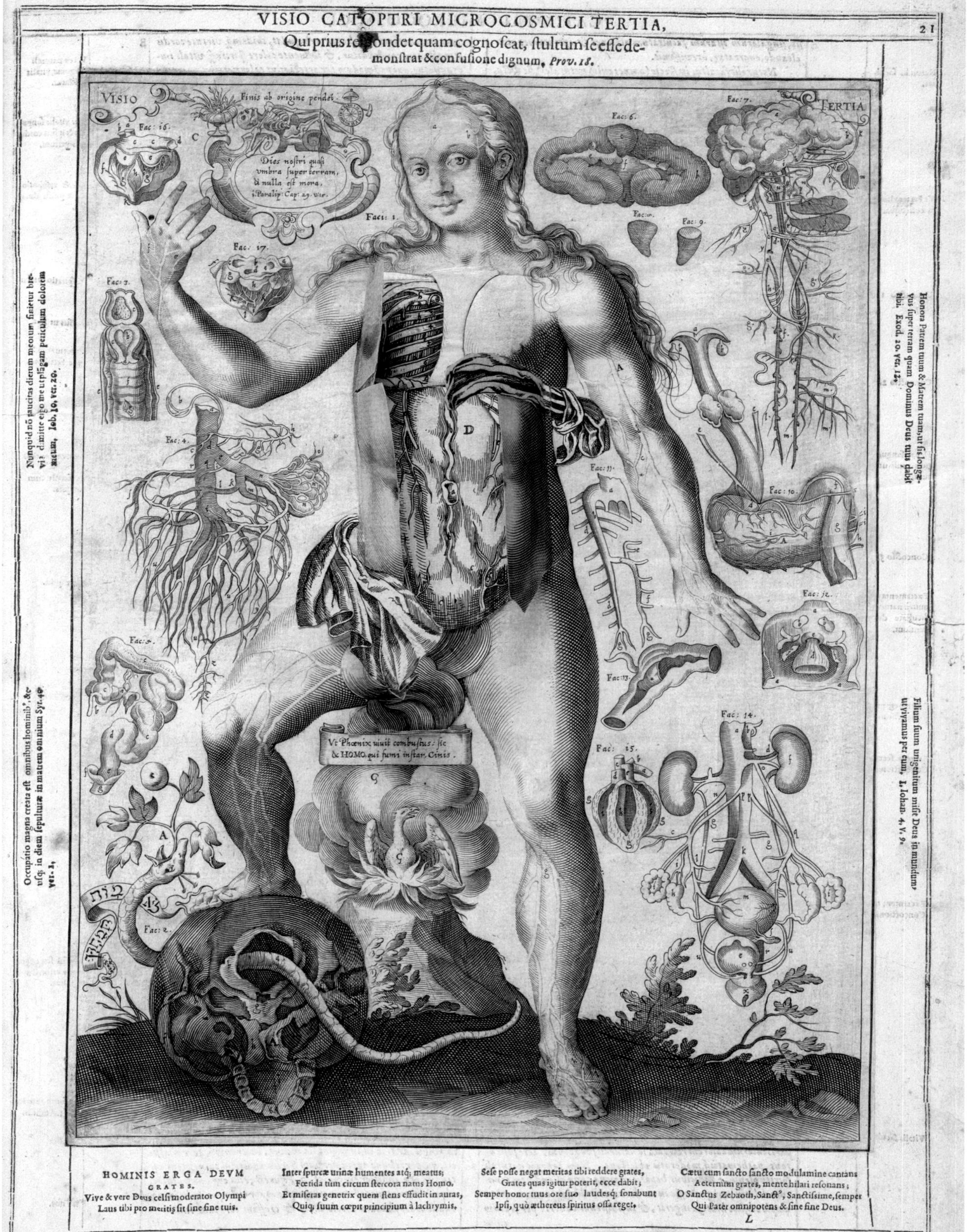
VISIO CATOPTRI MICROCOSMICI TERTIA,
Qui prius respondet quam cognoscat, stultum se esse demonstrat & confusione dignum, Prov. 18.
21
VISIO
Finis ab origine pendet.
TERTIA
Dies nostri quasi umbra super terram, & nulla est mora, i. Paralip. Cap. 29. ver.
Vt Phœnix uiuit combustus: sic & HOMO, qui fumi instar. Cinis.
Honora Patrem tuum & Matrem tuam, ut sis longæuus super terram quam Dominus Deus tuus dabit tibi. Exod. 20. ver. 12.
Filium suum unigenitum misit Deus in mundum, ut vivamus per eum, I. Iohan. 4. v. 9.
HOMINIS ERGA DEVM GRATES.
Vive & vere Deus celsi moderator Olympi
Laus tibi pro meritis sit sine fine tuis.
Inter spurcæ urinæ humentes atq; meatus;
Fœtida tùm circum stercora natus Homo.
Et miseras genetrix quem flens effudit in auras,
Quiq; suum cœpit principium à lachrymis.
Sese posse negat meritas tibi reddere grates,
Grates quas igitur poterit, ecce dabit;
Semper honor tuus ore suo laudesq; sonabunt
Ipsi, quò æthereus spiritus ossa reget.
Cætu cum sancto sancto modulamine cantans
Aeternum grates, mente hilari resonans;
O Sanctus Zebaoth, Sanct°, Sanctisime, semper
Qui Pater omnipotens & sine fine Deus.
L

52 Jules Galet

(French, active 1829–44), after Faustino Anderloni (Italian, 1766–1847), after Antonio Scarpa (Italian, 1752–1832); lithograph printed by Lemercier, Bernard & C.
View of the lungs and the vagus nerve, seen from the back
Hand-colored lithograph
From Jules Galet, *Le corps de l'homme: Traité complet d'anatomie et de physiologie humaines, contenant près de 200 planches dessinées d'après nature et lithographiées, et suivi d'un précis des systèmes de Lavater et de Gall* (Paris: Galet, Mansut fils, and Billet, 1844), vol. 4, facing p. 116, pl. 159
Los Angeles, Getty Research Institute, 92-B20039

BIBLIOGRAPHY

Richardson, B. W. 1901, 146–47; Choulant 1945, 298–99; Roberts and Tomlinson 1992, 375–81; Momesso 2007.

The popularizing aim of this four-volume work is denoted by the original title of the first edition of 1835–41, retained for some sections in this second edition: *Le corps de l'homme, ou l'anatomie et la physiologie humaines mises a la portée de toutes les classes de la société* (The body of man, or human anatomy and physiology, placed within the reach of all classes of society) (Galet 1844, 1:1). Although the illustrations were derived from a variety of sources, Jules Galet, the former head of clinic of the faculty of medicine in Montpellier, signed them as the designer and lithographer: "Galet D. M. del. et lith."

This posterior view of the lungs and the vagus nerve, displayed here with the ribs cut away and the spinal column removed, is adapted from an uncolored engraving that appeared as the fifth plate in Antonio Scarpa's illustrated book on the nerves of the heart, the *Tabulae neurologicae* (Scarpa 1794) (fig. Q). Scarpa's book was "the first in which the nerves of the heart itself were properly delineated. Other anatomists had already shown that the bloodvessels [*sic*] of the heart are accompanied by nerves; but to Scarpa is due the discovery that the muscular structure is also supplied with nerves" (Richardson, B. W. 1901, 146).

Scarpa, a professor of anatomy at the Università di Pavia, had a significant art collection (Momesso 2007). He also drew and supplied the drawings for the life-size engravings by Faustino Anderloni, which appeared with a separate outline plate to carry the lettering. In Galet's compilation, the image size has been reduced, the lettering has been curtailed and added directly to the image, and several details have been lost. For crispness of detail, neither Galet's lithograph nor the partially colored copy of the same plate in Jules Cloquet's *Manuel d'anatomie* could match Anderloni's original engraving (Cloquet 1825, pl. 89). In Galet's version, the outline of the form has been simplified. Gone is the framing drapery and the clavicles attached to the scapulae visible in the original. However, something has been gained with the addition of color, which adds to the sense of depth and helps greatly in differentiating the structures, bringing out the wandering course of the vagus nerve and its branches. —MK

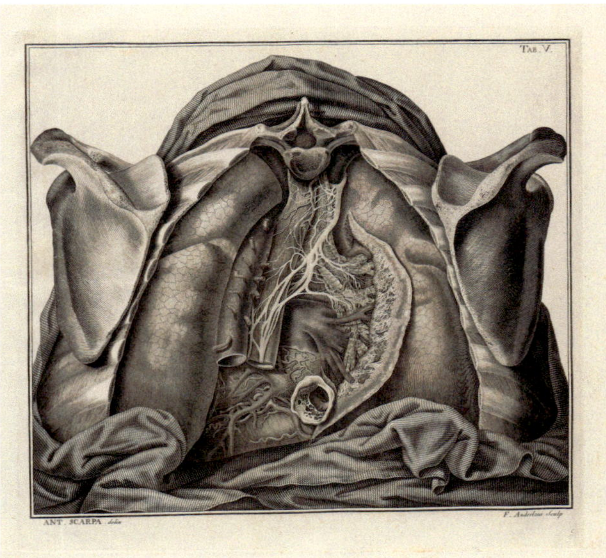

Fig. Q
Faustino Anderloni (Italian, 1766–1847), after Antonio Scarpa (Italian, 1752–1832).
Nerves of the posterior pulmonary plexus, engraving.
From Antonio Scarpa, *Tabulae neurologicae ad illustrandam historiam anatomicam cardiacorum nervorum, noni nervorum cerebri, glossopharyngaei, et pharyngaei ex octavo cerebri* (Pavia: Baldassare Comini, 1794), pl. 5.
Heidelberg, Universitätsbibliothek Heidelberg, P 1237-1 Gross RES.

PLANCHE 159

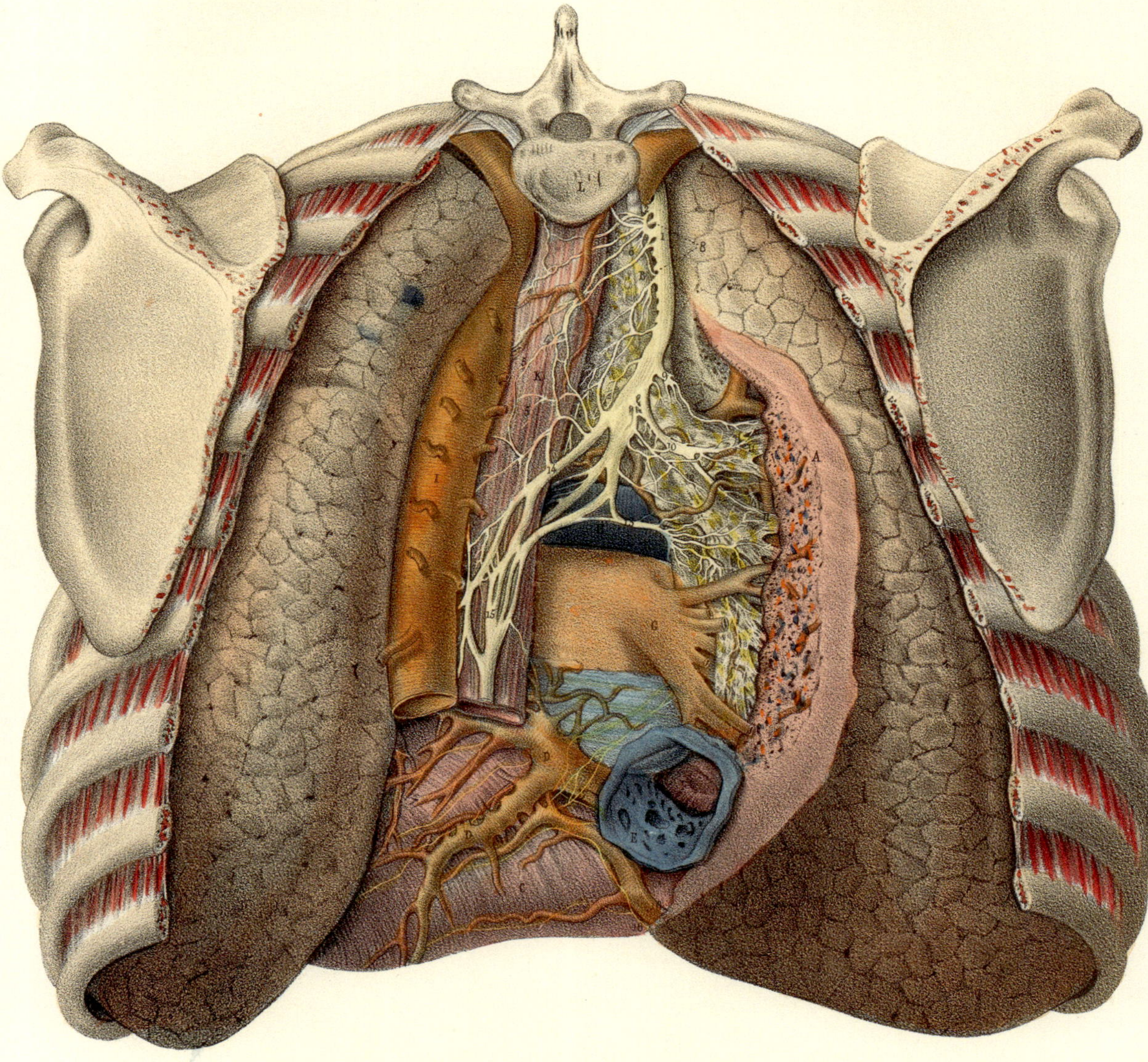

Galet D. M. del. et lith.

Imp. Lemercier, Benard et C.

53 **Thomas Sinclair**

(Scottish, ca. 1805–81), after Joseph Maclise
(Irish, ca. 1815–ca. 1880)
Dissection of chest and abdomen
Hand-colored lithograph
From Joseph Maclise, *Surgical Anatomy*
(Philadelphia: Blanchard and Lea, 1851), pl. 26
Los Angeles, Getty Research Institute, 1373-163

BIBLIOGRAPHY

Roberts and Tomlinson 1992, 562–65, 570–75, nos. 123–24; Coakley 1992, 74–76; Cazort in Cazort, Kornell, and Roberts 1996, 237–39, no. 126; Postle in Postle and Vaughan 1999, 26, no. 8; Caldwell 2006, 348–51; Kornell 2007, 212–13; *Plarr's Lives of the Fellows,* no. E002613.

The Irish-born surgeon Joseph Maclise considered the surface of the body as "a map to the surgeon, explanatory of the anatomy arranged beneath" (Maclise 1851, viii). For this reason, he purposely included intact areas of the bodily surface in the colored-lithograph illustrations of dissections in his well-received *Surgical Anatomy,* which appeared not long after the introduction of ether and chloroform for general anesthesia. In the twenty-sixth plate, the incision has been angled so that each layer from the surface, through muscle and ribs, can be clearly made out.

The lithographs were done after Maclise's own drawings of dissections at University College Hospital, London, where he studied under Richard Quain (1800–1887), Robert Liston (1794–1847), and Samuel Cooper (1780–1848) (the latter two, along with his fellow students, being the dedicatees of the book), and at the anatomy school connected with the Hôpital de la Pitié, Paris. The subject depicted in plate 26 may have been an unclaimed body made available for dissection by the Anatomy Act of 1832. Previously, cadavers for dissection in the United Kingdom had been restricted to those executed for crimes.

Maclise's figures occupy a liminal state between sleep and death; they glow with suffused lighting, an artistic effect also seen in the works of his elder brother, Daniel Maclise (1806–70), a painter who specialized in literary subjects and portraits. Maclise's corpses are on the whole more kempt, attractive, and alive than those he had depicted earlier for Quain's *The Anatomy of the Arteries of the Human Body* (1844). Three years later, Maclise published *Comparative Osteology* (1847), with lithographs after his own line drawings. In 1850, Maclise was admitted as a fellow of the Royal College of Surgeons and subsequently developed a successful surgical practice in London.

The first edition of *Surgical Anatomy* appeared in London in 1851 with thirty-five plates. The copy in the Getty Research Institute is the greatly expanded American edition of the same year, published in Philadelphia with sixty-eight plates. The positive reviews Maclise received cite the accuracy and artistry of the illustrations but also the aggressively low price, something that made the work accessible to students. The American publishers Blanchard and Lea promoted it in their catalog of July 1855 as "being one of the cheapest and best executed Surgica[l] works as yet issued in this country" (Blanchard and Lea 1855, 22). The run of one thousand copies of the first London edition sold out within months (*Edinburgh Medical and Surgical Journal* 1854, 580), and a second, revised English edition appeared in 1856. Three American editions followed in the next decade. Blanchard and Lea were also the American publishers of Henry Gray's *Anatomy, Descriptive and Surgical,* first published in London in 1858 (see fig. 1.5). It soon eclipsed Maclise's success and, in revised form, remains a medical best seller to this day. —MK

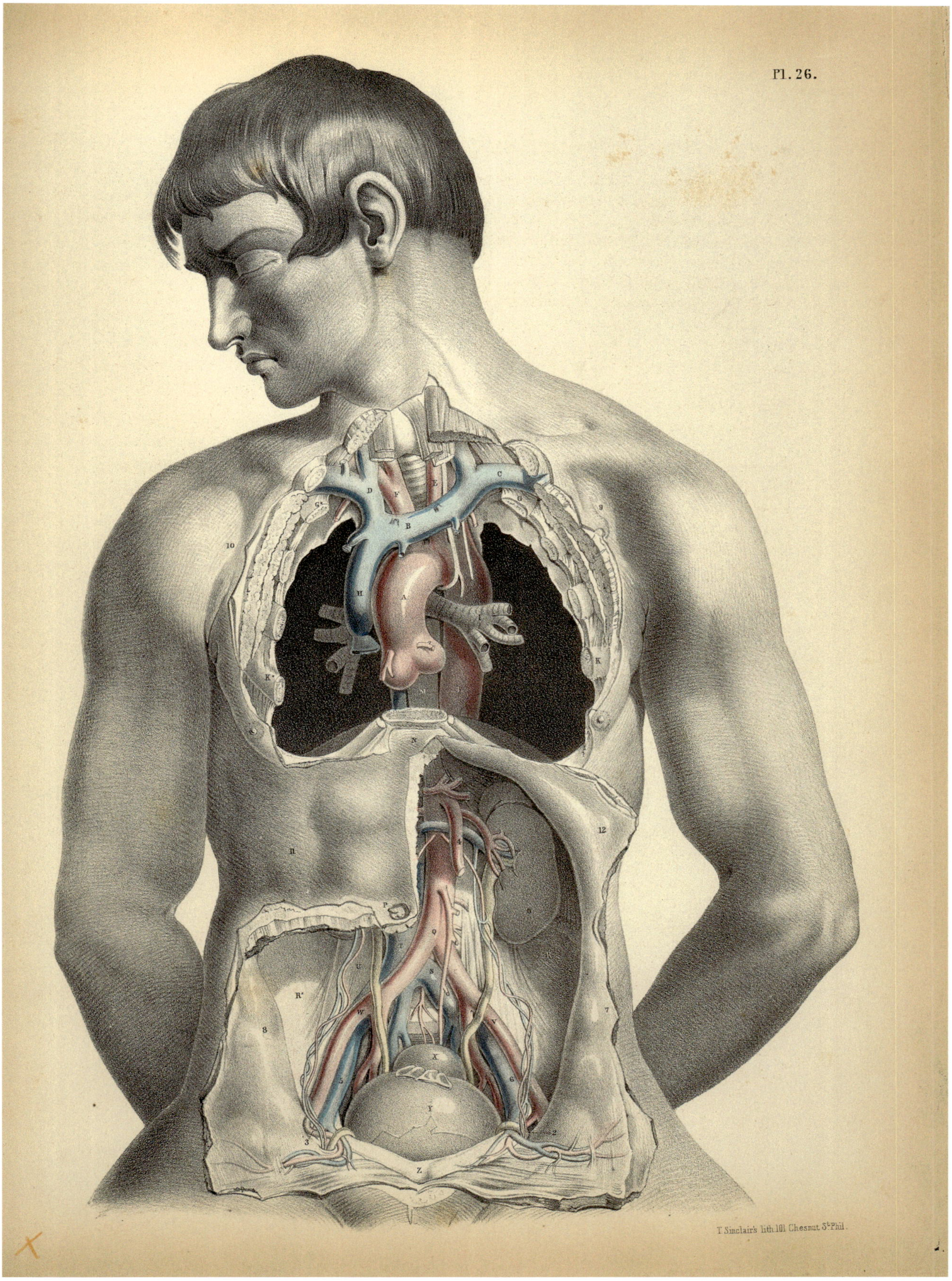
Pl. 26.
T. Sinclair's lith. 101 Chesnut St. Phil.

54 John Carbutt

(English, 1832–1905)
Left: *Mummified Hand of an Egyptian Princess*
Right: *Radiograph of Mummified Hand*
Half-tone
From William H. Meadowcroft, *The A B C of the X Rays* (London: Simpkin, Marshall, Hamilton, Kent & Co., [1896]), between pp. 112 [110] and 109 [111]
Los Angeles, Getty Research Institute, 91-B35082

BIBLIOGRAPHY
Wilson 1896; Brey 1984.

The hand was a popular subject of the earliest X-ray photographs, being easy to position in an X-ray machine and yielding an image relatively quickly. When Wilhelm Röntgen (1845–1923), the German physics professor who discovered X-rays, took the very first medical X-ray, it was of his wife's ringed hand. Meadowcroft's publication, which describes X-rays and gives instructions on the making of X-ray machines, has a preface dated 15 October 1896, less than a year after Röntgen had published the paper on his discovery in December 1895. It was part of the initial international enthusiasm for X-rays, before their adverse effects were known and when exposure times were dangerously long. *The A B C of the X Rays* includes an X-ray of the author's own hand, for which an exposure time of almost three minutes is recorded. Longer times were required for larger or denser subjects, with the torso requiring "fully ten times as much exposure" as a hand (Meadowcroft 1896, 155).

This X-ray of a mummified hand was taken by John Carbutt, a photographer and pioneer in photographic applications, such as X-ray plates (Meadowcroft 1896, 183–84) and celluloid film used in motion pictures. An early example of the use of X-rays for the scientific study of an ancient artifact, it was used to prove that the hand was an actual mummified hand and not a modern tourist fake. It had been published shortly before in *Wilson's Photographic Magazine* (Wilson 1896), accompanying "The Romance of Ro-ent-gen-ra" (a play on the name of Röntgen), in which the story of the hand's acquisition from an Egyptian grave robber by an American female tourist is related in a breathless, tongue-in-cheek fashion.

The English-born Meadowcroft (1853–1937) authored the similarly titled *Scholars' A B C of Electricity* (1896). He subsequently worked for over thirty years as a private secretary and assistant to the inventor Thomas Edison (1847–1931), who also briefly experimented with X-rays, until the effects of overexposure to radiation gradually became known. One of Edison's assistants, who was in the habit of taking test X-rays of his own hand, was an early fatality (Brey 1984, 171). —MK

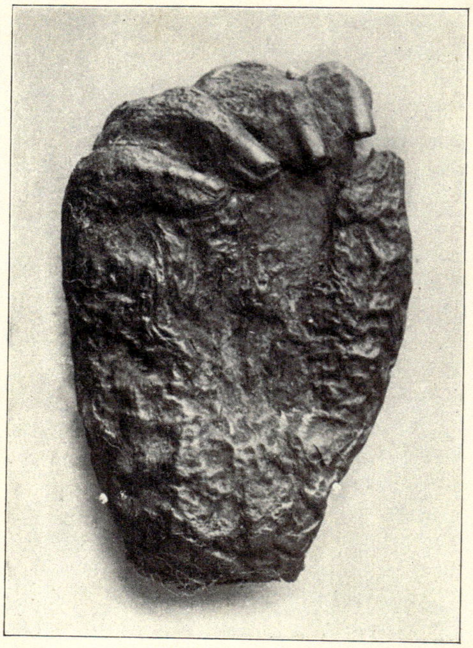

PHOTOGRAPH.

Mummified Hand of an Egyptian Princess, obtained near the Tombs of the Kings, Thebes, 1892.

The hand is believed to be between 3,000 and 4,000 years old.

Half-tone: Etched, using Carbutt Process plates.

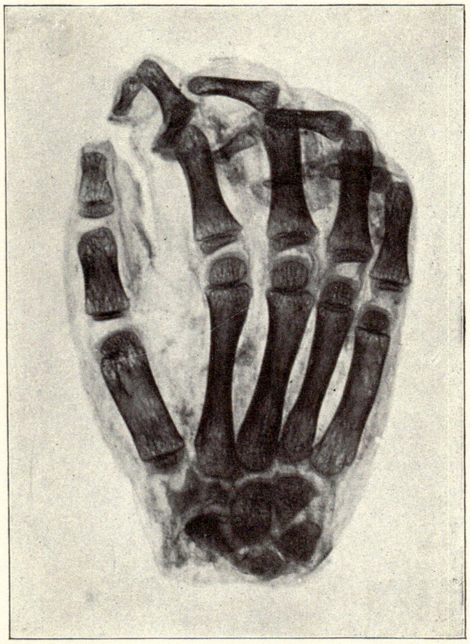

RADIOGRAPH

of Mummified hand on preceding page, made in the Laboratory of the Keystone Dry Plate Works, on a Carbutt X-Ray plate, by John Carbutt.

Half-tone: Etched, using Carbutt Process plates.

55 *INTERIOR OF THE RIGHT AURICLE AND VENTRICLE IN TOPOGRAPHICAL RELATION TO THE CHEST-WALL*

Silver gelatin stereographic photograph, mounted on card; length: 24 cm
From *Stereoscopic Studies of Anatomy,* prepared under authority of the University of Edinburgh by D. J. Cunningham, edited by David Waterston, Professor M. H. Cryer, and Frederick E. Neres, rev. ed. (New York: Imperial Publishing Company, [1909]), vol. 3, section III, no. 12
Los Angeles, Getty Research Institute, 91-B4743

BIBLIOGRAPHY

Crary 1990, 14–19, 59–62, 119–32; Kemp 1997, 144–45; Hallam 2016, 219–21; Pellerin 2019, 189.

The stereoscope, an optical device invented by Charles Wheatstone around 1832, originated as an investigation into the physiology of binocular vision, but by the 1860s had transformed into an object of mass entertainment, provoking an international stereoscopic mania. By presenting two planar images through the viewer, each representing an object or scene captured from slightly different angles, the stereoscope relies on depth perception to fuse the two images and conjure a single, unified image that appears in 3-D. The illusion of depth is the result of retinal convergence, producing a facsimile of natural perception.

The use of stereoscopy in medical illustration began as early as 1865. European and American physicians applied the medical gaze through this new binocular technology, largely in clinical and pathological case studies and to document surgical interventions. *Stereoscopic Studies of Anatomy* was the most successful application of this technology and represents an iconic example of early attempts to create three-dimensional imaging in medicine. First published in 1905, it evolved in subsequent editions over the next two decades into a behemoth of over three hundred stereographic cards that illustrate various parts of the human body in progressive detail. The edition of 1909 is composed of ten volumes and organized into regional sections. Each stereograph is mounted on a card and accompanied by a descriptive text and referential legend; the card could be inserted into the stereoscopic viewer to reveal the specimen in a striking illusion of solidity.

Marketed to professionals and students alike, *Stereoscopic Studies of Anatomy* served as a virtual and comprehensive anatomy lesson by re-creating the experience and process of dissection. Through successive images, the viewer sees the body unpeeled under the dissector's knife: first, the surface is demarcated with internal landmarks, followed by the removal of the skin, musculature, and tissues; then, the unseen dissector saws through the bones and finally reveals the interior organs. This mode of stratified excavation—unveiling the body to discover secrets beneath—structurally mimics the design of early flap anatomies. Contemporary reviewers hailed the stereoscope's immersive realism, accuracy, and educational powers, calling it "the nearest approach which has yet been found to the actual clinical or anatomical specimen" (M.A.G. 1910). While the photographer is unnamed, the dissections were prepared under the supervision of D. J. Cunningham (1850–1909), professor of anatomy at the University of Edinburgh (1903–9). Revised modern editions of his *Manual of Practical Anatomy,* first published in 1893–94, are still in use by students today. —TG

THE EDINBURGH STEREOSCOPIC ATLAS OF ANATOMY.

Section III.—No. 12

THORAX—No. 12.

INTERIOR OF THE RIGHT AURICLE AND VENTRICLE IN TOPOGRAPHICAL RELATION TO THE CHEST-WALL.

View from in front and slightly from the right side, to show the thoracic skeleton, and, at the same time, the parts which lie under cover of it.

The anterior auriculo-ventricular groove is seen to run from the middle of the sternum at the level of the third costal arch downwards and to the right, so that the cavity of the auricle lies almost entirely to the right of the sternum, while the ventricle lies behind the sternum and the inner end of the fifth right interspace.

Looking at the heart from the front, it may be said that the right auricle lies to the right of the sternum, the right ventricle occupies the area behind the sternum, and the left ventricle that to the left of the sternum.

The tip of the left auricular appendix is not seen here, but it usually appears on the left of the pulmonary artery.

The pulmonary valve lies behind the third left chondro-sternal junction, and the right auriculo-ventricular orifice behind the sternum opposite the extremity of the fourth intercostal space of the right side.

The figures indicate—

1-7. Corresponding costal arches.
8. Anterior interventricular furrow, with branch of left coronary artery.
9. Musculi pectinati of auricle.
10. Right coronary artery.
11. Anterior papillary muscle of ventricle.
12. Infundibular cusp of tricuspid valve.
13. Pulmonary valve.

THORAX No. 12.
(EDINBURGH STEREOSCOPIC ANATOMY.)

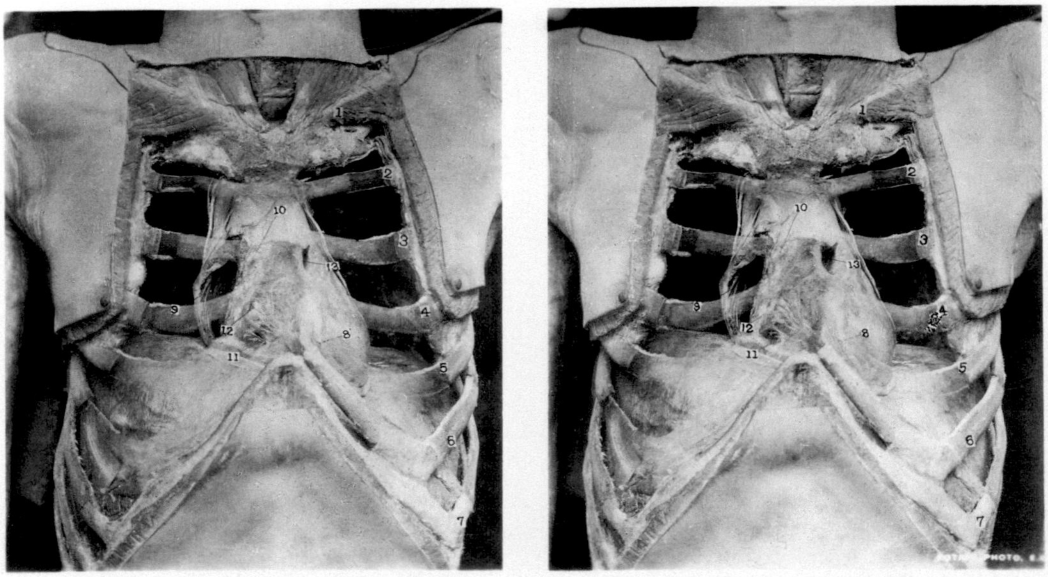

56 James Allen

(American, b. 1977)
The Body and Drugs
2014
Book excavation, edged in gold leaf, of two books from the *Life Science Library* series: Alan E. Nourse, *The Body* (New York: Time-Life, 1964), and Walter Modell and Alfred Lansing, *Drugs* (New York: Time-Life, 1967), unique, 23 × 23 × 30.5 cm
Los Angeles, Library Special Collections for Medicine and the Sciences, Louise M. Darling Biomedical Library, UCLA, BIOMED *N7433.4 A427 B668 2014

BIBLIOGRAPHY
Allen 2021; Heyenga 2013, 17, 24–29.

A book excavation transforms a book into a 3-D sculpture through selective cuts made at different layers. This one is composed of two books concerned with medicine intersecting each other: *The Body* (1964) and *Drugs* (1967). James Allen states, "As I cut I consider both narrative and compositional dynamics to create a new vision of the book's content. I enjoy how these Excavations turn the linear format of a book into a flat window through which to observe many pages at once" (Allen 2021). In carefully isolating different parts of a book while progressing through several layers, the artist evokes the painstaking process of dissection.

On *The Body* side, the shallower cuts juxtapose the diagrammatic system of medieval anatomical illustration with the naturalistic imagery of the Renaissance that would replace it. At the bottom, the recumbent cadaver and the disembodied arm and hand holding a dissection knife above it is from a woodcut illustration from the Venice 1493/94 edition of the *Fascicolo medicina,* a compilation of medical texts attributed to Johannes de Ketham, first published in Latin in 1491. The cadaver appears to float in space, since the trestle table that supports it in the original has been cut away. To the left is a medieval figure showing the nerves, one of a five-figure anatomical series in the Bodleian Library, Oxford (MS Ashmole 399, fol. 21r), dated to the late thirteenth century. As the cuts progress through the book, the modern era of photography is reached, and a photograph of the retina serves as a backdrop to a running écorché figure. On the *Drugs* side are medicinal plants and a medicinal jar of the type that was common in early apothecaries. —MK

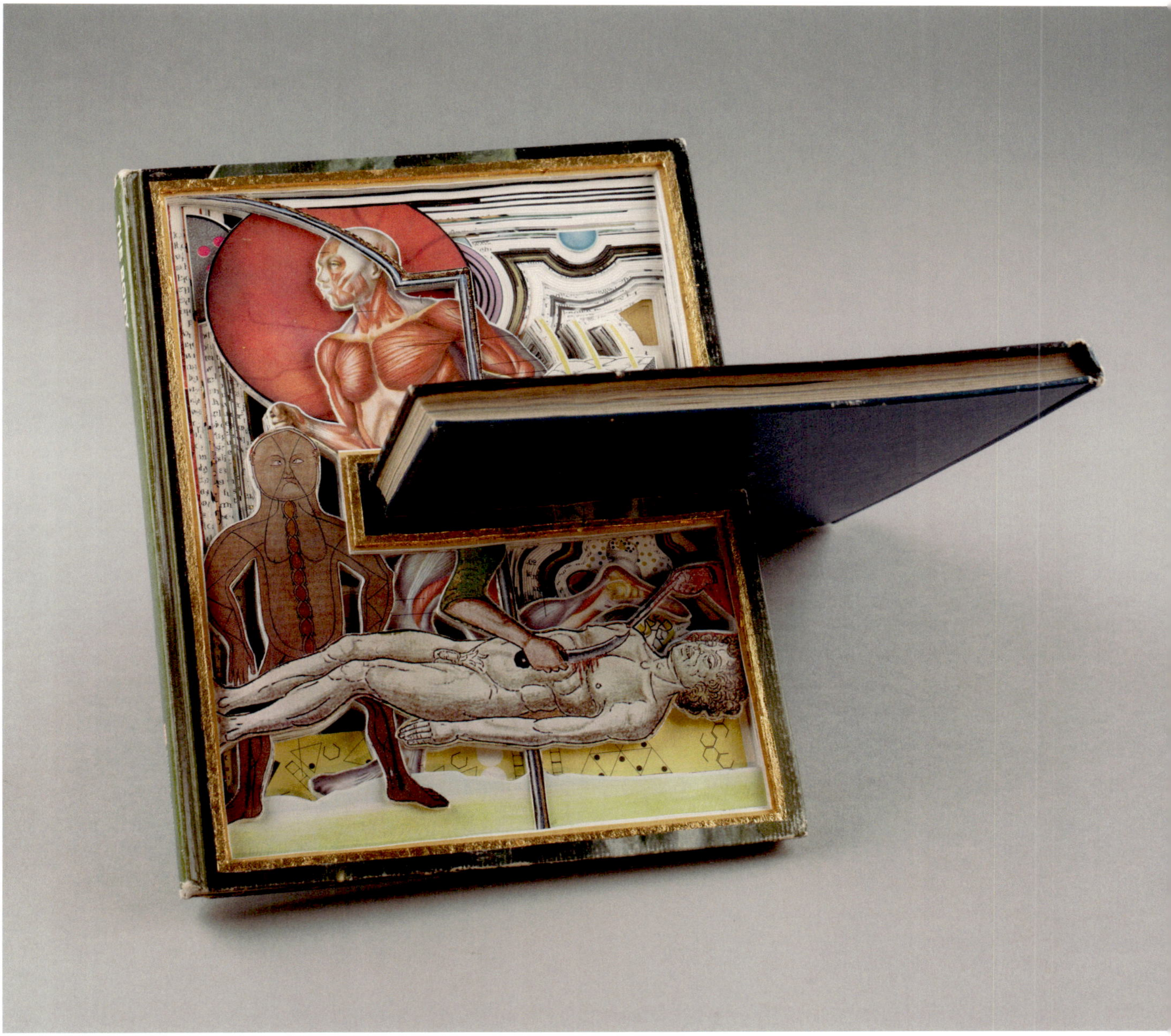

subscription announcements for anatomical prints by Antonio Cattani

Appendix A

Biblioteca Archiginnasio, Bologna, Inv. 37637, Fondo Spec Casali 31.2.2.003, SBN no. UBOE105135.

MANIFESTO.
Bologna lì 19. Marzo 1777.

Desiderando universalmente li Signori Pittori una Denominazione Anatomica stampata in Rame tanto per iscortare alli Studenti la metà del tempo abbisognevole per disegnare la Notomìa o da gessi, o da statue, quanto per procacciarli il comodo di tenerla nelle loro Cartelle così nella Scuola, come nell'Accademia in faccia al Nudo, affine d'illuminarli in occasione di qualche difficoltà, ricorrendo all'Anatomia, e così risolvere in un tratto quel tocco di contorno, o quel muscolo, che tante volte fa stranullare, per non dire perdere il tempo alli Studenti; oltre l'intelligenza, che per mezzo di questa possono procurarsi, imparando il nome appellativo delli Muscoli, o parti esteriori, e così farsi denomizzare dalli Maestri quelle parti, che perfettamente non sonosi disegnate, senza irritare la faticata pazienza col riffare di novo li sudati Lavori non per altro, che dal Maestro, e dallo Scolare viene ignorato il nome appellativo di quella parte, o Muscolo, che correggar si dovea, e così diventare Professori, che tutto eseguiscono a forza di pratica a testone dopo un gran stentato tempo, e così farsi maravigliosi, non che stimabili da que' veri Professori, che tutto hanno impegnato il loro talento non solo nella pratica, quanto nell'intelligenza, e teorica del vero: Massima in vero tanto stimata dalli Greci, quanto trascurata dalli Moderni.

L'infrascritto Incisore s'incoraggisce intraprendere tal Lavoro, che contiene Rami 18., cioè 4. Teste, 2. Mani, 3. Piedi, 4. Nudi, 4. Scheletri, ed il Frontispizio; avvertendo, che le estremità saranno grandi, e tratte dal vero scorticato, li 4. Nudi saranno tratti dalla picciol Statuetta anatomica di Ercole Lelio, e li 4. Scheletri saranno tratti dal vero, e grandi come li suddetti Nudi.

Si promette al Pubblico, che ad ogni Mese sortirà uno di questi Rami stampati in un mezzo foglio di Carta reale, e con decoro, incominciando dal Mese di Maggio.

Li Signori Associati non pagheranno che Soldi 10. moneta di Bologna per ogni Stampa, da sborsarsi subito ricevuta la Stampa.

Che non sarà Associato pagherà ogni Stampa Soldi 14. moneta di Bologna.

Antonio Cattani Piacentino Incisore.

Appendix B

From Antonio Cattani, *Osteografia e miografia della testa, mani, e piedi del corpo umano in misura naturale* (Bologna: Antonio Cattani and Antonio Nerozzi, 1780), at the bottom of tavola I, which is signed and dated "Ant. Cattani f. 16 Novembre 1778."

Si darà alli Signori Associati con il solito metodo di un Paolo per ogni Tavola, che sortirà ogni 40. giorni in circa un corso intero d'Anatomia di tutto il Corpo umano, avvertendo, che chi per tutto il mese di Marzo 1779: non si sarà associato pagherà le già stampate Tavole un Paolo e mezzo per ogniuna, e pagherà le altre avvenire un Paolo solo, se si associerà, dovendo innalterabilmente pagare due Paolo per ogni Tavola, chi Fra queste vorrà scegliere quelle, che più le piaceranno; dovendo i Signori Forastieri incaricarsi delle spese del porto.

Appendix C

Biblioteca Archiginnasio, Bologna, Inv. 37685, Fondo Spec Casali 31.2.2.015, SBN no. UBOE105352.

MANIFESTO.
Bologna 22. Settembre 1779.

Volendo lo sotto[s]critto Incisore continuare l'Opera sua Anatomica avanzata a quest'ora sino alle venti Tavole contenenti l'Osteografia, e Miografi della Testa, Mani e Piedi del Corpo umano in misura naturale co' nomi proprj di tutte le sue parti: Si è determinate d'incidere altre 4 Tavole Anatomiche in 20. fogli pure coll' indice della denominazione di tutte le sue parti, 10. de' quali rappresenteranno il Corpo umano virile denudato da' suoi integumenti di due figure grandi al naturale, una delle quali sarà veduta dalla parte anteriore, e l'altra dalla posteriore, dimostrando negli altri 10. fogli due Scheletri umani di grandezza parimenti naturale veduti anch'essi ne' sopraddetti aspetti.

Ognuna di queste 4. Tavole sarà composta di 5. fogli, che incollati un dopo l'altro formeranno una delle dette figure.

Chi si vorrà associare pagherà il solito Paolo, da sborsarsi ricevendo uno de' detti fogli che sortirà ogni 40. giorni in circa, avvertendo che a carico de' Forastieri saranno le spese del Porto, servendosi l'Autore della Posta ogni qual volta non disponessero qualche Persona costì in Bologna a prenderla.

BOLOGNA. Presso l'Autore
PARMA. Presso li Signori Fratelli Faure.
VENEZIA. Presso il Sig. Giuseppe Vagner.
Antonio Cattani Incisore Piacentino.

bibliography

Albinus 1747. Albinus, Bernhard Siegfried. *Tabulae sceleti et musculorum corporis humani.* Leiden: Johannes and Herman Verbeek, 1747.

Albinus 1753. Albinus, Bernhard Siegfried. *Tabulae ossium humanorum.* Leiden: Johannes and Herman Verbeek, 1753.

Albinus 1754. Albinus, Bernhard Siegfried. *The Explanation of Albinus's Anatomical Figures of the Human Skeleton and Muscles: With an Historical Account of the Work.* London: John and Paul Knapton, 1754.

Al-Gailani 2016. Al-Gailani, Salim. "The 'Ice Age' of Anatomy and Obstetrics: Hand and Eye in the Promotion of Frozen Sections around 1900." *Bulletin of the History of Medicine* 90, no. 4 (Winter 2016): 611–42.

Allen 2021. Allen, James. "Book Excavations." Accessed 13 March 2021. https://www.jamesallenstudio.com/book-excavations.

Amerson 1975. Amerson, Lawrence Price, Jr. "The Problem of the Écorché: A Catalogue Raisonné of Models and Statuettes from the Sixteenth Century and Later Periods." PhD diss., Pennsylvania State University, 1975.

Amerson 1976. Amerson, Lawrence Price, Jr. "Catalogue des moulages conservés à la Royal Academy of Arts de Londres." In *L'écorché: Expérience pédagogique à l'École régionale des beaux arts de Rouen,* edited by Bernard Venot, 13–34. Rouen: Musée des Beaux Arts de Rouen, 1976.

Antommarchi 1823–26. Antommarchi, Francesco. *Planches anatomiques du corps humain executees d'apres les dimensions naturelles.* Paris: Comte de Lasteyrie, 1823–26.

Arfe y Villafañe 1585–87. Arfe y Villafañe, Juan de. *De varia commensuración para la esculptura, y architectura.* Seville: Andrea Pescioni and Juan de León, 1585–87.

Arfe y Villafañe 1675. Arfe y Villafañe, Juan de. *Varia conmensuración para la escultura, y arquitectura.* Madrid: Francisco Sanz, 1675.

Armaroli 1981. Armaroli, Maurizio, ed. *Le cere anatomiche bolognesi del settecento.* Exh. cat. Bologna: Clueb, 1981.

Athens 2006. Athens, Elizabeth. "Knowledge and Authority in Thomas Eakins's *The Agnew Clinic.*" *Burlington Magazine* 148, no. 1240 (July 2006): 482–85.

Auclair 2011. Auclair, Valérie. "Usages et fonctions de la théorie des proportions dans le Livre de Portraiture de Jean Cousin fils." In *Proportions: Science, Musique, Peinture & Architecture; Actes du LIe Colloque International d'études Humanistes, 30 juin–4 juillet 2008,* edited by Sabine Rommevaux, Philippe Vendrix, and Vasco Zara, 373–94. Turnhout: Brepols, 2011.

Audubon 1840–44. Audubon, John James. *The Birds of America, from Drawings Made in the United States and Their Territories.* 7 vols. New York: J. J. Audubon and J. B. Chevalier, 1840–44.

Aymonino 2015. Aymonino, Adriano. "'Nature Perfected': The Theory & Practice of Drawing after the Antique." In Aymonino and Lauder 2015, 15–78.

Aymonino and Lauder 2015. Aymonino, Adriano, and Anne Varick Lauder, eds. *Drawn from the Antique: Artists & the Classical Ideal.* Exh. cat. London: Sir John Soane's Museum, 2015.

Azzolini 2006. Azzolini, Monica. "Leonardo da Vinci's Anatomical Studies in Milan: A Re-Examination of Sites and Sources." In *Visualizing Medieval Medicine and Natural History, 1200–1550,* edited by Jean A. Givens, Karen M. Reeds, and Alain Touwaide, 147–76. Aldershot, UK: Ashgate, 2006.

Baglione 1642. Baglione, Giovanni. *Le vite de' pittori scultori et architetti.* Rome: Andrea Fei, 1642.

Baldinucci 1974–75. Baldinucci, Filippo. *Notizie de' professori del disegno da Cimabue in qua.* 7 vols. Florence: S.P.E.S., 1974–75. First published 1681–1728 by Santi Franchi (Florence).

Ballestriero 2010. Ballestriero, Roberta. "Anatomical Models and Wax Venuses: Art Masterpieces or Scientific Craft Works?" *Journal of Anatomy* 216, no. 2 (February 2010): 223–34.

Bambach, Cox-Rearick, and Goldner 2010. Bambach, Carmen, Janet Cox-Rearick, and George Goldner. *The Drawings of Bronzino.* New York: Metropolitan Museum of Art, 2010.

Bambach 2019. Bambach, Carmen. *Leonardo da Vinci Rediscovered.* 4 vols. New Haven: Yale University Press, 2019.

Barclay 1819–20. Barclay, John. *A Series of Engravings, Representing the Bones of the Human Skeleton with the Skeletons of Some of the Lower Animals.* Edinburgh: Printed for E. Mitchell by Oliver and Boyd, 1819–20.

Barclay 1827. Barclay, John. *Introductory Lectures to a Course of Anatomy.* Edinburgh: Maclachlan and Stewart; London: Baldwin, Cradock, and Joy, 1827.

Baroni and Sellink 2012. Baroni, Alessandra, and Manfred Sellink, eds. *Stradanus (1523–1605): Court Artist of the Medici.* Turnhout: Brepols, 2012.

Barthe 1849. Barthe. "Biographie de Jacques Gamelin, peintre." *Mémoires de la Société des arts et des sciences de Carcassonne,* 1:388–426.

Bartholin 1651. Bartholin, Thomas. *Anatomia, ex Caspari Bartholini parentis institutionibus . . . reformata.* Leiden: Franciscum Hackium, 1651.

Bartholin 1678. Bartholin, Thomas. *De unicornu observationes novae.* Amsterdam: Hendrik Wetstein, 1678.

Bartholin 1686. Bartholin, Thomas. *Anatome ex omnium veterum recentiorumque observationibus.* 5th rev. ed. Leiden: Jacobum Hackium, 1686.

Basevi 2014. Basevi, Benedetta. In *Antico e moderno: Acquisizioni e donazioni per la storia di Bologna (2001–2013),* edited by Angela Mazza, 496–97. Exh. cat. Bologna: Bononia University Press, 2014.

Baudrier 1901. Baudrier, Henri-Louis. *Bibliographie lyonnaise: Recherches sur les imprimeurs, libraires, relieurs et fondeurs de lettres de Lyon au XVIe siècle.* Vol. 5. Lyon: Louis Brun, 1901.

Bayon 1935. Bayon, H. P. "The Authorship of Carlo Ruini's *Anatomia del cavallo.*" *Journal of Comparative Pathology and Therapeutics* 48 (June 1935): 138–48.

Bécane 1775. Bécane, Bertrand. *Abrégé des maladies qui attaquent la substance des os.* Toulouse: Dupleix, 1775.

Bécane 1778. Bécane, Bertrand. *Observations sur les effets du virus cancereux.* Toulouse: J. F. Desclassan, 1778.

Béguin 1996. Béguin, Sylvie. "A propos du chef-d'oeuvre de Bronzino: La déploration sur le Christ mort." In *Les Granvelle et l'Italie au XVIe siècle: Le mécénat d'une famille,* edited by Jacqueline Brunet and Gennaro Toscano, 129–51. Besançon: Cêtre, 1996.

Bell 1793. Bell, John. *The Anatomy of the Bones, Muscles, and Joints.* Edinburgh: Printed for G. Mudie and J. Johnson, 1793.

Bell 1794. Bell, John. *Engravings, Explaining the Anatomy of the Bones, Muscles, and Joints.* Edinburgh: John Paterson, 1794.

Bellini 1987. Bellini, Paolo. *The Illustrated Bartsch: Italian Masters of the Seventeenth Century.* Vol. 47, formerly vol. 21 (part 2). Edited by Walter L. Strauss. Translated by John Spike. New York: Abaris, 1987.

Bellori 1672. Bellori, Giovanni Pietro. *Le vite de' pittori, scultori e architetti moderni.* Rome: Mascardi, 1672.

Berendsen 1961. Berendsen, Olga. "The Italian Sixteenth and Seventeenth Century Catafalques." PhD diss., New York University, 1961.

Berengario 1521. Berengario da Carpi, Jacopo. *Carpi commentaria cum amplissimis additionibus super anatomia Mundini.* Bologna: H. de Benedictis, 1521.

Berengario 1523. Berengario da Carpi, Jacopo. *Isagogae breves perlucidae ac uberrimae in anatomiam humani corporis.* Bologna: B. Hectoris, 1523.

Berengario 1959. Berengario da Carpi, Jacopo. *A Short Introduction to Anatomy: (Isagogae Breves).* Translated and edited by L. R. Lind. Anatomical notes by Paul G. Roofe. Chicago: University of Chicago Press, 1959.

Berkowitz 2015. Berkowitz, Carin. "The Illustrious Anatomist: Authorship, Patronage, and Illustrative Style in Anatomy Folios, 1700–1840." *Bulletin of the History of Medicine* 89, no. 2 (Summer 2015): 171–208.

Bertinatti 1837–39. Bertinatti, Francesco. *Elementi di anatomia fisiologica applicata alle belle arti figurative.* 2 vols. Turin: Pietro Marietti, 1837–39.

Bertoloni Meli 2015. Bertoloni Meli, Domenico. "Visual Representations of Disease: The *Philosophical Transactions* and William Cheselden's *Osteographia.*" *Huntington Library Quarterly* 78, no. 2 (Summer 2015): 157–86.

Bertoloni Meli 2017. Bertoloni Meli, Domenico. *Visualizing Disease: The Art and History of Pathological Illustrations.* Chicago: University of Chicago Press, 2017.

Beukers 1989. Beukers, Harm. "Clinical Teaching in Leiden from Its Beginning until the End of the Eighteenth Century." In *Clinical Teaching, Past and Present,* edited by Harm Beukers and J. Moll, 139–52. Clio Medica, vol. 21. Amsterdam: Rodopi, 1989.

Bewick 1871. Bewick, William. *Life and Letters of William Bewick (Artist).* 2 vols. Edited by Thomas Landseer. London: Hurst and Blackett, 1871.

Bidloo 1685. Bidloo, Govard. *Anatomia humani corporis.* Amsterdam: Widow of J. à Someren, Heirs of J. à Dyk, H. & Widow of T. Boom, 1685.

Bignamini and Postle 1991. Bignamini, Ilaria, and Martin Postle. *The Artist's Model: Its Role in British Art from Lely to Etty.* Exh. cat. Nottingham: Nottingham University Art Gallery, 1991.

Birch 1757. Birch, Thomas. *The History of the Royal Society of London for Improving of Natural Knowledge, from Its First Rise.* Vol. 4. London: A. Millar, 1757.

Black 2007a. Black, Peter, ed. *"My Highest Pleasures": William Hunter's Art Collection.* Exh. cat. Glasgow: The Hunterian, University of Glasgow, 2007.

Black 2007b. Black, Peter. "Taste and the Anatomist: William Hunter's Art Collection; His Paintings, and the Representation of Art in His Library." In Black 2007a, 63–100.

Blanc 2008. Blanc, Jan. *Peindre et penser la peinture au XVIIe siècle: La théorie de l'art de Samuel van Hoogstraten.* Bern: Peter Lang, 2008.

Blanchard and Lea 1855. *Catalogue of Blanchard and Lea's Medical and Surgical Publications.* Philadelphia: Blanchard and Lea, 1855.

Bloemendal 2003. Bloemendal, Jan. *Spiegel van het dagelijks leven?: Latijnse school en toneel in de noordelijke Nederlanden in de zestiende en de zeventiende eeuw.* Hilversum: Verloren, 2003.

Bober and Rubenstein 2010. Bober, Phyllis Pray, and Ruth Rubenstein. *Renaissance Artists & Antique Sculpture: A Handbook of Sources.* 2nd ed. London: H. Miller, 2010.

Bonaveri ca. 1685–90. Bonaveri, Domenico. *Notomie di Titiano.* [Bologna]: n.p., ca. 1685–90.

Bolten 1985. Bolten, Jaap. *Method and Practice: Dutch and Flemish Drawing Books, 1600–1750.* Translated by Alexander Dietz. Landau: PVA, 1985.

Bosio 1800–1801. Bosio, Jean. *Traité élémentaire des règles du dessin.* Paris: Bosio, 1800–1801.

Bosse and Saint-Igny 1629. Bosse, Abraham, and Jean de Saint-Igny. *Le jardin de la noblesse françoise dans lequel ce peut ceuillir leur manierre de vettements.* Paris: François Langlois dit Chartres, 1629.

Bouchardon 1741. Bouchardon, Edme. *L'anatomie nécessaire pour l'usage du dessein.* Paris: Huquier, [1741].

Bouchardon [1787–89]. Bouchardon, Edme. *L'anatomie nécessaire pour l'usage du dessein.* Paris: Chéreau & Joubert, [1787–89].

Bowen and Imhof 2008. Bowen, Karen Lee, and Dirk Imhof. *Christopher Plantin and Engraved Book Illustrations in Sixteenth-Century Europe.* Cambridge: Cambridge University Press, 2008.

Brey 1984. Brey, William. *John Carbutt, on the Frontiers of Photography.* Cherry Hill, NJ: Willowdale Press, 1984.

Bridson and White 1990. Bridson, Gavin D. R., and James J. White. *Plant, Animal and Anatomical Illustration in Art and Science: A Bibliographical Guide from the 16th Century to the Present Day.* Winchester, UK: St. Paul's Bibliographies, 1990.

Brink 2012. Brink, Sonja. "Ein Teilskelettierter Kadaver Passarottis in Düsseldorf, das Grabmal der Elena Savelli von Jacopo Del Duca und der Anatomieatlas des Girolamo Genga." *Rivista d'arte* 2 (2012): 173–98.

Brisbane 1769. Brisbane, John. *The Anatomy of Painting; or, A Short and Easy Introduction to Anatomy: Being a New Edition, on a Smaller Scale, of Six Tables of Albinus, with Their Linear Figures; Also, a New Translation of Albinus's History of That Work, and of His Index to the Six Tables; To Which Are Added the Anatomy of Celsus, with Notes, and the Physiology of Cicero; With an Introduction, Giving a Short View of Picturesque Anatomy.* London: George Scott, 1769.

Brock 1983. Brock, C. Helen, ed. *William Hunter 1718–1783: A Memoir by Samuel Foart Simmons and John Hunter.* Glasgow: University of Glasgow Press, 1983.

Brown, M. 2013. Brown, Meg. "Flip, Flap, and Crack: The Conservation and Exhibition of 400+ Years of Flap Anatomies." *Book and Paper Group Annual* 32 (2013): 6–14.

Brown, Sienna 2010. Brown, Sienna. "The Lithographs of Robert Rauschenberg." PhD diss., Emory University, 2010.

Brown, Sue 2009. Brown, Sue. *Joseph Severn, A Life: The Rewards of Friendship.* Oxford: Oxford University Press, 2009.

Brownell 1879. Brownell, William C. "The Art Schools of Philadelphia." *Scribner's Monthly* 18, no. 5 (September 1879): 737–50.

Brusati 1995. Brusati, Celeste. *Artifice and Illusion: The Art and Writing of Samuel van Hoogstraten.* Chicago: University of Chicago Press, 1995.

Buckley 2013. Buckley, Cali. "Johann Remmelin's *Catoptrum Microcosmicum* and the End of an Era." *Bodleian Library Record* 26, no. 1 (April 2013): 18–35.

Burris 1966. Burris, George P. "Estienne's *De Dissectione* (1545), an Example of Sixteenth Century Anatomical Illustration." *Bios* 37, no. 4 (December 1966): 147–56.

Bylebyl 2008. Bylebyl, Jerome J. "Colombo, Realdo." In *Complete Dictionary of Scientific Biography,* 3:354–57. Detroit: Charles Scribner's Sons, 2008. http://www.encyclopedia.com/science/dictionaries-thesauruses-pictures-and-press-releases/colombo-realdo.

Caldwell 2006. Caldwell, Janis McLarren. "The Strange Death of the Animated Cadaver: Changing Conventions in Nineteenth-Century British Anatomical Illustration." *Literature and Medicine* 25, no. 2 (Fall 2006): 325–57.

Callen 2018. Callen, Anthea. *Looking at Men: Anatomy, Masculinity, and the Modern Male Body.* New Haven: Yale University Press, 2018.

Calvesi et al. 1987. Calvesi, Maurizio, Vincenzo di Flavio, Andrea di Nicola, Paola Ferraris, and Antonio Vannugli, eds. *Carlo Cesi: Pittore e incisore del Seicento tra ambiente cortonesco e classicismo marattiano 1622–1682.* Rieti: Secit, 1987.

Campbell and Flis 2018. Campbell, Mungo, and Nathan Flis, eds. *William Hunter and the Anatomy of the Modern Museum.* Exh. cat. New Haven: Yale Center for British Art, 2018.

Campi 1585. Campi, Antonio. *Cremona fedelissima città.* Cremona: In casa dell'auttore, per Hippolito Tromba & Hercoliano Battoli, 1585.

Campori 1870. Campori, Giuseppe. *Raccolta di cataloghi ed inventarii inedita.* Modena: Carlo Vincenzi, 1870.

Canadelli 2011. Canadelli, Elena. "'Scientific Peep Show': The Human Body in Contemporary Science Museums." *Nuncius* 26, no. 1 (2011): 159–84.

Canalis 2018. Canalis, Rinaldo Fernando. "Vesalius's Methods in the Production of the *Fabrica* with Emphasis on the Neuroanatomy Chapters." In Canalis and Ciavolella 2018, 131–65.

Canalis and Ciavolella 2018. Canalis, Rinaldo Fernando, and Massimo Ciavolella, eds. *Andreas Vesalius and the "Fabrica" in the Age of Printing: Art, Anatomy, and Printing in the Italian Renaissance.* Cursor Mundi, vol. 33. Turnhout: Brepols, 2018.

Canani 1925. Canani, Giovanni Battista. *Musculorum humani corporis picturata dissectio (Ferrara 1541?).* Facsimile annotated by Harvey Cushing and Edward C. Streeter. Monumenta Medica, vol. 4. Florence: R. Lier, 1925.

Carducho 1633. Carducho, Vincente. *Dialogos de la pintura.* Madrid: Francisco Martinez, 1633.

Carlino 1999a. Carlino, Andrea. *Books of the Body: Anatomical Ritual and Renaissance Learning.* Translated by John Tedeschi and Anne C. Tedeschi. Chicago: University of Chicago Press, 1999.

Carlino 1999b. Carlino, Andrea. *Paper Bodies: A Catalogue of Anatomical Fugitive Sheets, 1538–1687.* Translated by Noga Arikha. London: Wellcome Institute for the History of Medicine, 1999.

Carlino 2004. Carlino, Andrea. "Vesalio e la cultura visiva delle anatomie a stampa del Rinascimento." In Olmi 2004, 75–91.

Carlisle 1807. Carlisle, Anthony. "The Artist: No. XVII." In *The Artist: A Collection of Essays, Relative to Painting, Poetry, Sculpture, Architecture, the Drama, Discoveries of Science, and Various Other Subjects,* edited by Prince Hoare, 1–12. London: John Murray, John White, W. Miller, J. Harding, and J. Asperne, 1807.

Carradori 1802. Carradori, Francesco. *Istruzione elementare per gli studiosi della scultura.* Florence: n.p., 1802.

Carradori 1979. Carradori, Francesco. *Istruzione elementare per gli studiosi della scultura (1802).* Edited by Gianni Carlo Sciolla. Treviso: Canova, 1979.

Carradori 2002. Carradori, Francesco. *Elementary Instructions for Students of Sculpture.* Translated by Matti Kalevi Auvinen. Los Angeles: J. Paul Getty Museum, 2002.

Casseri 1600–1601. Casseri, Giulio. *De vocis auditusque organis historia anatomica.* Ferrara: Victorius Baldinus, 1600–1601.

Casseri 1609. Casseri, Giulio. *Pentaestheseion, hoc est de quinque sensibus liber.* Venice: Nicolo Misserino, 1609.

Casseri 1627. Casseri, Giulio. *Tabulae anatomicae.* Venice: Evangelista Deuchinus, 1627.

Cattani 1780. Cattani, Antonio. *Osteografia e miografia della testa, mani, e piedi del corpo umano in misura naturale con il catalogo de' nomi proprj di tutte le sue parti incise in rame da Antonio Cattani incisore Piacentino.* Bologna: Antonio Cattani and Antonio Nerozzi, 1780.

Cattani 1787. Cattani, Antonio. *Dell'arte ostetrizia fogli periodici con rami colorati: Trimestre primo.* Bologna: San Tommaso d'Aquino, 1787.

Caturla 1968–69. Caturla, María Luisa. "Documentos en torno a Vicencio Carducho." *Arte español* 26, no. 3 (1968–69): 145–221.

Cavalcanti 1634. Cavalcanti, Andrea. *Esequie del serenissimo principe Francesco, celebrate in Fiorenza dal serenissimo Ferdinando II, granduca di Toscana, suo fratello, nell'insigne collegiata di S. Lorenzo il dì 30 d'agosto 1634.* Florence: Giovanni Batista Landini, 1634.

Cavanagh 1983. Cavanagh, G. S. T. "A New View of the Vesalian Landscape." *Medical History* 27, no. 1 (1983): 77–79.

Cazes 2004. Cazes, Hélène. "Théâtres imaginaires du livre et de l'anatomie: *La Dissection des parties du corps humain,* Charles Estienne, 1545–1546." *Fabula/Les colloques, Fiction du savoir à la Renaissance,* 11 June 2004. http://www.fabula.org/colloques/document103.php.

Cazort 1996. Cazort, Mimi. "The Theatre of the Body." In Cazort, Kornell, and Roberts 1996, 11–42.

Cazort, Kornell, and Roberts 1996. Cazort, Mimi, Monique Kornell, and K. B. Roberts. *The Ingenious Machine of Nature: Four Centuries of Art and Anatomy.* Exh. cat. Ottawa: National Gallery of Canada, 1996.

Cellini 1901. Cellini, Benvenuto. *Vita di Benvenuto Cellini: Testo critico con introduzione e note storiche.* Edited by Orazio Bacci. Florence: G. C. Sansoni, 1901.

Cellini 1979. Cellini, Benvenuto. *Sopra i principii e 'l modo d'imparare l'arte del disegno.* In *Trattati d'arte del Cinquecento.* Edited by Paolo Barocchi, 8: 1933–40. Turin: Giulio Einaudi, 1979.

Cesi 1679. Cesi, Carlo. *Cognitione de muscoli del corpo humano per il disegno.* Rome: Francesco Collignon, 1679.

Chatelain 2018. Chatelain, Antoine. "À l'ombre de Robert et Fragonard, catalogue des dessins français de la seconde moitié du XVIIIe siècle du musée de Besançon." MA thesis, École du Louvre, 2018.

Cheselden 1713. Cheselden, William. *The Anatomy of the Humane Body.* London: N. Cliff, D. Jackson, and W. Innys, 1713.

Cheselden 1722. Cheselden, William. *The Anatomy of the Humane Body.* 2nd ed. London: S. Collins, 1722.

Cheselden 1733. Cheselden, William. *Osteographia; or, the Anatomy of the Bones.* London: n.p., 1733.

Cheselden 1740. Cheselden, William. *The Anatomy of the Human Body.* 5th ed. London: William Bowyer, 1740.

Choulant 1945. Choulant, Ludwig. *History and Bibliography of Anatomic Illustration.* Rev. ed. Translated by Mortimer Frank. New York: Schuman's, 1945.

Christie's, London 2020. *Valuable Books and Manuscripts.* Christie's, London, 9 December 2020.

Christie's, New York 2007. *Anatomy as Art: The Dean Edell Collection.* Christie's, New York, 5 October 2007.

Ciardi 1984. Ciardi, Roberto Paolo. "Michelangelo come Galeno: Un'ipotesi iconologica." In *Studi in onore di Giulio Carlo Argan,* edited by Silvana Macchioni and Bianca Tavassi La Greca, 1:173–81. Rome: Multigrafica editrice, 1984.

Ciardi 2003. Ciardi, Roberto Paolo. "Le corps transparent: Modalités d'iconographie anatomique chez Michel-Ange." In *L'anatomie chez Michel-Ange de la réalité à l'idéalité,* edited by Chiara Rabbi-Bernard, 51–75. Paris: Hermann, 2003.

Ciardi 2009. Ciardi, Roberto Paolo. "Anatomia: Esercizio e mito." In *La bella anatomia: Il disegno del corpo fra arte e scienza nel Rinascimento,* edited by Andrea Carlino, Roberto P. Ciardi, and Annamari Petrioli Tofani, 25–39. Milan: Silvano, 2009.

Ciardi and Tongiorgi Tomasi 1984. Ciardi, Roberto Paolo, and Lucia Tongiorgi Tomasi. *Immagini anatomiche e naturalistiche nei disegni degli Uffizi: Secc. XVI e XVII.* Exh. cat. Gabinetto disegni e stampe degli Uffizi, vol. 60. Florence: Leo S. Olschki, 1984.

Cicognara 1821. Cicognara, Leopoldo. *Catalogo ragionato dei libri d'arte e d'antichità posseduti dal Conte Cicognara.* 2 vols. in one. Pisa: Niccolò Capurro, 1821.

Clark 1981. Clark, Harry. "Foiling the Pirates: The Preparation and Publication of Andreas Vesalius's *De humani corporis fabrica.*" *Library Quarterly* 51, no. 3 (July 1981): 301–11.

Clayton and Philo 2012. Clayton, Martin, and Ron Philo. *Leonardo da Vinci: Anatomist.* London: Royal Collection Publications, 2012.

Cloquet 1825. Cloquet, Jules. *Manuel d'anatomie descriptive du corps humain: Représentée en planches lithographiées.* Paris: Béchet Jeune, 1825.

Coakley 1992. Coakley, Davis. "Anatomy and Art: Irish Dimensions." In Kennedy and Coakley 1992, 57–87.

Cobb 2006. Cobb, Matthew. *Generation: The Seventeenth-Century Scientists Who Unraveled the Secrets of Sex, Life, and Growth.* New York: Bloomsbury, 2006.

Cobb 2012. Cobb, Matthew. "An Amazing 10 Years: The Discovery of Egg and Sperm in the 17th Century." *Reproduction in Domestic Animals* 47, suppl. 4 (2012): 2–6.

Coiter 1955. Coiter, Volcher. *Lectiones Gabrielis Fallopii de partibus similaribus humani corporis, ex diversis exemplaribus a Volchero Coiter . . . collectae* [1575]. In *Opuscula selecta Neerlandicorum de arte medica.* Edited and translated by Araham Shierbeek and B. W. Th. Nuyens, 18:162–259. Amsterdam: Sumptibus Societatis, 1955.

Cole 1949. Cole, F. J. *A History of Comparative Anatomy: From Aristotle to the Eighteenth Century.* London: Macmillan Press, 1949.

Collins 1978. Collins, Marcia. *The Dance of Death in Book Illustration.* Exh. cat. Columbia: University of Missouri, 1978.

Colombo 1559. Colombo, Realdo. *De re anatomica libri XV.* Venice: Nicolò Bevilacqua, 1559.

Colombo 2014. Colombo, Realdo. *De re anatomica libri XV = Anatomia.* Translated and edited by Gianluigi Baldo. Paris: Les Belles lettres, 2014.

Comar 2008a. Comar, Philippe. "Une leçon d'anatomie a l'Ecole des Beaux-Arts." In Comar 2008b, 19–65.

Comar 2008b. Comar, Philippe, ed. *Figures du corps: Une leçon d'anatomie à l'École des Beaux-Arts.* Exh. cat. Paris: Beaux-Arts de Paris, 2008.

Comparini 1996. Comparini, Leonetto. "L'opera anatomica di Paolo Mascagni dall'*Istoria dei Vasi Linfatici* alla *Grande Anatomia.*" In *La scienza illuminata: Paolo Mascagni nel suo tempo (1755–1815),* edited by Francesca Vannozzi, 47–57. Exh. cat. Siena: Nuova immagine, 1996.

Condivi 1553. Condivi, Ascanio. *Vita di Michelangelo Buonarroti.* Rome: Antonio Blado, 1553.

Conring 1687. Conring, Hermann. *In universam artem medicam singulasque ejus partes.* Helmstedt: Georg Wolfgang Hamm, 1687.

Cook 2002. Cook, Harold. "Time's Bodies: Crafting the Preparation and Preservation of Naturalia." In *Merchants and Marvels: Commerce, Science, and Art in Early Modern Europe,* edited by

Pamela H. Smith and Paula Findlen, 237–47. New York: Routledge, 2002.

Cook 2007. Cook, Harold J. *Matters of Exchange: Commerce, Medicine, and Science in the Dutch Golden Age.* New Haven: Yale University Press, 2007.

Cope 1953. Cope, Sir Zachary. *William Cheselden, 1688–1752.* Edinburgh: E. & S. Livingstone, 1953.

Coppola 1957. Coppola, Edward D. "The Discovery of the Pulmonary Circulation: A New Approach." *Bulletin of the History of Medicine* 31 (1957): 44–77.

Coquery 2002. Coquery, Emmanuel. "L'anatomie d'une Académie." In *L'idéal classique: Les échanges artistiques entre Rome et Paris au temps de Bellori (1640–1700),* edited by Olivier Bonfait and Anne-Lise Desmas, 140–60. Paris: Somogy, 2002.

Coquery 2013. Coquery, Emmanuel. *Charles Errard, ca. 1601–1689: La noblesse du décor.* Paris: Arthena, 2013.

Cousin 1595. Cousin, Jean (the Younger). *Livre de pourtraicture.* Paris: David Leclerc, 1595.

Cousin 1671. Cousin, Jean (the Younger). *Livre de pourtraicture; La vraye science de la pourtraicture descrite et demontrée par maistre Jean Cousin, peintre & geometrien tres-excellent.* Paris: Guillaume le Bé, 1671.

Cousin ca. 1685. Cousin, Jean (the Younger). *L'art de dessiner de Maistre Jean Cousin.* Paris: François Jollain, ca. 1685.

Cowper 1694. Cowper, William. *Myotomia reformata; or, A New Administration of All the Muscles of Humane Bodies.* London: Samuel Smith and Benjamin Walford, 1694.

Cowper 1698. Cowper, William. *The Anatomy of Humane Bodies.* Oxford: Samuel Smith and Benjamin Walford, 1698.

Cowper 1724. Cowper, William. *Myotomia reformata; or, An Anatomical Treatise on the Muscles of the Human Body, Illustrated with Figures after the Life.* 2nd ed. London: Robert Knaplock, William and John Innys, and Jacob Tonson, 1724.

Crary 1990. Crary, Jonathan. *Techniques of the Observer: On Vision and Modernity in the Nineteenth Century.* Cambridge, MA: MIT Press, 1990.

Crespi 1769. Crespi, Luigi. *Vite de' pittori bolognesi non descritte nella Felsina pittrice.* Rome: Marco Pagliarini, 1769.

Croke 1992. Croke, Fionnuala. "Bernard Siegfried Albinus and His Ideal Man." In Kennedy and Coakley 1992, 37–55.

Crutchfield and Hobbs 2013. Crutchfield, Jean, and Robert Hobbs. *Seen/Unseen: Tavares Strachan.* Hong Kong: ArtAsiaPacific, 2013.

Cuir 2009. Cuir, Raphaël. *The Development of the Study of Anatomy from the Renaissance to Cartesianism: Da Carpi, Vesalius, Estienne, Bidloo.* Lewiston, NY: Edwin Mellen Press, 2009.

Cummings 1963. Cummings, Frederick. "B. R. Haydon and His School." *Journal of the Warburg and Courtauld Institutes* 26 (1963): 367–80.

Cunningham, A. 1997. Cunningham, Andrew. *The Anatomical Renaissance: The Resurrection of the Anatomical Projects of the Ancients.* Aldershot, UK: Scolar Press, 1997.

Cunningham, A. 2016. Cunningham, Andrew. *The Anatomist Anatomis'd: An Experimental Discipline in Enlightenment Europe.* London: Routledge, 2016.

Cunningham, D. J. [1909]. Cunningham, D. J. *Stereoscopic Studies of Anatomy.* Rev. ed. 10 vols. Edited by David Waterston, Professor M. H. Cryer, and Frederick E. Neres. New York: Imperial Publishing Company, [1909].

Cunsolo 2005. Cunsolo, Elisabetta. "Tra metodo e arte: Giulio Casserio e la rinascita della trattatistica anatomica all'inizio del Seicento." In Pigozzi 2005, 90–110.

Cushing 1937. Cushing, Harvey. "Ercole Lelli and his Écorché." *Yale Journal of Biology and Medicine* 9, no. 3 (1937): 199–213.

Cushing 1962. Cushing, Harvey. *A Bio-Bibliography of Andreas Vesalius.* 2nd ed. Hamden, CT: Archon, 1962.

Cuzin, Gaborit, and Pasquier 2000. Cuzin, Jean-Pierre, Jean-René Gaborit, and Alain Pasquier. *D'après l'antique.* Exh. cat. Paris: Réunion des Musées Nationaux, 2000.

Dackerman 2011a. Dackerman, Susan. "Introduction: Prints as Instruments." In Dackerman 2011b, 19–35.

Dackerman 2011b. Dackerman, Susan, ed. *Prints and the Pursuit of Knowledge in Early Modern Europe.* Exh. cat. Cambridge, MA: Harvard Art Museums, 2011.

Dacome 2017. Dacome, Lucia. *Malleable Anatomies: Models, Makers, and Material Culture in Eighteenth-Century Italy.* Oxford: Oxford University Press, 2017.

Daly Davis 1982. Daly Davis, Margaret. "Beyond the 'Primo libro' of Vincenzo Danti's *Trattato delle perfette proporzioni.*" *Mitteilungen des Kunsthistorischen Institutes in Florenz* 26, no. 1 (1982): 63–84.

Darlington 1990. Darlington, Anne. "The Royal Academy of Arts and Its Anatomical Teachings: With an Examination of the Art-Anatomy Practices during the Eighteenth and Early Nineteenth Centuries in Britain." PhD diss., University of London, 1990.

Daston and Galison 1992. Daston, Lorraine, and Peter Galison. "The Image of Objectivity." *Representations* 40 (1992): 81–128.

De Caro 2018. De Caro, Raffaele. "Anatomy in Colour: Girolamo Fabrici d'Acquapendente (c. 1537–1619)." In Canalis and Ciavolella 2018, 201–23.

De Feyfer 1914. De Feyfer, F. M. G. "Die Schriften des Andreas Vesalius." *Janus* 19 (1914): 435–507.

De Graaf 1686. De Graaf, Reinier. *Alle de wercken, so in de ontleedkunde, als andere deelen der medicyne.* Amsterdam: Abraham Abrahamse, 1686.

De Lairesse 1701. De Lairesse, Gérard. *Grondlegginge ter Teekenkonst.* Amsterdam: Willem de Coup, 1701.

De Lairesse 1712. De Lairesse, Gérard. *Het Groot Schilderboek.* 2 vols. Amsterdam: David Mortier, 1712.

Del Medico 1811. Del Medico, Giuseppe. *Anatomia per uso dei pittori e scultori.* Rome: Vincenzo Poggioli, 1811.

Delsenbach 1733. Delsenbach, Johann Adam. *Kurtzer Begriff der Anatomie, worinnen hauptsächlich die nöthigsten Stücke der Osteologie und Myologie in XIX kupfer Tabellen enthalten.* Nuremberg: Lorenz Bieling, 1733.

de Piles 1708. de Piles, Roger. *Cours de peinture par principes.* Paris: Jacques Estienne, 1708.

Desmas et al. 2016. Desmas, Anne-Lise, Édouard Kopp, Guilhem Scherf, and Juliette Trey. *Bouchardon: Royal Artist of the Enlightenment.* Exh. cat. Los Angeles: J. Paul Getty Museum, 2016.

De Vesme and Massar 1971. De Vesme, Alexandre Baudi, and Phyllis Dearborn Massar. *Stefano della Bella: Catalogue raisonné.* 2 vols. New York: Collectors Editions, 1971.

De Vries 1998. De Vries, Lyckle. *Gerard de Lairesse: An Artist between Stage and Studio.* Amsterdam: Amsterdam University Press, 1998.

Douglas 1707. Douglas, James. *Myographiae comparatae specimen; or, A Comparative Description of All the Muscles in a Man and in a Quadruped.* London: W. B., 1707.

Douglas 1735. Douglas, John. *Animadversions on a Late Pompous Book, Intituled, Osteographia; or, The Anatomy of the Bones by William Cheselden Esq.* London: John Douglas, 1735.

Doyle 1999. Doyle, Jennifer. "Sex, Scandal and Thomas Eakins's *The Gross Clinic.*" *Representations* 68 (Fall 1999): 1–33.

Drake 1707. Drake, James. *Anthropologia nova; or, a New System of Anatomy.* London: Samuel Smith and Benjamin Walford, 1707.

Dryander 1541. Dryander, Joannes. *Anatomia Mundini.* Marburg: Christian Egenolff, 1541.

Dufresnoy 1668. Dufresnoy, Charles. *L'art de peinture.* Paris: Nicolas L'Anglois, 1668.

Duhme 1980. Duhme, Lüdike. *Die "Tabulae Anatomicae" des Pietro Berrettini da Cortona.* Arbeiten der Forschungsstelle des Instituts für Geschichte der Medizin der Universität zu Köln, vol. 18. Cologne: Forschungsstelle Robert-Koch-Straße, 1980.

Dumaître 1982. Dumaître, Paule, in collaboration with Janine Samion-Contet. *La curieuse destinée des planches anatomiques de Gérard de Lairesse, peintre en Holland: Lairesse, Bidloo, Cowper.* Amsterdam: Rodopi, 1982.

Düring, Didi-Huberman, and Poggesi 1999. Düring, Monika von, Georges Didi-Huberman, and Marta Poggesi. *Encyclopedia Anatomica: Museo La Speccola Florence.* Cologne: Taschen, 1999.

Dürr 1879. Dürr, Alphons. *Adam Friedrich Oeser: Ein Beitrag zur Kunstgeschichte des 18. Jahrhunderts.* Leipzig: Alphons Dürr, 1879.

Duval and Cuyer 1898. Duval, Mathias, and Édouard Cuyer. *Histoire de l'anatomie plastique: Les maîtres, les livres et les écorchés.* Paris: Société française d'éditions d'art, 1898.

***Edinburgh Medical and Surgical Journal* 1854.** Unsigned review of *Surgical Anatomy,* by Joseph Maclise, first four fascicules of 2nd ed. *Edinburgh Medical and Surgical Journal* 81, no. 198 (1854): 580–81.

Egmond 2017. Egmond, Florike. *Eye for Detail: Images of Plants and Animals in Art and Science, 1500–1630.* London: Reaktion, 2017.

Engelskirchen 2014. Engelskirchen, Maria. "Jean Cousin." In Heilmann et al. 2014, 66–67, cat. no. 3.1.

Ercolani 1873. Ercolani, Giovanni Battista. *Carlo Ruini: Curiosità storiche e bibliografiche intorno alla scoperta della circolazione del sangue.* Bologna: Nicola Zanichelli, 1873.

Esterquest 1962. Esterquest, Ralph T. "Library Notes: The Lelli Ecorche." *Harvard Medical Alumni Bulletin* 36, no. 4 (Spring 1962): 11.

Estienne and La Rivière 1545. Estienne, Charles, and Estienne de la Rivière. *De dissectione partium corporis humani libri tres, à Carolo Stephano, doctore Medico, editi: Unà cum figuris, & incisionum declarationibus, à Stephano Riverio Chirurgo compositis.* Paris: Simon de Colines, 1545.

Fabricius ab' Acquapendente 1600. Fabricius ab' Acquapendente, Hieronymus. *De visione, voce, auditu.* Venice: Francesco Bolzetta, 1600.

Fabricius ab' Acquapendente 1933. Fabricius ab' Acquapendente, Hieronymus. *De venarum ostiolis 1603 of Hieronymus Fabricius of Aquapendente (1533?–1619).* Facsimile edition with introduction, translation, and notes by Kenneth J. Franklin. Springfield, IL: Charles C. Thomas, 1933.

Fanti and Chiossi 1984. Fanti, Mario, and Rosa Chiossi. *Richerche su Carlo Ruini (1530–1598).* Bologna: Li Causi, 1984.

Fantuzzi 1781–94. Fantuzzi, Giovanni. *Notizie degli scrittori bolognesi.* 9 vols. Bologna: San Tommaso D'Aquino, 1781–94.

Fau 1845. Fau, Julien. *Anatomie des formes extérieures du corps humain, à l'usage des peintres et des sculpteurs.* Paris: Méquignon-Marvis, 1845.

Felfe 2013. Felfe, Robert. "Naer het leven: Eine sprachliche Formel zwischen bildgenerieren Ubertrangungsvorgängen und ästhetischer Vermittlung." In *Ad Fontes! Niederländische Kunst des 17. Jahrhunderts in Quellen,* edited by Claudia Fritzche, Karin Leonhard, and Gregor J. M. Weber, 165–95. Petersberg: Michael Imhof, 2013.

Felsen 2014. Felsen, Sidney B. Interview by James McElhinney, 9 November 2013 and 13 December 2013, New York, New York. *Rauschenberg Oral History Project: The Reminiscences of Sidney B. Felsen.* Columbia Center for Oral History, Columbia University, 2014. https://www.rauschenbergfoundation.org/artist/oral-history/sidney-felsen.

Fend 2017. Fend, Mechthild. *Fleshing Out Surfaces: Skin in French Art and Medicine, 1650–1850.* Manchester: Manchester University Press, 2017.

Fend 2019. Fend, Mechthild. "Drawing the Cadaver *ad vivum:* Gérard de Lairesse's Illustrations for Govard Bidloo's *Anatomia Humani Corporis.*" In *Ad vivum? Visual Materials and the Vocabulary of Life-Likeness in Europe before 1800,* edited by Thomas Balfe, Joanna Woodall, and Claus Zittel, 294–327. Leiden: Brill, 2019.

Ferraris and Vannugli 1987. Ferraris, Paola, and Antonio Vannugli. "La fase Romana di Carlo Cesi." In Calvesi et al. 1987, 21–34.

Findlen 1994. Findlen, Paula. *Possessing Nature: Museums, Collecting, and Scientific Culture in Early Modern Italy.* Berkeley: University of California Press, 1994.

Fine 1984. Fine, Ruth E. *Gemini G.E.L.: Art and Collaboration.* Exh. cat. Washington, D.C.: National Gallery of Art, 1984.

Fine 1997. Fine, Ruth E. "Writing on Rocks, Rubbing on Silk, Layering on Paper." In *Robert Rauschenberg: A Retrospective,* edited by Walter Hopps and Susan Davidson, 376–89. Exh. cat. New York: Solomon R. Guggenheim Museum, 1997.

Firmin-Didot 1872. Firmin-Didot, Ambroise. *Étude sur Jean Cousin: Suivie de notices sur Jean Leclerc et Pierre Woeirot.* Paris: Ambroise Firmin-Didot, 1872.

Flaxman 1829. Flaxman, John. *Lectures on Sculpture.* London: J. Murray, 1829.

Flis 2018. Flis, Nathan. "Skeletons in Hunter's Closet: James Douglas and the Fashioning of William Hunter." In Campbell and Flis 2018, 49–71.

Forlani Tempesti 1973. Forlani Tempesti, Anna. *Mostra di incisioni di Stefano della Bella.* Exh. cat. Gabinetto disegni e stampe degli Uffizi, vol. 39. Florence: Leo S. Olschki, 1973.

Fortgens 1958. Fortgens, H. W. *Schola latina: Uit het verleden van ons voorbereidend hoger onderwijs.* Zwolle: Tjeenk Willink, 1958.

Foster 1970. Foster, Edward A. *Robert Rauschenberg: Prints 1948/1970.* Exh. cat. Minneapolis: Minneapolis Institute of Arts, 1970.

French 1985. French, R. K. "Berengario da Carpi and the Use of Commentary in Anatomical Teaching." In *The Medical Renaissance of the Sixteenth Century,* edited by A. Wear, R. K. French, and I. M. Lonie, 42–74. Cambridge: Cambridge University Press, 1985.

Fuchs 1542. Fuchs, Leonhart. *De historia stirpium commentarii insignes, maximis impensis et vigiliis elaborati, adjectis earundem vivis plusquam quingentis imaginibus.* Basel: Michael Isingrin, 1542.

Gaeta Bertelà and Petrioli Tofani 1969. Gaeta Bertelà, Giovanna, and Annamaria Petrioli Tofani. *Feste e apparati medicei da Cosimo I a Cosimo III.* Exh. cat. Gabinetto disegni e stampe degli Uffizi, vol. 31. Florence: Leo S. Olschki, 1969.

Gale and Dewdney 2010. Gale, Peggy, and Christopher Dewdney. *Shelagh Keeley.* Exh. cat. Oshawa: Robert McLaughlin Gallery, 2010.

Galet 1844. Galet, Jules. *Le corps de l'homme: Traité complet d'anatomie et de physiologie humaines.* Vol. 4. Paris: Galet, Mansut fils, and Billet, 1844.

Galison 1998. Galison, Peter. "Judgment against Objectivity." In *Picturing Science, Producing Art,* edited by Caroline A. Jones and Peter Galison, 327–59. New York: Routledge, 1998.

Gallay 2019. Gallay, Antoine. "When Sight Penetrates the Body: The Use and Promotion of Stereoscopic Radiography in Britain, 1896–1918." In Graciano 2019, 105–28.

Gamberini 2005. Gamberini, Giliola. "Raffigurare per identificare il disegno anatomico a Bologna nei primi anni del Cinquecento." In Pigozzi 2005, 63–90.

Gamelin 1779. Gamelin, Jacques. *Nouveau recueil d'ostéologie et de myologie, dessiné d'après nature . . . pour l'utilité des sciences et des arts.* Toulouse: J. F. Desclassan, 1779.

Gamer 2018. Gamer, Meredith. "Scalpel to Burin: A Material History of William Hunter's *Anatomy of the Human Gravid Uterus.*" In Campbell and Flis 2018, 109–25.

Gamer 2019. Gamer, Meredith. "The *Smugglerius,* Re-Viewed," *Sculpture Journal* 28, no. 3 (2019): 331–44.

Gasparotto 2018. Gasparotto, Davide. "The Renaissance Nude and the Study of the Antique." In *The Renaissance Nude,* edited by Thomas Kren with Jill Burke and Stephen J. Campbell, 247–59. Exh. cat. Los Angeles: J. Paul Getty Museum, 2018.

Gautier Dagoty, A. É. 1773. Gautier Dagoty, Arnauld Éloi. *Cours complet d'anatomie: Peint et gravé en couleurs naturelles.* Nancy: Jean-Baptiste-Hyacinthe Leclerc, 1773.

Gautier Dagoty, J. F. [1752]. Gautier Dagoty, Jacques Fabien. *Anatomie générale des viscères en situation, de grandeur et couleur naturelle.* Paris: s.n., [1752].

Gautier Dagoty, J. F. 1773. Gautier Dagoty, Jacques Fabien. *Anatomie des parties de la génération de l'homme et de la femme . . . jointe a l'angéologie de tout le corps humain, et a ce qui concerne la grossesse et les accouchemens.* Paris: J. B. Brunet & Demonville, 1773.

Geminus 1545. Geminus, Thomas. *Compendiosa totius anatomie delineatio, aere exarate.* London: Thomas Geminus, 1545.

Geminus 1959. Geminus, Thomas. *Compendiosa totius anatomie delineatio.* Facsimile of the first English edition of 1553 translated by Nicholas Udall, with an introduction by C. D. O'Malley. London: Dawson's of Pall Mall, 1959.

Genga 1672. Genga, Bernardino. *Anatomia chirurgica, cioè istoria anatomica dell'ossa, e muscoli del corpo humano.* Rome: Nicolò Angelo Tinassi, 1672.

Genga 1694. Genga, Bernardino. *In Hippocratis Aphorismos ad chirurgiam spectantes.* Rome: Typis Rev. Cam. Apost., 1694.

Genga and Lancisi 1691. Genga, Bernardino, and Giovanni Maria Lancisi. *Anatomia per uso et intelligenza del disegno, ricercato non solo su gl'ossi, e muscoli del corpo humano; ma dimostrata ancora su le statue antiche più insigni.* Rome: Domenico de Rossi, 1691.

Genga and Lancisi 1723. Genga, Bernardino, and Giovanni Maria Lancisi. *Anatomy Improv'd and Illustrated with Regard to the Uses Thereof in Designing: Not Only Laid Down from an Examen of the Bones and Muscles of the Human Body, but Also Demonstrated and Exemplified from the Most Celebrated Antique Statues in Rome.* London: John Senex, 1723.

Gerdy 1829. Gerdy, Pierre-Nicolas. *Anatomie des formes extérieures du corps humain: Appliquée à la peinture, à la sculpture et à la chirurgie.* Paris: Béchet Jeune, 1829.

Gilhofer & Ranschburg 1994. *Medicine.* Luzern: Gilhofer & Ranschburg, 1994.

Gilmour 1986. Gilmour, Pat. *Ken Tyler, Master Printer, and the American Print Renaissance.* New York: Hudson Hills Press, 1986.

Gogelein 1975. Gogelein, A. J. F., et al. *Leidse universiteit 400: Stichting en eerste bloei 1575–ca.1650.* Exh. cat. Amsterdam: Rijksmuseum, 1975.

Goldner 1990. Goldner, George. "A New Bronzino Drawing." *Master Drawings* 28, no. 3 (Autumn 1990): 262–64.

Gombrich 1960. Gombrich, Ernst Hans. *Art and Illusion: A Study in the Psychology of Pictorial Representation.* New York: Pantheon, 1960.

Goodrich 1933. Goodrich, Lloyd. *Thomas Eakins: His Life and Work.* New York: Whitney Museum of American Art, 1933.

Goree 2014. Goree, John A. "The Woodblocks of Vesalius and the Printings: From the Renaissance to the Modern Era," *Printing History* 16 (July 2014): 3–41.

Graciano 2019. Graciano, Andrew, ed. *Visualizing the Body in Art, Anatomy, and Medicine since 1800: Models and Modeling.* New York: Routledge, 2019.

Gray 1858. Gray, Henry. *Anatomy, Descriptive and Surgical.* London: John W. Parker and Son, 1858.

Green and Moretti 2019. Green, Daryl, and Laura Moretti, eds. *Thinking 3D: Books, Images and Ideas from Leonardo to the Present.* Oxford: Oxford Bodleian Library, 2019.

Gregory 2018. Gregory, Sharon. "Michelangelo, St Bartholomew, and Northern Italy." *Renaissance Studies* 32, no. 5 (2018): 778–805.

Griffin 2019. Griffin, Jonathan. "Tavares Strachan." In *May You Live in Interesting Times: Biennale Arte 2019*, by Ralph Rugoff et al., 1:330–31. 2 vols. Exh. cat. Venice: La Biennale di Venezia, 2019.

Griffiths 2016. Griffiths, Antony. *The Print before Photography: An Introduction to European Printmaking 1550–1820.* London: British Museum Press, 2016.

Grivel 2013. Grivel, Marianne. "Les publications de Jean Cousin Fils." In Scailliérez 2013, 256–59.

Grodecki 1986. Grodecki, Catherine. *Documents du minutier central des notaires de Paris: Histoire de l'art au XVIe siècle. II: Sculpture, peinture, broderie, émail et faïence, orfèvrerie, armures.* Paris: Archives Nationales, 1986.

Groenendijk 2008. Groenendijk, Pieter. *Beknopt biografisch lexicon van Zuid- en Noord-Nederlandse schilders, graveurs, glasschilders, tapijtwevers et cetera van ca. 1350 tot ca. 1720.* Utrecht: Groenendijk, 2008.

Gründler 2011. Gründler, Hana. "Against 'the Fatigue in Mind': Leonardo's Anatomical Drawings as Multiperspectival Epistemic Spaces." In Nova and Laurenza 2011, 131–53.

Guest 2014. Guest, Clare E. L. "Art, Antiquarianism and Early Anatomy." *Medical Humanities* 40, no. 2 (2014): 97–104.

Guillemeau 1586. Guillemeau, Jacques. *Tables anatomiques, avec les pourtraicts et declaration d'iceulx: Ensemble un denombrement de cinq cens maladies diverses.* Paris: Jean Charron, 1586.

Haddjeri 2018. Haddjeri, Maëlyss. "Le recueil d'anatomie de Bartolomeo Passerotti (1529–1592)." MA thesis, École du Louvre, 2018.

Hainy 2019. Hainy, Josh. "Grecian Theory at the Royal Academy: John Flaxman and the Pedagogy of Corporeal Representation." In Graciano 2019, 61–83.

Hallam 2016. Hallam, Elizabeth. *Anatomy Museum: Death and Body Displayed.* London: Reaktion, 2016.

Haller 1753. Haller, Albrecht von. *Icones anatomicae quibus praecipuae aliquae partes corporis humani delineatae proponuntur & arteriarum potissimum historia continetur.* Part 6, *Arteriae pectoris & brachii.* Göttingen: Widow of Abraham Vandenhoeck, 1753.

Haller 1774–77. Haller, Albrecht von. *Bibliotheca anatomica.* 2 vols. Zürich: Orell, Gessner, Fuessli, et Socc., 1774–77.

Hammelmann and Boase 1975. Hammelmann, Hanns, and T. S. R. Boase. *Book Illustrators in Eighteenth-Century England.* New Haven: Yale University Press, 1975.

Hansen 1996. Hansen, Julie. "Resurrecting Death: Anatomical Art in the Cabinet of Dr. Frederik Ruysch." *Art Bulletin* 78, no. 4 (1996): 663–79.

Hanson 2012. Hanson, Craig Ashley. "Anatomy, Newtonian Physiology and Learned Culture: The *Myotomia reformata* and Its Context within Georgian Scholarship." In *Anatomy and the Organization of Knowledge, 1500–1850,* edited by Matthew Landers and Brian Muñoz, 157–70. London: Pickering & Chatto, 2012.

Harcourt 1987. Harcourt, Glenn. "Andreas Vesalius and the Anatomy of Antique Sculpture." *Representations* 17 (1987): 28–61.

Haskell and Penny 1981. Haskell, Francis, and Nicholas Penny. *Taste and the Antique: The Lure of Classical Sculpture, 1500–1900.* New Haven: Yale University Press, 1981.

Hébert et. al. 1968. Hébert, Michèle, Edmond Pognon, and Yves Bruand, with the collaboration of Yves Sjöberg. *Inventaire du fonds français, graveurs du XVIIIe siècle.* Vol 10. Paris: Bibliothèque Nationale, 1968.

Heilmann et al. 2014. Heilmann, Maria, Nino Nanobashvili, Ulrich Pfisterer, and Tobias Teutenberg, eds. *Punkt, Punkt, Komma, Strich: Zeichenbücher in Europa, ca. 1525–1925.* Exh. cat. Passau: Dietmar Klinger, 2014.

Heilmann et al. 2015. Heilmann, Maria, Nino Nanobashvili, Ulrich Pfisterer, and Tobias Teutenberg, eds. *Lernt Zeichnen! Techniken zwischen Kunst und Wissenschaft 1525–1925.* Passau: Dietmar Klinger, 2015.

Heinen and Jaffé 2005. Heinen, Ulrich, and David Jaffé. "Écorché Figures and Anatomical Studies." In *Rubens: A Master in the Making,* edited by David Jaffé and Elizabeth McGrath, 102–10, cat. nos. 33–44. Exh. cat. London: National Gallery, 2005.

Hendrix 2015. Hendrix, Lee, ed. *Noir: The Romance of Black in Nineteenth-Century French Drawings and Prints.* Los Angeles: J. Paul Getty Museum, 2015.

Herrlinger 1970. Herrlinger, Robert. *History of Medical Illustration from Antiquity to 1600.* Translated by Graham Fulton-Smith. New York: Editions Medicina Rara, 1970.

Heseler 1959. Heseler, Baldasar. *Andreas Vesalius' First Public Anatomy at Bologna, 1540: An Eyewitness Report.* Translated and edited by Ruben Eriksson. Uppsala: Almqvist & Wiksells, 1959.

Heyenga 2013. Heyenga, Laura, ed. *Art Made from Books: Altered, Sculpted, Carved, Transformed.* San Francisco: Chronicle, 2013.

Hildebrand 1992. Hildebrand, Reinhard. "Anatomie und Revolutions des Menschenbildes." *Sudhoffs Archiv* 76, no. 1 (1992): 1–27.

Hillard 2007. Hillard, Caroline. "Michelangelo and Realdo Colombo: A Dialogue on Art and Anatomy." In *Italian Art, Society, and Politics: A Festschrift in Honor of Rab Hatfield,* edited by Barbara Deimling, Jonathan K. Nelson, and Gary M. Radke, 163–77. Florence: Syracuse University in Florence, 2007.

Hilloowala 1986. Hilloowala, Rumy. "Bartolomeo Eustachio: His Influence on Albinus and the Anatomical Models at La Specola, Florence." *Journal of the History of Medicine and Allied Sciences* 41 (October 1986), 442–48.

Hirst 1969. Hirst, Michael. "Salviati illustrateur de Vidus Vidius." *Revue de l'Art* 6 (1969): 19–28.

Hoboken 1675. Hoboken, Nicolaas. *Anatomia secundinae humanae.* Utrecht: Johannes Ribbius, 1675.

Hoffman 1939. Hoffman, Malvina. *Sculpture Inside and Out.* London: George Allen & Unwin, 1939.

Hoffman 1965. Hoffman, Malvina. *Yesterday Is Tomorrow: A Personal History.* New York: Crown Publishers, 1965.

Hollstein 1988. Hollstein, F. W. H. *Hollstein's German Engravings, Etchings and Woodcuts, 1400–1700. Vol. 14a, Hans Holbein the Younger.* Edited by Tilman Falk and compiled by Robert Zijlma. Roosendaal: Koninklijke van Poll, 1988.

Hollstein 2010. Hollstein, F. W. H. *The New Hollstein: German Engravings, Etchings and Woodcuts, 1400–1700. Wenceslaus Hollar: Part 4.* Edited by Giulia Bartrum and compiled by Simon Turner. Ouderkerk aan den Ijssel, Netherlands: Sound & Vision, 2010.

Hook and Norman 1991. Hook, Diana H., and Jeremy M. Norman. *The Haskell F. Norman Library of Science and Medicine.* 2 vols. San Francisco: J. Norman, 1991.

Hooper 1807. Hooper, Robert. *Anatomical Plates of the Bones and Muscles: Diminished from Albinus, for the Use of Students in Anatomy, and Artists.* 3rd ed. London: J. Murray, 1807.

Houtzager 2014. Houtzager, Hans. "Editions, Copies and Replicas of *Fabrica* and *Epitome,* Published between 1543 and 1725." In *Art of Vesalius,* edited by Robrecht van Hee, 191–220. Antwerp: Garant, 2014.

Hoyle and Miedema 1996. Hoyle, Michael, and Hessel Miedema. "Philips Angel. *Praise of Painting.*" *Simiolus: Netherlands Quarterly for the History of Art* 24, no. 2/3 (1996): 227–58.

Huisman 1992. Huisman, Tim. "Squares and Diopter: The Drawing System of a Famous Anatomical Atlas." *Tractrix* 4 (1992): 1–11.

Huisman 2009. Huisman, Tim. *The Finger of God: Anatomical Practice in 17th-Century Leiden.* Leiden: Primavera Pers, 2009.

Huisman 2015. Huisman, Tim. "Resilient Collections: The Long Life of Leiden's Earliest Anatomical Collections." In *The Fate of Anatomical Collections,* edited by Rina Knoeff and Robert Zwijnenberg, 58–70. Burlington, VT: Ashgate, 2015.

Hunter 1774. Hunter, William. *Anatomia uteri humani gravidi tabulis illustrata = The Anatomy of the Human Gravid Uterus Exhibited in Figures.* Birmingham: John Baskerville, 1774.

Hüttel 2008. Hüttel, Richard. *Das Evangelium des Schönen: Zeichnungen von Adam Friedrich Oeser (1717–1799).* Exh. cat. Leipzig: Museum der bildenden Künste, 2008.

Ittmann 2017. Ittmann, John, ed. *The Enchanted World of German Romantic Prints 1770–1850.* Exh. cat. Philadelphia: Philadelphia Museum of Art, 2017.

Jacobs 2005. Jacobs, Fredrika H. *The Living Image in Renaissance Art.* Cambridge: Cambridge University Press, 2005.

Jaffé 1987. Jaffé, Michael. "Rubens's Anatomy Book." In *Old Master Drawings,* Christie's, London, 6–7 July 1987, 58–61, lots 57–67.

Janson 1946. Janson, H. W. "Titian's Laocoon Caricature and the Vesalius-Galenist Controversy." *Art Bulletin* 28, no. 1 (March 1946): 49–53.

Jenty 1758. Jenty, Charles Nicholas. *The Demonstrations of a Pregnant Uterus of a Woman at Her Full Time. In Six Tables, as Large as Nature. Done from the Pictures Painted, after Dissections, by Mr. Van Riemsdyk.* London: C. N. Jenty, 1758.

Johns 1983. Johns, Elizabeth. *Thomas Eakins: Heroism of Modern Life.* Princeton: Princeton University Press, 1983.

Joly 2008. Joly, Morwena. *La leçon d'anatomie: Le corps des artistes de la Renaissance au romantisme.* Paris: Hazan, 2008.

Jordanova 1985. Jordanova, Ludmilla. "Gender, Generation and Science: William Hunter's Obstetrical Atlas." In *William Hunter and the Eighteenth-Century Medical World,* edited by W. F. Bynum and Roy Porter, 385–411. Cambridge: Cambridge University Press, 1985.

Jordanova 1989. Jordanova, Ludmilla. *Sexual Visions: Images of Gender in Science and Medicine between the Eighteenth and Twentieth Centuries.* Madison: University of Wisconsin Press, 1989.

Jorink, Lehmann, and Ramakers 2019. Jorink, Eric, Ann-Sophie Lehmann, and Bart Ramakers, eds. *Nederlands Kunsthistorisch Jaarboek: Lessons in Art: Art, Education and Modes of Instruction since 1500* 68 (2019).

Jurin 1996. Jurin, James. *The Correspondence of James Jurin (1684–1750): Physician and Secretary to the Royal Society.* Edited by Andrea A. Rusnock. Clio Medica 39, The Wellcome Institute Series in the History of Medicine. Amsterdam: Rodopi, 1996.

Karp 1985. Karp, Diane R., ed. *Ars Medica: Art, Medicine, and the Human Condition; Prints, Drawings, and Photographs from the Collection of the Philadelphia Museum of Art.* Exh. cat. Philadelphia: Philadelphia Museum of Art, 1985.

Keele and Pedretti 1978–80. Keele, Kenneth D., and Carlo Pedretti, eds. *Leonardo da Vinci: Corpus of the Anatomical Studies in the Collection of Her Majesty the Queen at Windsor Castle.* 3 vols. London: Johnson Reprint, 1978–80.

Keeley 1991. Keeley, Shelagh. *Notes on the Body.* New York: Granary Books, 1991.

Keil 1849. Keil, Georg. *Catalog des Kupferstichwerkes von Johann Friedrich Bause: Mit einigen biographischen notizen.* Leipzig: Rudoph Weigel, 1849.

Kellett 1964. Kellett, C. E. "Two Anatomies." *Medical History* 8, no. 4 (1964): 342–59.

Kemp 1970. Kemp, Martin. "A Drawing for the *Fabrica;* And Some Thoughts upon the Vesalius Muscle-Men." *Medical History* 14, no. 3 (1970): 277–88.

Kemp 1975. Kemp, Martin, ed. *Dr William Hunter at the Royal Academy of Arts.* Glasgow: University of Glasgow Press, 1975.

Kemp 1976. Kemp, Martin. "Dr William Hunter on the Windsor Leonardos and His Volume of Drawings Attributed to Pietro da Cortona." *Burlington Magazine* 118, no. 876 (March 1976): 144–48.

Kemp 1993. Kemp, Martin. "'The Mark of Truth': Looking and Learning in Some Anatomical Illustrations from the Renaissance and Eighteenth Century." In *Medicine and the Five Senses,* edited by W. F. Bynum and Roy Porter, 85–121. Cambridge: Cambridge University Press, 1993.

Kemp 1997. Kemp, Martin. "'A Perfect and Faithful Record': Mind and Body in Medical Photography before 1900." In *Beauty of Another Order: Photography in Science,* edited by Ann Thomas, 120–49. Exh. cat. New Haven: Yale University Press, 1997.

Kemp 2000. Kemp, Martin. "Babes in Bottles." In *Visualizations: The Nature Book of Art and Science,* 24–25. Berkeley: University of California Press, 2000.

Kemp 2004. Kemp, Martin. "'Il mio bell'ingenio': L'anatomia visive nel *Theatrum totius animalis fabricate* di Fabrici." In Rippa Bonati and Pardo-Tomás 2004, 83–107.

Kemp 2006. Kemp, Martin. *Leonardo da Vinci: The Marvellous Works of Nature and Man.* Rev. ed. Oxford: Oxford University Press, 2006.

Kemp 2010. Kemp, Martin. "Style and Non-Style in Anatomical Illustration: From Renaissance Humanism to Henry Gray." *Journal of Anatomy* 216 (2010): 192–208.

Kemp and Wallace 2000. Kemp, Martin, and Marina Wallace. *Spectacular Bodies: The Art and Science of the Human Body from Leonardo to Now.* Exh. cat. London: Hayward Gallery, 2000.

Kennedy and Coakley 1992. Kennedy, Brian P., and Davis Coakley, eds. *The Anatomy Lesson: Art and Medicine.* Dublin: National Gallery of Ireland, 1992.

Klestinec 2010. Klestinec, Cynthia. "Practical Experience in Anatomy." In *The Body as Object and Instrument of Knowledge: Embodied Empiricism in Early Modern Science,* edited by Charles T. Wolfe and Ofer Gal, 33–57. Dordrecht: Springer, 2010.

Knoeff 2011. Knoeff, Rina. "The Visitor's View: Early Modern Tourism and the Polyvalence of Anatomical Exhibits." In *Centres and Cycles of Accumulation in and around the Netherlands during the Early Modern Period,* edited by Lissa Roberts, 155–75. Berlin: Lit, 2011.

Knoeff 2012. Knoeff, Rina. "Sex in Public: On the Spectacle of Female Anatomy in Amsterdam around 1700." *L'homme: Europäische Zeitschrift für Feministische Geschichtswissenschaft* 23, no. 1 (2012): 43–58.

Knoeff 2015. Knoeff, Rina. "Touching Anatomy: On the Handling of Preparations in the Anatomical Cabinets of Frederik Ruysch (1638–1731)." *Studies in History and Philosophy of Biological and Biomedical Sciences* 49 (2015): 32–44.

Knox 1852. Knox, Robert. *A Manual of Artistic Anatomy: For the Use of Sculptors, Painters, and Amateurs.* London: Henry Renshaw, 1852.

Kooijmans 2011. Kooijmans, Luuc. *Death Defied: The Anatomy Lessons of Frederik Ruysch.* Translated by Diane Webb. Leiden: Brill, 2011.

Kopp 2017. Kopp, Édouard. *The Learned Draftsman: Edme Bouchardon.* Los Angeles: J. Paul Getty Museum, 2017.

Kornell 1989a. Kornell, Monique. "Anatomical Drawings by Battista Franco." *Bulletin of the Cleveland Museum of Art* 76, no. 9 (November 1989): 302–25.

Kornell 1989b. Kornell, Monique. "Rosso Fiorentino and the Anatomical Text." *Burlington Magazine* 131, no. 1041 (December 1989): 843–47.

Kornell 1993. Kornell, Monique. "Artists and the Study of Anatomy in Sixteenth-Century Italy." PhD diss., University of London, 1993.

Kornell 1996. Kornell, Monique. "The Study of the Human Machine: Books of Anatomy for Artists." In Cazort, Kornell, and Roberts 1996, 43–70.

Kornell 1998. Kornell, Monique. "Drawings for Bartolomeo Passarotti's Book on Anatomy." In *Drawing 1400–1600: Invention and Innovation,* edited by Stuart Currie, 172–88. Aldershot, UK: Ashgate, 1998.

Kornell 2000. Kornell, Monique. "Vesalius's Method of Articulating the Skeleton and a Drawing in the Collection of the Wellcome Library." *Medical History* 44, no. 1 (January 2000): 97–110.

Kornell 2007. Kornell, Monique. "The Eternal Cadaver: Anatomy and Its Representation." In *Remember Me: Constructing Immortality; Beliefs on Immortality, Life, and Death,* edited by Margaret Mitchell, 209–26. New York: Routledge, 2007.

Kornell 2011. Kornell, Monique. "Accuracy and Elegance in Cheselden's *Osteographia* (1733)." *Public Domain Review.* 22 August 2011. http://publicdomainreview.org/2011/08/22/accuracy-and-elegance-in-cheseldens-osteographia-1733/.

Kornell 2016a. Kornell, Monique. "A Dating for Domenico Bonaveri's *Notomie di Titiano.*" *Print Quarterly* 33, no. 4 (December 2016): 379–90.

Kornell 2016b. Kornell, Monique. "Edme Bouchardon's Anatomy Book for Artists." *Getty Research Journal* 8 (2016): 39–54.

Kornell 2018. Kornell, Monique. "Jan Steven van Calcar (c. 1515–c. 1546), Vesalius's Illustrator." In Canalis and Ciavolella 2018, 99–130.

Kornell 2019. Kornell, Monique. "Drawings by William Cowper (c. 1666/7–1710) for His *Myotomia reformata* (London, 1724)." *Master Drawings* 57, no. 4 (2019): 489–510.

Kotz 2004. Kotz, Mary Lynn. *Rauschenberg: Art and Life.* 2nd rev. ed. New York: Abrams, 2004.

Kroon 1911. Kroon, Just Emile. *Bijdragen tot de Geschiedenis van het Geneeskundig Onderwijs aan de Leidsche Universiteit 1575–1625.* Leiden: S. C. Van Doesburgh, 1911.

Kusukawa 2012. Kusukawa, Sachiko. *Picturing the Book of Nature: Image, Text, and Argument in Sixteenth-Century Human Anatomy and Medical Botany.* Chicago: University of Chicago Press, 2012.

Lancisi 1714. Lancisi, Giovanni Maria. *Tabulae anatomicae clarissimi viri Bartholomaei Eustachii quas è tenebris tandem vindicatas . . . praefatione, notisque illustravit, ac . . . publici juris fecit Jo. Maria Lancisius.* Rome: Francesco Gonzaga, 1714.

Landau and Parshall 1994. Landau, David, and Peter Parshall. *The Renaissance Print, 1470–1550.* New Haven: Yale University Press, 1994.

Landrus 2019. Landrus, Matthew. "Leonardo da Vinci, 500 Years On." In Green and Moretti 2019, 70–91.

Laurenza 2003. Laurenza, Domenico. *La ricerca dell'armonia: Rappresentazioni anatomiche nel Rinascimento.* Florence: Leo S. Olschki, 2003.

Laurenza 2009. Laurenza, Domenico. *Leonardo: L'anatomia.* Florence: Giunti, 2009.

Laurenza 2012. Laurenza, Domenico. *Art and Anatomy in Renaissance Italy: Images from a Scientific Revolution.* New York: Metropolitan Museum of Art, 2012.

Lautensack 1564. Lautensack, Heinrich. *Des Circkels unnd Richtscheyts auch der Perspectiva und Proportion des Menschen und Rosse, kurtze doch gründtliche underweisung deß rechten gebrauchs.* Frankfurt: Sigmund Feyerabend and Heinrich Lautensack, 1564.

Lavater 1790. Lavater, Johann Heinrich. *Anleitung zur Anatomischen Kenntniss des menschlichen Körpers für Zeichner und Bildhauer: Mit vielen Kupfertafeln, grösstentheils nach den Albinschen des Herrn Ploos von Amstel.* Zurich: Ziegler und Söhne, 1790.

Lazzaro 2016. Lazzaro, Claudia. "Michelangelo's Medici Chapel and Its Aftermath: Scattered Bodies and Florentine Identities under the Duchy." *California Italian Studies* 6, no. 1 (2016): 1–35. https://escholarship.org/uc/item/4gv6c9b5.

Leonardo 1965. Leonardo da Vinci. *Leonardo da Vinci on Painting: A Lost Book (Libro A); Reassembled from the Codex Vaticanus Urbinas 1270 and from the Codex Leicester.* Edited by Carlo Pedretti. London: P. Owen, 1965.

Leonardo 1970. Leonardo da Vinci. *The Literary Works of Leonardo* *da Vinci.* 3rd ed. 2 vols. Edited by Jean Paul Richter. Commentary by Carlo Pedretti. London: Phaidon, 1970.

Leonardo 1977. Leonardo da Vinci. *The Literary Works of Leonardo da Vinci.* 2 vols. Edited by Jean Paul Richter with commentary by Carlo Pedretti. London: Phaidon, 1977.

Leproux 2013. Leproux, Guy-Michel. "Jean Cousin Père et Fils d'après les textes." In Scaillíérez 2013, 14–23.

Lichtenstein and Michel 2006. Lichtenstein, Jacqueline, and Christian Michel, eds. *Conférences de l'Académie royale de peinture et de sculpture.* Vol. 2 of *Les conférences au temps d'Henry Testelin, 1648–1681.* Paris: École Nationale Supérieure des Beaux-Arts, 2006.

Liebert 2016. Liebert, Emily. "'Looking Also Had to Happen in Time': The Printed Trace." In *Robert Rauschenberg,* edited by Leah Dickerman and Achim Borchardt-Hume, 361–68. New York: Museum of Modern Art, 2016.

Lifchez 2009. Lifchez, Raymond. "Jean-Galbert Salvage and His *Anatomie du gladiateur combattant:* Art and Patronage in Post-Revolutionary France." *Metropolitan Museum Journal* 44 (2009): 163–84.

Lind 1975. Lind, L. R. *Studies in Pre-Vesalian Anatomy: Biography, Translations, Documents.* Philadelphia: American Philosophical Society, 1975.

Lippard 1967. Lippard, Lucy. "Robert Rauschenberg." In *Robert Rauschenberg: Booster and 7 Studies,* by Robert Rauschenberg. Los Angeles: Gemini G.E.L., 1967.

Lo 2018. Lo, Melissa. "Cut, Copy, and English Anatomy: Thomas Geminus and the Reordering of Vesalius's Canonical Body." In Canalis and Ciavolella 2018, 225–56.

Lo Bianco 1997. Lo Bianco, Anna, ed. *Pietro da Cortona: 1597–1669.* Exh. cat. Milan: Electa, 1997.

Lockhart 1974. Lockhart, R. D. *Living Anatomy: A Photographic Atlas of Muscles in Action and Surface Contours.* 7th ed. London: Faber and Faber, 1974.

Loire 1992. Loire, Stéphane. "François Tortebat." In *Simon Vouet: Actes du colloque international; Galeries nationales du Grand Palais, 5–6–7 février 1991,* edited by Stéphane Loire, 435–54. Paris: Documentation Française, 1992.

Longo and Reynolds 2016. Longo, Lawrence D., and Lawrence P. Reynolds. *Wombs with a View: Illustrations of the Gravid Uterus from the Renaissance through the Nineteenth Century.* Switzerland: Springer, 2016.

López Martínez 1928. López Martínez, Celestino. *Arquitectos, escultores y pintores vecinos de Sevilla.* Seville: Rodrigues, Giménez y C., 1928.

Lowengard 2006. Lowengard, Sarah. *The Creation of Color in Eighteenth-Century Europe.* New York: Columbia University Press, 2006.

Lukehart 2015. Lukehart, Peter M. "The Practice and Pedagogy of Drawing in the Accademia di San Luca." In Heilmann et al. 2015, 45–57.

Maclise 1851. Maclise, Joseph. *Surgical Anatomy.* Philadelphia: Blanchard and Lea, 1851.

Maclise 1856. Maclise, Joseph. *Surgical Anatomy.* 2nd ed. London: J. Churchill, 1856.

Maerker 2011. Maerker, Anna. *Model Experts: Wax Anatomies and Enlightenment in Florence and Vienna, 1775–1815.* Manchester: Manchester University Press, 2011.

Maerker 2013. Maerker, Anna. "Anatomizing the Trade: Designing and Marketing Anatomical Models as Medical Technologies, ca. 1700–1900." *Technology and Culture* 54, no. 3 (July 2013): 531–62.

Maerker 2015. Maerker, Anna. "Towards a Comparative History of Touch and Spaces of Display: The Body as Epistemic Object." *Historical Social Research/Historische Sozialforschung* 40, no. 1 (2015): 284–300.

M. A. G. 1910. M. A. G. Review of *Stereoscopic Anatomy of the Head and Neck* by D. J. Cunningham, David Waterston, Matthew H. Cryer, and Frederick E. Neres. New York: Imperial Publishing Company, 1910. *Laryngoscope* 20 (December 1910): 1185–86.

Măgureanu 2013–14. Măgureanu, Ioana. "Questions of Authorship and Authority in Some Early Modern Anatomical Images: The *Tabulae Anatomicae* of Pietro da Cortona." In Irina Vainovski-Mihai, ed., *New Europe College "Ştefan Odobleja" Program Yearbook* (2013–14): 251–92.

Malvasia 1678. Malvasia, Cesare Carlo. *Felsina pittrice: Vite de pittori bolognesi.* 2 vols. Bologna: Heirs of Domenico Barbieri, 1678.

Maral 2003. Maral, Alexandre. *Sculptures: La galerie du musée Granet.* Paris: Somogy, 2003.

Margócsy 2014. Margócsy, Dániel. *Commercial Visions: Science, Trade, and Visual Culture in the Dutch Golden Age.* Chicago: University of Chicago Press, 2014.

Margócsy 2019. Margócsy, Dániel. "From Vesalius through Ivins to Latour: Imitation, Emulation and Exactly Repeatable Pictorial Statements in the *Fabrica.*" *Word & Image* 35, no. 3 (2019): 315–33.

Margócsy, Somos, and Joffe 2018. Margócsy, Dániel, Márk Somos, and Stephen N. Joffe. *The "Fabrica" of Andreas Vesalius: A Worldwide Descriptive Census, Ownership, and Annotations of the 1543 and 1555 Editions.* Medieval and Early Modern Philosophy and Science, vol. 28. Leiden: Brill, 2018.

Marr 2014. Marr, Alexander. "Walther Ryff, Plagiarism and Imitation in Sixteenth-Century Germany." *Print Quarterly* 31, no. 2 (2014): 131–43.

Martens 1803. Martens, Franz Heinrich. *Leichtfassliche Darstellung der Theorie des Gehirn- und Schädelbaues und der daraus entspringenden physiognomischen und psychologischen Folgerungen des Herrn Dr. Gall in Wien.* Leipzig: Friedrich August Leo, 1803.

Martínez 1740. Martínez, Crisóstomo. *Nouvelle exposition de deux grandes planches gravées et dessinées d'après nature, par Chrysostome Martinez, Espagnol: Représentant des figures très singulières de proportions & d'anatomie. Ouvrage important, & utile non seulement aux médecins & aux chirurgiens mais encore à tous les peintres, sculpteurs, graveurs, dessinateurs, & généralement toutes les personnes sçavantes & curieuses de connoître exactement la structure du corps humain.* Paris: Veuve D'Houry, 1740.

Martineau and Salvesen 2016. Martineau, Paul, and Britt Salvesen. *Robert Mapplethorpe: The Photographs.* Los Angeles: J. Paul Getty Museum, 2016.

Mascagni 1787. Mascagni, Paolo. *Vasorum lymphaticorum corporis humani historia et ichnographia.* Siena: Pazzini Carli, 1787.

Mascagni 1816. Mascagni, Paolo. *Anatomia per uso degli studiosi di scultura et pittura.* Florence: Giovanni Marenigh, 1816.

Mascagni 1823–31. Mascagni, Paolo. *Anatomia universa: XLIV tabulis aeneis juxta archetypum hominis adulti accuratissime repraesentata.* Edited by Andrea Vaccà Berlinghieri, Giacomo Barzellotti, and Giovanni Rosini. Pisa: Nicola Capurro, 1823–31.

Mascagni 1833. Mascagni, Paolo. *Anatomia universale del professore Paolo Mascagni: Rappresentata con tavole in rame ridotte a minori forme di quelle della grande edizione Pisana per Antonio Serantoni disegnatore, incisore e modellatore in cera.* Florence: V. Batelli e figli, 1833.

Massari 1983. Massari, Stefania. *Giulio Bonasone.* Exh. cat. 2 vols. Rome: Quasar, 1983.

Massart, Passeri, and Ciardi 1976. Massart, Curzio, Pietro Passeri, and Roberto P. Ciardi. *Le tavole anatomiche di Paolo Mascagni.* Exh. cat. Pisa: Università degli Studi, 1976.

Massey 2005. Massey, Lyle. "Pregnancy and Pathology: Picturing Childbirth in Eighteenth-Century Obstetric Atlases." *Art Bulletin* 87, no. 1 (March 2005): 73–91.

Massey 2008. Massey, Lyle. "On Waxes and Wombs: Eighteenth-Century Representations of the Gravid Uterus." In Panzanelli 2008, 83–105.

Massey 2013. Massey, Lyle. "The Alchemical Womb: Johann Remmelin's *Catoptrum microcosmicum.*" In *Visual Cultures of Secrecy in Early Modern Europe,* edited by Timothy McCall, Sean Roberts, and Giancarlo Fiorenza, 208–28. Kirksville, MO: Truman State University Press, 2013.

Massey 2017. Massey, Lyle. "Against the 'Statue Anatomized': The 'Art' of Eighteenth-Century Anatomy on Trial." *Art History* 40, no. 1 (February 2017): 68–103.

Mattison 2003. Mattison, Robert S. *Robert Rauschenberg: Breaking Boundaries.* New Haven: Yale University Press, 2003.

Maugeri 1997. Maugeri, Vincenza. "Fialetti, Odoardo." In vol. 47 of *Dizionario biografico degli italiani.* Rome: Istituto della Enciclopedia italiana, 1997. http://www.treccani.it/enciclopedia/odoardo-fialetti_(Dizionario-Biografico)/.

Mayor 1984. Mayor, A. Hyatt. *Artists and Anatomists.* New York: Metropolitan Museum of Art, 1984.

Mazzolini 2004. Mazzolini, Renato G. "Plastic Anatomies and Artificial Dissections." In *Models: The Third Dimension of Science,* edited by Soraya de Chadaverian and Nick Hopwood, 43–70. Stanford: Stanford University Press, 2004.

McCormack 2018. McCormack, Helen. *William Hunter and His Eighteenth-Century Cultural Worlds: The Anatomist and the Fine Arts.* London: Routledge, 2018.

McGrath 2002. McGrath, Roberta. *Seeing Her Sex: Medical Archives and the Female Body.* Manchester: Manchester University Press, 2002.

Meadowcroft 1896. Meadowcroft, William H. *The A B C of the X Rays.* London: Simpkin, Marshall, Hamilton, Kent & Co., [1896].

Medici 1856. Medici, Michele. "Elogio d'Ercole Lelli." *Memorie della Accademia delle scienze dell'Istituto di Bologna,* 7 (1856):156–86.

***Mercure de France* October 1741.** "Estampes nouvelles." *Mercure de France* (October 1741): 2279–80.

***Mercure de France* July 1774.** "Arts gravures." *Mercure de France* (July 1774): 173–74.

Merz 1991. Merz, Jörg Martin. *Pietro da Cortona: Der Aufstieg zum führenden Maler in barocken Rom.* Tübingen: Ernst Wasmuth, 1991.

Merz 1997. Merz, Jörg Martin. "Cortona giovane." In Lo Bianco 1997, 55–66.

Messbarger 2010. Messbarger, Rebecca. *The Lady Anatomist: The Life and Work of Anna Morandi Manzolini.* Chicago: University of Chicago Press, 2010.

Meyer and Wirt 1943. Meyer, A. W., and Sheldon K. Wirt. "The Amuscan Illustrations." *Bulletin of the History of Medicine* 14, no. 5 (1943): 667–87.

Michel 2003. Michel, Olivier. "Gamelin, Jacques." Grove Art Online, 2003. https://doi.org/10.1093/gao/9781884446054.article.T030575.

Missirini 1823. Missirini, Melchior. *Memorie per servire alla storia della Romana Accademia di S. Luca fino alla morte di Antonio Canova.* Rome: Stamperia de Romanis, 1823.

Mittman and Sciacca 2017. Mittman, Asa Simon, and Christine Sciacca. "Robed in Martyrdom: The Flaying of St. Bartholomew in the Laudario of Sant'Agnese." In *Flaying in the Pre-Modern World: Practice and Representation,* edited by Larissa Tracy, 140–72. Cambridge: D. S. Brewer, 2017.

Moehsen 1771. Moehsen, J. C. W. *Verzeichnis einer Samlung von Bildnissen, grösstentheils berühmter Aerzte.* Berlin: Friederich Wilhelm Birnstiel, 1771.

Molhuysen 1918. Molhuysen, P. C. *Bronnen tot de Geschiedenis der Leidsche Universiteit.* The Hague: Martinus Nijhoff, 1918.

Molhuysen 1920. Molhuysen, P. C. *Bronnen tot de Geschiedenis der Leidsche Universiteit.* The Hague: Martinus Nijhoff, 1920.

Molhuysen et al. 1930. Molhuysen, P. C., P. J. Blok, and Fr. K. H. Kossmann, eds. *Nieuw Nederlandsch Biografisch Woordenboek.* Vol. 8. Leiden: A. W. Sijthoff's Uitgevers-Maatschappij N.V., 1930. http://resources.huygens.knaw.nl/retroboeken/nnbw/#source=8&page=59&view=imagePane.

Momesso 2007. Momesso, Sergio. *La collezione di Antonio Scarpa, 1752–1832.* Cittadella: Bertoncello, 2007.

Monnet 1774. Monnet, Charles. *Études d'anatomie à l'usage des peintres.* Paris: Demarteau, 1774.

Montaiglon 1887. Montaiglon, Anatole de. *Correspondance des directeurs de l'Académie de France à Rome avec les surintendants des batiments publiée d'après les manuscrits des Archives nationales.* Vol. 1, 1666–94. Paris: Charavay Frères, 1887.

Moretti 2014. Moretti, Lino. "Marco Boschini: 'Compendiosa informazione.'" *Ricche minere* 1, no. 1 (March 2014): 7–19.

Moro 1679. Moro, Giacopo. *Anotomia ridotta all'uso de' pittori, e scultori.* Venice: Giovanni Francesco Valvasense, 1679.

Morrison 2017. Morrison, Elizabeth. "The Light at the End of the Tunnel: Manuscript Illumination and the Concept of Death." In *The Ivory Mirror: The Art of Mortality in Renaissance Europe,* edited by Stephen Perkinson, 83–105. Brunswick, ME: Bowdoin College Museum of Art, 2017.

Morrisroe 1995. Morrisroe, Patricia. *Mapplethorpe: A Biography.* New York: Random House, 1995.

Mortimer 1964. Mortimer, Ruth. *Harvard College Library Department of Printing and Graphic Art Catalogue of Books and Manuscripts: Part I; French 16th Century Books.* Vol. 1. Cambridge, MA: Belknap Press of Harvard University Press, 1964.

Mozziconacci and Michel 1990. Mozziconacci, Jean-François, and Olivier Michel. *Jacques Gamelin, 1783–1803.* Collections du Musée des Beaux-Arts de Carcassonne, vol. 2. Carcassonne: Musée des Beaux-Arts, 1990.

Mulder and Beukers 1991. Mulder, W. J., and H. Beukers. "Injected Specimens in the Anatomy Museum of Leiden." In *Acte du 5e colloque des conservateurs des musées d'histoire des sciences medicales, 5 au 8 septembre 1990: Médecine en musées; Aujourd'hui, demain,* 9–17. Lyon: Collection Fondation Marcel Mérieux, 1991.

Nanobashvili 2018. Nanobashvili, Nino. *Die Ausbildung von Künstlern und Dilettanti: Das ABC des Zeichnens.* Petersberg: Michael Imhof, 2018.

Nichols 1812. Nichols, John. *Literary Anecdotes of the Eighteenth Century.* Vol. 1. London: Nichols, Son, and Bentley, 1812.

Norman 1986. Norman, Jeremy M. *The Anatomical Plates of Pietro da Cortona.* New York: Dover Publications, 1986.

Nova and Laurenza 2011. Nova, Alessandro, and Domenico Laurenza, eds. *Leonardo da Vinci's Anatomical World: Language, Context and "Disegno."* Venice: Marsilio, 2011.

Nutton 2001. Nutton, Vivian. "Representation and Memory in Renaissance Anatomical Illustration." In *Immagini per conoscere: Dal Rinascimento alla rivoluzione scientifica,* edited by Fabrizio Meroi and Claudio Pogliano, 61–79. Istituto Nazionale di Studi sul Rinascimento: Atti di Convegni, vol. 21. Florence: Leo S. Olschki, 2001.

Nutton 2013. Nutton, Vivian. *Ancient Medicine.* 2nd ed. London: Routledge, 2013.

Nutton 2014. Nutton, Vivian. "Introduction." In Vesalius 2014, 1: lxxv–cix.

Nutton 2016. Nutton, Vivian. "Vesalius and His Publishers." In *La "Fabrique" de Vésale: La mémoire d'un livre,* edited by Jacqueline Vons, 27–36. Paris: Bibliothèque Interuniversitaire de Santé, 2016. http://www.biusante.parisdescartes.fr/ressources/pdf/histmed-vesale-actes2014.pdf.

Ogilvie 2006. Ogilvie, Brian. *The Science of Describing: Natural History in Renaissance Europe.* Chicago: University of Chicago Press, 2006.

Olmi 2004. Olmi, Giuseppe, ed. *Rappresentare il corpo: Arte e anatomia da Leonardo all'Illuminismo.* Exh. cat. Bologna: Bononia University Press, 2004.

O'Malley 1958. O'Malley, Charles D. "The Anatomical Sketches of Vitus Tritonius Athesinus and Their Relationship to Vesalius's *Tabulae Anatomicae.*" *Journal of the History of Medicine and Allied Sciences* 13, no. 3 (1958): 395–97.

O'Malley 1964. O'Malley, Charles D. *Andreas Vesalius of Brussels, 1514–1564.* Berkeley: University of California Press, 1964.

O'Malley 2008. O'Malley, Charles D. "Bartholin, Thomas." In *Complete Dictionary of Scientific Biography,* 1: 482–82. Detroit: Charles Scribner's Sons, 2008.

Owen 2016. Owen, Harry. *Simulation in Healthcare Education: An Extensive History.* Cham, Germany: Springer, 2016.

Panofsky 1969. Panofsky, Erwin. *Problems in Titian, Mostly Iconographic.* London: Phaidon, 1969.

Pantin 2014. Pantin, Isabelle. "Analogy and Difference: A Comparative Study of Medical and Astronomical Images in Books, 1470–1500." In *Observing the World through Images: Diagrams and Figures in the Early-Modern Arts and Sciences,* edited by Nicholas Jardine and Isla Fay, 9–44. Leiden: Brill, 2014.

Panzanelli 2008. Panzanelli, Roberta, ed. *Ephemeral Bodies: Wax Sculptures and the Human Figure.* Los Angeles: Getty Research Institute, 2008.

Park 1994. Park, Katharine. "The Criminal and the Saintly Body: Autopsy and Dissection in Renaissance Italy." *Renaissance Quarterly* 47, no. 1 (Spring 1994): 1–33.

Park 2006. Park, Katharine. *Secrets of Women: Gender, Generation, and the Origins of Human Dissection.* New York: Zone, 2006.

Parshall 1993. Parshall, Peter. "Imago Contrafacta: Images and Facts in the Northern Renaissance." *Art History* 16, no. 4 (December 1993): 554–79.

Pellerin 2019. Pellerin, Denis. "Two Eyes Are Better Than One." In Green and Moretti 2019, 186–89.

Pennington 1982. Pennington, Richard. *A Descriptive Catalogue of the Etched Work of Wenceslaus Hollar, 1607–1677.* Cambridge: Cambridge University Press, 1982.

Petherbridge and Jordanova 1997. Petherbridge, Deanna, and Ludmilla Jordanova. *The Quick and the Dead: Artists and Anatomy.* Exh. cat. London: Royal College of Art, 1997.

Petherbridge, Ritschard, and Carlino 1998. Petherbridge, Deanna, Claude Ritschard, and Andrea Carlino, eds. *Corps à vif: Art et anatomie.* Exh. cat. Geneva: Département Municipal des Affaires Culturelles, 1998.

Petraglia 1788. Petraglia, Francisco. *Tabulae anatomicae ex archetypis egregii pictoris Petri Berrettini Cortonensis expressae et in aes incisae: Opus chirurgis et pictoribus apprime necessarium; Alteram hanc editionem recensuit nothas iconas expunxit perpetuas explicationes adjecit.* Rome: Venanzio Monaldini, 1788.

Petrioli 1740. Petrioli, Gaetano. *Riflessioni anatomiche sulle note di Monsignor Gio: Maria Lancisi fatte sopra le tavole del celebre Bartolomeo Eustachio.* Rome: Giovanni Zempel, 1740 [1741].

Petrioli 1741. Petrioli, Gaetano. *Tabulae anatomicae a celeberrimo pictore Petro Berrettino Cortonensi delineatae, & egregiè aeri incisae nunc primum prodeunt, et a Cajetano Petrioli Romano doctore, regis Sardiniae chirurgo, publico anatomico, & inter arcades Erasistrato Coo notis illustratae.* Rome: Antonio de' Rossi, 1741.

Petrioli 1750. Petrioli, Gaetano. *Le otto tavole anatomiche con cinquanta figure in foglio delineate per compimento dell'opera sublime, et imperfetta del celebre Bartolomeo Eustachio.* Rome: Antonio de'Rossi, 1750.

Phillips 2008. Phillips, Glenn, ed. *California Video: Artists and Histories.* Los Angeles: Getty Research Institute / J. Paul Getty Museum, 2008.

Pierguidi 2017. Pierguidi, Stefano. "'Tanto che basti': La 'notomia' nelle arti figurative di età barocca e una polemica tra Carlo Cesi e Carlo Maratti." *RIHA Journal* 178 (August 2017): 1–21. https://www.riha-journal.org/articles/2017/0178-pierguidi.

Pigeaud 1990. Pigeaud, Jackie. "Formes et normes dans le *De fabrica* de Vésale." In *Le corps à la Renaissance: Actes du XXXe* *coloque de Tours 1987,* edited by Jean Céard, Marie-Madeleine Fontaine, and Jean-Claude Margolin, 399–421. Paris: Aux Amateurs de Livres, 1990.

Pigozzi 2005. Pigozzi, Marinella, ed. *Il corpo in scena: I trattati di anatomia della Biblioteca Comunale Passerini-Landi.* Piacenza: Tip.Le.Co., 2005.

***Plarr's Lives of the Fellows,* no. E002613.** s.v. "Maclise, Joseph." Obituary. https://livesonline.rcseng.ac.uk/client/en_GB/default/search/results?qu=%22RCS: E002613%22&rt=false|||IDENTIFIER|||Resource+Identifier.

Platter 1583. Platter, Felix. *De corporis humani structura et usu libri III.* Basel: H. Froben, 1583.

Platter 1961. Platter, Felix. *Beloved Son Felix: The Journal of Felix Platter, a Medical Student in Montpellier in the Sixteenth Century.* Translated by Seán Jennett with a foreword by Jack Lindsay. London: Frederick Muller, 1961.

Pliny 1968. Pliny. *Natural History.* Translated by H. Rackham. Loeb Classical Library, vol. 9. London: William Heinemann, 1968.

Ploos van Amstel 1783. Ploos van Amstel, Cornelis. *Aanleiding tot de kennis der anatomie, in de tekenkunst, betreklyk tot het menschbeeld: Met eenige plaaten, en daar bygevoegde verklaaringen, opgehelderd.* Amsterdam: J. Yntema, 1783.

Portal 1770–73. Portal, Antoine. *Histoire de l'anatomie et de la chirurgie.* 6 vols. Paris: P. Fr. Didot le jeune, 1770–73.

Portmann 2014. Portmann, Maria. *L'image du corps dans l'art espagnol aux XVIe et XVIIe siècles: Autour du "Libro segundo" de Juan de Arfe y Villafañe (1585).* Natur, Wissenschaft und die Künste, vol. 14. Bern: Peter Lang, 2014.

Postle 2004. Postle, Martin. "Flayed for Art: The Écorché Figure in the English Art Academy." *British Art Journal* 5, no. 1 (Spring/Summer 2004): 55–63.

Postle and Vaughan 1999. Postle, Martin, and William Vaughan. *The Artist's Model: From Etty to Spencer.* Exh. cat. London: Merrell Holberton, 1999.

Poulet 2003. Poulet, Anne L. *Jean-Antoine Houdon: Sculptor of the Enlightenment.* Exh. cat. Washington, D.C.: National Gallery of Art, 2003.

***Print Collector's Newsletter* 1992.** Unsigned review of "Shelagh Keeley." Vol. 23, no. 2 (May–June 1992): 64.

Punt 1983. Punt, H. *Bernard Siegfried Albinus (1697–1770) on "Human Nature": Anatomical and Physiological Ideas in Eighteenth Century Leiden.* Amsterdam: B. M. Israël, 1983.

Putscher 1972. Putscher, Marielene. *Geschichte der medizinischen Abbildung: Von 1600 bis zur Gegenwart.* Munich: Heinz Moos, 1972.

Putti 1937. Putti, Vittorio. *Berengario da Carpi: Saggio biografico e bibliografico seguito dalla traduzione del "De fractura calvae sive cranei."* Bologna: L. Cappelli, 1937.

Quesnay 1749. Quesnay, François. *Histoire de l'origine et des progrès de la chirurgie en France.* Paris: Nyon, 1749.

Rees 1819–20. Rees, Abraham. *The Cyclopaedia; or, Universal Dictionary of Arts, Sciences, and Literature.* Vol. 2. London: Longman, Hurst, Rees, Orme, & Brown, 1819. s.v. "Anatomy, Picturesque."

Reilly 2004. Reilly, Patricia L. "Benvenuto Cellini's *On the Principles and Method of Learning the Art of Drawing* and the Question of Amateur Drawing Education." In *Benvenuto Cellini: Sculptor, Goldsmith, Writer,* edited by Margaret A. Gallucci and Paolo L. Rossi, 26–50. Cambridge: Cambridge University Press, 2004.

Remmelin 1619. Remmelin, Johann. *Catoptrum microcosmicum: Suis aere incisis visionibus splendens, cum historia, & pinace, de novo prodit.* Augsburg: David Franck, 1619.

Remond 2019. Remond, Jaya. "'Draw Everything That Exists in the World': *'T Light der Teken en Schilderkonst* and the Shaping of Art Education in Early Modern Northern Europe." In Jorink, Lehmann, and Ramakers 2019, 287–321.

Resta 2014. Resta, Duke. "Resurrecting Vivisection: Michelangelo among the Anatomists." *Sixteenth Century Journal* 45, no. 4 (2014): 921–50.

Riccòmini 1965. Riccòmini, Eugenio. *Mostra della scultura bolognese del Settecento.* Exh. cat. Bologna: Tamari, 1965.

Richardson, B. W. 1901. Richardson, Benjamin Ward. *Disciples of Aesculapius.* Vol. 1. New York: E. P. Dutton, 1901.

Richardson, R. 1987. Richardson, Ruth. *Death, Dissection and the Destitute.* 2nd ed. London: Phoenix Press, 2001.

Richardson, R. 2008. Richardson, Ruth. *The Making of Mr. Gray's Anatomy.* Oxford: Oxford University Press, 2008.

Richer 1895. Richer, Paul. *Physiologie artistique de l'homme en mouvement.* Paris: Octave Doin, 1895.

Rifkin, Ackerman, and Folkenberg 2006. Rifkin, Benjamin A., Michael J. Ackerman, and Judith Folkenberg. *Human Anatomy: Depicting the Body from the Renaissance to Today.* London: Thames & Hudson, 2006.

Rinaldi 2016. Rinaldi, Furio. "Denys Calvaert in Rome." *Burlington Magazine* 158, no. 1356 (March 2016): 182–88.

Rippa Bonati and Pardo-Tomás 2004. Rippa Bonati, Maurizio, and José Pardo-Tomás, eds. *Il teatro dei corpi: Le "Pitture colorate d'anatomia" di Girolamo Fabrici d'Acquapendente.* Exh. cat. Milan: Mediamed, 2004.

Riva et al. 2001. Riva, Alessandro, Beniamino Orrù, Alessio Pirino, and Francesca Testa Riva. "Iulius Casserius (1552–1616): The Self-Made Anatomist of Padua's Golden Age." *Anatomical Record* 265 (2001): 168–75.

Roani 2016. Roani, Roberta. "' . . . una Produzione che porterebbe il titolo d'Istruzione per i Giovani apprendisti, che vogliano abilitarsi nell'Arte della Scultura': Breve nota di presentazione dell'*Istruzione elementare per gli studiosi della scultura,* di Francesco Carradori." In *Accademia di belle arti di Firenze: Scultura 1784–1915,* edited by Sandro Bellesi, 123–29. Pisa: Pisa University Press, 2016.

Roberts, A 1981. Roberts, Andrew. "Hoxton House." *The Lunacy Commission: A Study of Its Origin, Emergence and Character.* 1981. http://studymore.org.uk/3_06.htm#HoxtonHouseSite.

Roberts, J. 2016. Roberts, Jennifer L. "Screenprinting and the Art of the 1960s." In *Three Centuries of American Prints from the National Gallery of Art,* edited by Judith Brodie et al., 240–44. Washington, D.C.: National Gallery of Art, 2016.

Roberts and Tomlinson 1992. Roberts, K. B., and J. D. W. Tomlinson. *The Fabric of the Body: European Traditions of Anatomical Illustration.* Oxford: Clarendon, 1992.

Roccasecca 2009. Roccasecca, Pietro. "Teaching in the Studio of the 'Accademia del Disegno dei Pittori, Scultori e Architetti di Roma' (1594–1636)." In *The Accademia Seminars: The*

Accademia di San Luca in Rome, c. 1590–1635, edited by Peter M. Lukehart, 123–59. Washington, D.C.: National Gallery of Art, 2009.

Rodari 1996. Rodari, Florian, ed. *Anatomie de la couleur: L'invention de l'estampe en couleurs.* Exh. cat. Paris: Bibliothèque Nationale de France, 1996.

Röhrl 2000. Röhrl, Boris. *History and Bibliography of Artistic Anatomy: Didactics for Depicting the Human Figure.* Hildesheim: Olms, 2000.

Roncuzzi 2016. Roncuzzi, Valeria, ed. *Le regole del corpo norma e arbitrio: Dalla tradizione di studio nell'Accademia di Belle Arti di Bologna ai linguaggi contemporanei della video arte e della performance.* Exh. cat. Bologna: Minerva, 2016.

Ronot 1970. Ronot, Henry. "Le traité d'anatomie d'Edme Bouchardon." *Bulletin de la Société de l'histoire de l'art français, année 1968* (1970): 93–100.

Rosenberg 2000. Rosenberg, Raphael. *Beschreibungen und Nachzeichnungen der Skulpturen Michelangelos: Eine Geschichte der Kunstbetrachtung.* Kunstwissenschaftliche Studien, vol. 82. Munich: Deutscher Kunstverlag, 2000.

Røstvik 2019. Røstvik, Camilla. "Remarkable Reproductions." In Green and Moretti 2019, 176–81.

Roth 1889. Roth, Moritz. "Plutonis et Harpagi dissecti dialogus, anatomisches Gedicht aus dem Anfang des 16. Jahrhunderts." *Archiv für pathologische Anatomie und Physiologie und für klinische Medicin* 117, no. 1 (July 1889): 192–98.

Roth 1892. Roth, Moritz. *Andreas Vesalius Bruxellensis.* Berlin: Georg Reimer, 1892.

Roux 1949. Roux, Marcel. *Inventaire du fonds français, graveurs du XVIIIe siècle.* Vol 6. Paris: Bibliothèque Nationale, 1949.

Royal Academy of Arts Archive, SA/34/18. s.v. "Brisbane, John." https://www.royalacademy.org.uk/art-artists/archive/dr-brisbane-great-titchfield-street-to-joshua-kirby.

Rudy 2013. Rudy, Elizabeth M. "On the Market: Selling Etchings in Eighteenth-Century France." In *Artists and Amateurs: Etching in Eighteenth-Century France,* edited by Perrin Stein, 40–67. Exh. cat. New York: Metropolitan Museum of Art, 2013.

Ruini 1599. Ruini, Carlo. *Anatomia del cavallo, infermità, et suoi rimedii.* Venice: Gasparo Bindoni the Younger, 1599.

Ruiz-Gomez 2017. Ruiz-Gomez, Natasha. "Shaking the Tyranny of the Cadaver: Doctor Paul Richer and the 'Living Écorché.'" In *Bodies beyond Borders: Moving Anatomies, 1750–1950,* edited by Kaat Wils, Raf de Bont, and Sokhieng Au, 231–57. Leuven: Leuven University Press, 2017.

Rupp 1990. Rupp, Jan. "Matters of Life and Death: The Social and Cultural Conditions of the Rise of Anatomical Theatres, with Special Reference to Seventeenth Century Holland." *History of Science* 28, no. 3 (1990): 263–87.

Russell 1952. Russell, K. F. "The *De dissectione* of Charles Estienne." *Australian and New Zealand Journal of Surgery* 22, no. 2 (November 1952): 147–48.

Russell 1953. Russell, K. F. "The *De re anatomica* of Realdo Colombo." *Australian and New Zealand Journal of Surgery* 22, no. 3 (February 1953): 225–26.

Russell 1954. Russell, K. F. "The *Osteographia* of William Cheselden." *Bulletin of the History of Medicine* 28, no. 1 (1954): 32–49.

Russell 1987. Russell, K. F. *British Anatomy 1525–1800: A Bibliography of Works Published in Britain, America and on the Continent.* 2nd ed. Winchester: St Paul's Bibliographies, 1987.

Ruysch 1703. Ruysch, Frederik. *Thesaurus anatomicus tertius cum figuris aeneis = Het derde anatomisch cabinet.* Amsterdam: Johannes Wolters, 1703.

Ruysch 1744. Ruysch, Frederik. *Alle de Ontleed- Genees- en Heelkundige Werken.* Amsterdam: Janssoons van Waesberge, 1744.

Saint-Igny [1630]. Saint-Igny, Jean de. *Elemens de pourtraiture; ou, La metode de representer et pourtraire toutes les parties du corps humain.* Paris: François Langlois dit Chartres, [1630].

Sanders 2005. Sanders, Mark A. "William Cowper and His Decorated Copperplate Initials." *Anatomical Record (Part B: The New Anatomist)* 282B, no. 1 (2005): 5–12.

Salvage 1812. Salvage, Jean-Galbert. *Anatomie du gladiateur combattant, applicable aux beaux arts; ou, Traité des os, des muscles, du mécanisme des mouvemens, des proportions et des caractères du corps humain.* Paris: Jean-Galbert Salvage & l'imprimerie de Mame, 1812.

San Juan 2008. San Juan, Rose Marie. "Restoration and Translation in Juan de Valverde's *Historia de la composision del cuerpo humano.*" In *The Virtual Tourist in Renaissance Rome: Printing and Collecting the "Speculum Romanae Magnificentiae,"* edited by Rebecca Zorach, 53–61. Chicago: University of Chicago, 2008.

San Juan 2012. San Juan, Rose Marie. "The Turn of the Skull: Andreas Vesalius and the Early Modern Memento Mori." *Art History* 35, no. 5 (2012): 959–75.

Sappol 2006. Sappol, Michael. *Dream Anatomy.* Exh. cat. Bethesda, MD: National Library of Medicine, 2006.

Sappol 2017. Sappol, Michael. "Anatomy's Photography: Objectivity, Showmanship, and the Reinvention of the Anatomical Image, 1860–1950." *Remedia,* 23 January 2017. https://remedianetwork.net/2017/01/23/anatomys-photography-objectivity-showmanship-and-the-reinvention-of-the-anatomical-image-1860–1950/.

Saunders and O'Malley 1950. Saunders, J. B. de C. M., and Charles D. O'Malley. *The Illustrations from the Works of Andreas Vesalius of Brussels: With Annotations and Translations, a Discussion of the Plates and Their Background, Authorship and Influence, and a Biographical Sketch of Vesalius.* Cleveland: World Publishing, 1950.

Savio 1968. Savio, Pietro. "Ricerche sull'anatomico Guglielmo Riva." *Bollettino storico-bibliografico subalpino* 66 (1968): 229–67.

Savio 1971. Savio, Pietro. "Ricerche sui medici e chirurghi dell'Ospedale di Santo Spirito in Sassia sec. XVI–XVII." *Archivio della Società Romana di Storia Patria* 94 (1971): 145–68.

Sawday 1995. Sawday, Jonathan. *The Body Emblazoned: Dissection and the Human Body in Renaissance Culture.* London: Routledge, 1995.

Scailliérez 2013. Scailliérez, Cécile, ed. *Jean Cousin, père et fils: Une famille de peintres au XVIe siècle.* Exh. cat. Paris: Musée du Louvre: Somogy, 2013.

Scarpa 1794. Scarpa, Antonio. *Tabulae neurologicae ad illustrandam historiam anatomicam cardiacorum nervorum, noni nervorum cerebri, glossopharyngaei, et pharyngaei ex octavo cerebri.* Pavia: Baldassare Comini, 1794.

Scarpa 1832. Scarpa, Antonio. *Engravings of the Cardiac Nerves, the Nerves of the Ninth Pair, the Glosso-Pharyngeal, and the Pharyngeal Branch of the Pneumo-Gastric.* Translated by Robert Knox. 3rd ed. Edinburgh: Maclachlan & Stewart, 1832.

Scheurleer 1975. Scheurleer, Th. H. Lunsingh. "Un Amphithéâtre d'Anatomie Moralisée." *Leiden University in the Seventeenth Century: An Exchange of Learning,* edited by Th. H. Lunsingh Scheurleer and G. H. M. Posthumus Meyjes, 217–77. Leiden: Universitaire Pers Leiden, 1975.

Schiebinger 1986. Schiebinger, Londa. "Skeletons in the Closet: The First Illustrations of the Female Skeleton in Eighteenth-Century Anatomy." *Representations* 14 (1986): 42–82.

Schiebinger 1989. Schiebinger, Londa. *The Mind Has No Sex? Women in the Origins of Modern Science.* Cambridge, MA: Harvard University Press, 1989. Reprint, Cambridge, MA: Harvard University Press, 1993.

Schmidt 2011. Schmidt, Suzanne Karr, with Kimberly Nichols. *Altered and Adorned: Using Renaissance Prints in Daily Life.* Exh. cat. Chicago: Art Institute of Chicago, 2011.

Schnalke 1995. Schnalke, Thomas. "Anatomie fur alle! Trew und sein Projekt eines anatomischen Tafelwerks." In *Natur im Bild: Anatomie und Botanik in der Sammlung des Nürnberger Arztes Christoph Jacob Trew,* edited by Thomas Schnalke, 99–129. Exh. cat. Erlangen: Universitätsbibliothek, 1995.

Schraven 2014. Schraven, Minou. *Festive Funerals in Early Modern Italy: The Art and Culture of Conspicuous Commemoration.* Farnham, Surrey: Ashgate, 2014.

Schwartz 2008. Emmanuel Schwartz. "L'anatomie face à l'antique: De l'usage du moulage dans l'enseignement académique." In Comar 2008b, 82–87.

Scott 2003. Scott, Katie. "Invention and Privilege in Early Modern France: The Case of Color." In *Les chemins de la nouveauté: Innover, inventer au regard de l'histoire,* edited by Liliane Hilaire-Pérez and Anne-Françoise Garçon, 167–86. Paris: Comité des Travaux Historiques et Scientifiques, 2003.

Sellink 1992. Sellink, Manfred. "'As a Guide to the Highest Learning': An Antwerp Drawing Book Dated 1589." *Simiolus: Netherlands Quarterly for the History of Art* 21, no. 1/2 (1992): 40–56.

Sénéchal 1998. Sénéchal, Philippe. "*L'Anatomie du Gladiateur combatant* de Jean-Galbert Salvage: Science et art à Paris sous l'Empire." In *Curiosité: Études d'histoire de l'art en l'honneur d'Antoine Schnapper,* edited by Olivier Bonfait, Véronique Gerard Powell, and Philippe Sénéchal, 219–28. Paris: Flammarion, 1998.

Sewell 1982. Sewell, Darrel. *Thomas Eakins: Artist of Philadelphia.* Exh. cat. Philadelphia: Philadelphia Museum of Art, 1982.

Sharpe 1818. Sharpe, James Birch. *Elements of Anatomy; Designed for the Use of Students in the Fine Arts.* London: R. Hunter, 1818.

Shedd 1991. Shedd, Meredith. "Prometheus the Primeval Sculptor: Archaeology and Anatomy in Émeric-David's *Recherches sur l'art statuaire.*" *Zeitschrift für Kunstgeschichte* 54, no. 1 (1991): 88–106.

Shotwell 2018. Shotwell, Allen. "The Illustrated Dissection: Drawings in the Extant Student Notes from Vesalius's Dissections." In *Towards the Authority of Vesalius: Studies on Medicine and the Human Body from Antiquity to the Renaissance and Beyond,* edited by Erika Gielen and Michèle Goyens, 383–402. Turnhout: Brepols, 2018.

Silver and Wyckoff 2008. Silver, Larry, and Elizabeth Wyckoff, eds. *Grand Scale: Monumental Prints in the Age of Dürer and Titian.* Exh. cat. Wellesley, MA: Davis Museum and Cultural Center, Wellesley College, 2008.

Simblet 2001. Simblet, Sarah. *Anatomy for the Artist.* New York: Dorling Kindersley, 2001.

Simons and Kornell 2008. Simons, Patricia, and Monique Kornell. "Annibal Caro's After-Dinner Speech (1536) and the Question of Titian as Vesalius's Illustrator." *Renaissance Quarterly* 61, no. 4 (2008): 1069–97.

Simpson and Burnet 1908–9. Simpson, G. F. Barbour, and Edward Burnet, eds. *The Edinburgh Stereoscopic Atlas of Obstetrics.* 4 vols. London: Caxton Publishing, 1908–9.

Siraisi 1994. Siraisi, Nancy G. "Vesalius and Human Diversity in *De humani corporis fabrica.*" *Journal of the Warburg and Courtauld Institutes* 57 (1994): 60–88.

Sjöberg 1974. Sjöberg, Yves. *Inventaire du fonds français, graveurs du XVIIIe siècle.* Vol. 13. Paris: Bibliothèque Nationale, 1974.

Skaarup 2015. Skaarup, Bjørn Okholm. *Anatomy and Anatomists in Early Modern Spain.* London: Routledge, 2015.

Sørenson 2002. Sørensen, Bent. "The Enduring Vitality of a Flayed Horse: Carlo Ruini, Bouchardon and Others." *Apollo* 155, no. 481 (2002): 30–39.

Squanquerillo 1841. Squanquerillo, Costantino. *Trattato di anatomia pittorica.* Rome: Tipografia delle Belle Arti, 1841.

Staden 1992. Staden, Heinrich von. "The Discovery of the Body: Human Dissection and Its Cultural Contexts in Ancient Greece." *Yale Journal of Biology and Medicine* 65, no. 3 (1992): 223–41.

Stafford 1991. Stafford, Barbara Maria. *Body Criticism: Imaging the Unseen in Enlightenment Art and Medicine.* Cambridge, MA: MIT Press, 1991.

Steinberg 1982. Steinberg, Leo. "Michelangelo and the Doctors." *Bulletin of the History of Medicine* 56, no. 4 (1982): 543–53.

Sterzi 1910. Sterzi, Giuseppe. "Giulio Casseri: Anatomico e Chirurgo (c. 1552–1616), parte II." *Nuovo Archivio Veneto* 19 (1910): 25–111.

Stetskevich 2019. Stetskevich, Elena S. "Drawing Education at the Russian Academy of Sciences in the First Half of the Eighteenth Century." In *Drawing Education: Worldwide! Continuities—Transfers—Mixtures,* edited by Nino Nanobashvili and Tobias Teutenberg, 147–66. Heidelberg: Heidelberg University Publishing, 2019. https://doi.org/10.17885/heiup.457.

Stoichita 2013. Stoichita, Victor I., ed. *Le corps transparent.* Rome: "L'Erma" di Bretschneider, 2013.

Swammerdam 1737. Swammerdam, Jan. *Biblia naturae; Sive Historia insectorum.* Leiden: Isaak Severinus, Boudewyn Vander Aa, Pieter Vander Aa, 1737.

Swan 1995. Swan, Claudia. "*Ad vivum, naer het leven,* From the Life: Defining a Mode of Representation." *Word & Image* 11, no. 4 (1995): 353–72.

Syson et al. 2018. Syson, Luke, Sheena Wagstaff, Emerson Bowyer, and Brinda Kumar, with contributions by Bharti Kher et al. *Like Life: Sculpture, Color, and the Body.* New York: Metropolitan Museum of Art, 2018.

Talvacchia 1999. Talvacchia, Bette. *Taking Positions: On the Erotic in Renaissance Culture.* Princeton: Princeton University Press, 1999.

Teeuwisse 2011. Teeuwisse, Nicolaas. "Skull of a Seven-Month-Old Child." *Selected Works XI.* https://www.teeuwisse.de/catalogues/adam-friedrich-oeser-skull-of-a-seven-month-old-child.

Thieme, Becker, and Vollmer 1925. Thieme, Ulrich, Felix Becker, and Hans Vollmer, eds. *Allgemeines Lexikon der bildenden Künstler: Von der Antike bis zur Gegenwart,* vol. 18. Leipzig: Seemann, 1925.

Tinney 1743. Tinney, John. *Compendium Anatomicum; or A Compendious Treatise of Anatomy Adapted to the Arts of Painting and Sculpture. . . .* London: John Tinney, 1743.

Tortebat and de Piles 1668. Tortebat, François [and Roger de Piles]. *Abregé d'anatomie, accommodé aux arts de peinture et de sculpture, et mis dans un order nouveau, dont la methode est tres-facile, & débarassée de toutes les difficultez & choses inutiles, qui ont toûjours esté un grand obstacle aux peintres, pour arriver à la perfection de leur art.* Paris: Tortebat, 1668.

Travers 2019. Travers, Erin. "Jacob van der Gracht's *Anatomie* for Artists." In Jorink, Lehmann, and Ramakers 2019, 250–84.

Trey 2016. Trey, Juliette, with Hélène Grollemund. *Inventaire general des dessins du Musée du Louvre; École Française: Edme Bouchardon (1698–1762): Catalogue raisonné.* Paris: Mare & Martin; Musée du Louvre, 2016.

Trew 1767. Trew, Christoph Jacob. *Tabulae osteologicae.* Nuremberg: n.p., 1767.

Turner 1878. Turner, Edouard. "Ce que sont devenues les planches de Vésale, publiées en 1543 dans le grand ouvrage d'anatomie et dans l'Epitome." *Gazette Hebdomadaire de Médecine et de Chirurgie* 25 (1878): 49–58, 65–78, 113–19, 129–41, 161–66, 177–94.

Turner, Hendrix, and Plazzotta 1997. Turner, Nicholas, Lee Hendrix, and Carol Plazzotta. *European Drawings 3: Catalogue of the Collections.* Los Angeles: J. Paul Getty Museum, 1997.

Valverde 1556. Valverde de Amusco, Juan. *Historia de la composicion del cuerpo humano.* Rome: Antonio Salamanca and Antonio Lafrerij, 1556.

Van de Passe 1643–44. Van de Passe, Crispijn. *La prima-[quinta] parte della luce del dipingere et disegnare.* Amsterdam: Crispijn van de Passe, 1643–44.

Van den Spiegel 1645. Van den Spiegel, Adriaan. *Opera quae extant, omnia.* Edited by Johannes Antonides van der Linden. Amsterdam: Joan Blaeu, 1645.

Van der Aa 1712. Van der Aa, Pieter. *Les delices de Leide, une des célébres villes de l'Europe.* Leiden: Pieter van der Aa, 1712.

Van der Gracht 1634. Van der Gracht, Jacob. *Anatomie der wtterlicke deelen van het menschelick lichaem.* The Hague: Jacob van der Gracht, 1634.

Van de Roemer 2010. Van de Roemer, Gijsbert M. "From *Vanitas* to Veneration: The Embellishments in the Anatomical Cabinet of Frederik Ruysch." *Journal of the History of Collections* 22, no. 2 (2010): 169–86.

Van de Roemer 2011. Van de Roemer, Gijsbert M. "Regulating the Arts: Willem Goeree versus Samuel van Hoogstraten." In Erik Jorink and Bart Ramakers, eds., *Nederlands Kunsthistorisch Jaarboek: Art and Science in the Early Modern Netherlands* 61 (2011): 185–203.

Van Glabbeek and Biesbrouck 2020. Van Glabbeek, Francis, and Maurits Biesbrouck. "Giovanni Battista Canani (1515–1579ns) and Andreas Vesalius (1514–1564)." In Van Hee 2020b, 101–48.

Van Hee 2020a. Van Hee, Robrecht. "In the Footsteps of Vesalius: Plantin's Anatomy Editions of Juan Valverde and David van Mauden." In Van Hee 2020b, 149–203.

Van Hee 2020b. Van Hee, Robrecht, ed. *In the Shadow of Vesalius.* Antwerp: Garant, 2020

Van Hoogstraten 1678. Van Hoogstraten, Samuel. *Inleyding tot de Hooge Schoole der Schilderkonst, anders de Zichtbaere Werelt.* Rotterdam: François van Hoogstraten, 1678.

Van Hoogstraten 2021. Van Hoogstraten, Samuel. *Samuel van Hoogstraten's "Introduction to the Academy of Painting; or, The Visible World."* Edited by Celeste Brusati. Translated by Jaap Jacobs. Los Angeles: Getty Research Institute, 2021.

Van Leeuwenhoek 1677. van Leeuwenhoek, Antonie. "Observationes D. Anthonii Lewenhoeck, de natis è semine genitali animalculis." *Philosophical Transactions of the Royal Society* 12, no. 142 (1677): 1040–43.

Van Mander 1994–99. Van Mander, Karel. *The Lives of the Illustrious Netherlandish and German Painters, from the First Edition of the Schilder-boeck (1603–1604).* Edited by Hessel Miedema 6 vols. Doornspijk: Davaco, 1994–99.

Vannoni 1838. Vannoni, Pietro. *Biografia di Antonio Serantoni: Disegnatore, incisore e lavoratore di cere anatomiche.* Florence: V. Batelli e figli, 1838.

Vannugli 1997. Vannugli, Antonio. "Carlo Cesi." In Lo Bianco 1997, 257–64.

Vasari 1568. Vasari, Giorgio. *Le vite de' più eccellenti pittori, scultori, et architettori.* 3 parts in 2 vols. Florence: Giunti, 1568.

Venturoli 1977. Venturoli, Paolo. "Carradori, Francesco." *Dizionario Biografico degli Italiani* 20 (1977). http://www.treccani.it/enciclopedia/francesco-carradori_(Dizionario-Biografico)/.

Vesalius 1538. Vesalius, Andreas. *Tabulae anatomicae sex.* Venice: Bernardino dei Vitali, 1538.

Vesalius 1543a. Vesalius, Andreas. *De humani corporis fabrica libri septem.* Basel: J. Oporinus, 1543.

Vesalius 1543b. Vesalius, Andreas. *De humani corporis fabrica librorum epitome.* Basel: J. Oporinus, 1543.

Vesalius 1555. Vesalius, Andreas. *De humani corporis fabrica libri septem.* 2nd rev. ed. Basel: J. Oporinus, 1555.

Vesalius 1559. Vesalius, Andreas. *Beschreibung und Antzeigung Mannes unnd Weibes innerlicher Glider in xij cupffere figuren warafftichlich unnd gar kunstlich verfaßt, und getzogen auß die große Anatomey des hochgelertten Philosophen un Artzen Doctor Andree Vesalij; nitt allein tzo nutz den Artzen und Wundartzen, aber auch einen yeden der die kunst lieb hatt.* N.p., 1559.

Vesalius 1560. Vesalius, Andreas. *Description et demonstration des membres interieurs de l'homme & de la femme, en douze Tables, tirées au naturel, selon la vraye Anatomie de André Wesal, Philosophe, & Docteur en Medicine. Oeuvre util & necessaire non seulement aux Medicins & Chirurgiens, ains aussi aux Portrayeurs, & Architectes.* Lyon: Clément Baudin, 1560.

Vesalius 1566. Vesalius, Andreas. *Vivae imagines partium corporis humani aereis formis expressae.* Antwerp: Christophe Plantin, 1566.

Vesalius 1934. Vesalius, Andreas. *Andreae Vesalii Bruxellensis Icones anatomicae.* Munich: Bremer, 1934.

Vesalius 2014. Vesalius, Andreas. *The Fabric of the Human Body: An Annotated Translation of the 1543 and 1555 Editions of "De Humani Corporis Fabrica Libri Septem."* Edited and translated by Daniel H. Garrison and Malcolm H. Hast. 2 vols. Basel: Karger, 2014.

Vesalius 2015. Vesalius, Andreas. *Vesalius: "The China Root Epistle."* Edited and translated by Daniel H. Garrison. New York: Cambridge University Press, 2015.

Vidius 1544. Vidius, Vidus. *Chirurgia è Graeco in Latinum conversa.* Paris: Pierre Gaultier, 1544.

Vincent and Perrot 2016. Vincent, Jean-François, and Chloé Perrot. "Johannes Van Horne and Marten Sagemolen's Myology: Four Volumes of Anatomical Drawings of the Golden Age Rediscovered at the Bibliothèque Interuniversitaire de Santé (Paris)." BIU Santé, 2016. https://www.biusante.parisdescartes.fr/ressources/pdf/van-horne_en.pdf.

Vogel 1999. Vogel, Klaus. "The Transparent Man—Some Comments on the History of a Symbol." In *Manifesting Medicine,* edited by Robert Bud, Bernard Finn, and Helmuth Trischler, 31–61. London: Science Museum, 2004.

Vogel 2017. Vogel, Gerd-Helge. *Adam Friedrich Oeser: Götterhimmel und Idylle; Zum 300. Geburtstag des Künstlers.* Exh. cat. Berlin: Lukas, 2017.

Vrolijk 2017. Vrolijk, Arnoud. "Arabic Studies in the Netherlands and the Prerequisite of Social Impact—a Survey." In *The Teaching and Learning of Arabic in Early Modern Europe,* edited by Jan Loop, Alastair Hamilton, and Charles Burnett, 13–32. Leiden: Brill, 2017.

Wagner 2017. Wagner, Corinna. "Replicating Venus: Art, Anatomy, Wax Models, and Automata." *19: Interdisciplinary Studies in the Long Nineteenth Century* 24 (2017): 1–27.

Walker 1787. Walker, John. *The Artist's Pocket Companion: Being a Compendious Treatise of Anatomy, Adapted to the Arts of Designing, Painting, and Sculpture. . . .* 2nd ed. London: John Walker, 1787.

Walters 2009. Walters, Laura M. "Odoardo Fialetti (1573–c. 1638): The Interrelation of Venetian Art and Anatomy, and His Importance in England." PhD diss., 2 vols., University of St. Andrews, 2009.

Warthin 1931. Warthin, Aldred Scott. *The Physician of the Dance of Death: A Historical Study of the Evolution of the Dance of Death Mythus in Art.* New York: Paul B. Hoeber, 1931.

Webster 2011. Webster, Mary. *Johan Zoffany, 1733–1810.* New Haven: Yale University Press, 2011.

Weems 2008. Weems, Jason. "Scanning Bob's Body: Rauschenberg's *Booster,* Science, and the Techno-Visual Subject." *Bulletin of the Detroit Institute of Arts* 82, no. 1/2 (2008): 4–15.

Werbel 2007. Werbel, Amy Beth. *Thomas Eakins: Art, Medicine, and Sexuality in Nineteenth-Century Philadelphia.* New Haven: Yale University Press, 2007.

Weststeijn 2008. Weststeijn, Thijs. *The Visible World: Samuel van Hoogstraten's Art Theory and the Legitimation of Painting in the Dutch Golden Age.* Amsterdam: Amsterdam University Press, 2008.

Wilson 1896. Wilson, Edward L. "The Romance of Ro-ent-gen-ra." *Wilson's Photographic Magazine* 33, no. 477 (September 1896): 416–19.

Wilton 2012. Wilton, Andrew. "Full-length Écorché Figure ?1791 by Joseph Mallord William Turner." In *J. M. W. Turner: Sketchbooks, Drawings and Watercolours,* edited by David Blayney Brown. London: Tate Research Publications, 2012. https://www.tate.org.uk/art/research-publications/jmw-turner/joseph-mallord-william-turner-full-length-corche-figure-r1140308.

Winckelmann 2006. Winckelmann, Johann Joachim. *History of the Art of Antiquity.* Translated by Harry Francis Mallgrave. Los Angeles: Getty Research Institute, 2006.

Witkam 1970. Witkam, Henricus Joannes. *De Dagelijkse Zaken van de Leidse Universiteit van 1582 tot 1596.* Vol. 1. Leiden: H. J. Witkam, 1970.

Witkam 1980. Witkam, Henricus Joannes. *Catalogues of All the Chiefest Rarities in the Publick Anatomie Hall of the University of Leyden.* Leiden: H. J. Witkam, 1980.

Witkowski 1878–88. Witkowski, Gustave-Joseph. *Human Anatomy and Physiology. Part IV. A Movable Atlas Showing the Mechanism of Vision; The Eye; The Organs of Vision.* Translated and edited by Henry Power. London: Ballière, Tindall and Cox, 1878–88.

Wolf-Heidegger and Cetto 1976. Wolf-Heidegger, Gerhard, and Anna Maria Cetto. *Die anatomische Sektion in bildlicher Darstellung.* Basel: Karger, 1967.

Woollett 2021. Woollett, Anne T. "Visualizing an Epic Past: Rubens, Antiquity, and the Art of Painting." In Anne T. Woollett, Davide Gasparotto, and Jeffrey Spier, eds. *Rubens: Picturing Antiquity,* 5–27. Exh. cat. Los Angeles: J. Paul Getty Museum, 2021.

Wragge-Morley 2016. Wragge-Morley, Alexander. "Connoisseurship and the Communication of Anatomical Knowledge: The Case of William Cheselden's *Osteographia* (1733)." In *The Material Cultures of Enlightenment Arts and Sciences,* edited by Adriana Craciun and Simon Schaffer, 271–89. London: Palgrave Macmillan, 2016.

Zanotti 1739. Zanotti, Giampietro. *Storia dell'Accademia Clementina di Bologna Aggregata all'Instituto delle scienze e dell'arti.* 2 vols. Bologna: Lelio dalla Volpe, 1739.

Zeitlin & Ver Brugge 1961. Zeitlin & Ver Brugge, Booksellers. *The Medical Sciences: Rare and Important Books Illustrating the History of Medicine and Related Sciences.* Pasadena, CA: Castle Press, 1961.

contributors

Monique Kornell specializes in the history of anatomical book illustration and the study of anatomy by artists. She has published articles on anatomical drawings and prints, and on illustrated anatomy books from the sixteenth to the eighteenth century, with a particular interest in anatomy books for artists. Dr. Kornell was a cocurator of *The Ingenious Machine of Nature: Four Centuries of Art and Anatomy,* an exhibition of prints and drawings organized by the National Gallery of Canada in 1996. She is a guest assistant professor in the Program in the History of Medicine, Graduate School of Biomedical Sciences, Cedars-Sinai Medical Center, Los Angeles.

Thisbe Gensler is a research associate at the Getty Research Institute. She is a historian of nineteenth-century visual culture, focusing specifically on the intersections of art and medicine. Her research examines the biomedical body, with a special interest in representations of sexuality and reproduction in both aesthetic and clinical contexts. Her work considers the ethics and epistemology of vision, realism and the real, and political propaganda and feminist theory. She has written on the history of gynecology, anatomical illustration, and stereoscopic imagery in medicine.

Naoko Takahatake is curator of prints and drawings at the Getty Research Institute. Previously, she was curator of prints and drawings at the Los Angeles County Museum of Art, where she coorganized the exhibition *The Serial Impulse at Gemini G.E.L.* (2016) and curated the exhibition *The Chiaroscuro Woodcut in Renaissance Italy* (2018), for which she also edited the catalog.

Erin Travers is a specialist of seventeenth-century Dutch art and visual culture. Her interests include the early modern body, works on paper, knowledge production, and the history of collecting. These topics were explored in her PhD dissertation, "Boundaries of the Body: The Art of Anatomy in the Seventeenth-Century Netherlands" (2018), which was generously supported by the Samuel H. Kress Foundation, the University of California Humanities Research Institute, the Albert and Elaine Borchard Foundation, and the Social Sciences and Humanities Research Council of Canada.

illustration credits

Photographs of items in the holdings of the Getty Research Institute are courtesy the Research Institute. The following sources have granted additional permission to reproduce illustrations in this volume.

Chapter One

Page xii: Thomas Sinclair (Scottish, ca. 1805–81), after Joseph Maclise (Irish, ca. 1815–ca. 1880), Dissection of chest and abdomen (detail). See cat. no. 53.

Figs. 1.2, 1.8. Library Special Collections for Medicine and the Sciences, Louise M. Darling Biomedical Library, UCLA.

Figs. 1.4, 1.7. Wellcome Collection.

Chapter Two

Page 14: Cornelis Huyberts (Dutch, 1669/70–ca. 1712), Fetal skeletons in an anatomical assemblage (detail). See cat. no. 6.

Figs. 2.4, 2.6, 2.7. J. Paul Getty Museum.

Fig. 2.10. The Huntington Library, San Marino, California.

Fig. 2.11. Bibliothèque de Institut National d'Histoire de l'Art, Collections Jacques Doucet, 12 Res 2100 (1-2).

Chapter Three

Page 24: Title page showing dissection scene (detail), from Realdo Colombo, *De re anatomica libri XV* (1559). See cat. no. 16.

Fig. 3.7. Österreichische Nationalbibliothek.

Chapter Four

Page 34: After Charles Errard (French, ca. 1606/9–89), *Statue of the Gladiator in the Villa Borghese, . . . Considered Anatomically* (detail). See cat. no. 32.

Figs. 4.4, 4.6. Wellcome Collection.

Fig. 4.9. Elaine Lesser / Alamy Stock Photo.

Chapter Five

Page 46: Antonio Cattani (Italian, active 1777–ca. 1790), after Ercole Lelli (Italian, 1702–66), Écorché figure, seen from the back (detail), 1781. See cat. no. 37.

Fig. 5.2. bpk-Bildagentur / Herzog Anton Ulrich Museum / Art Resource, NY.

Fig. 5.5. Schloss Friedenstein Gotha Foundation.

Fig. 5.6. © Photo Josse / Bridgeman Images.

Chapter Six

Page 58: Nikolaj Utkin (Russian, 1780–1868), after Jean-Galbert Salvage (French, 1770–1813), The head of the Apollo Belvedere, rendered anatomically (detail). See cat. no. 34.

Fig. 6.1. Michele Falzone / Alamy Stock Photo.

Fig. 6.2. UCLA Grunwald Center for the Graphic Arts, Hammer Museum. Bequest of Walter Otto Schneider.

Fig. 6.3. Marcel / Alamy Stock Photo.

Figs. 6.4, 6.8, 6.10. J. Paul Getty Museum.

Fig. 6.5. © National Gallery, London / Art Resource, NY.

Fig. 6.6. © RMN-Grand Palais / Art Resource, NY / Photo: Thierry Le Mage.

Fig. 6.7. Scala / Art Resource, NY.

Chapter Seven

Page 70: The Leiden anatomy theater (detail), from Pieter van der Aa, *Les delices de Leide, une des célébres villes de l'Europe* (1712). See cat. no. 7.

Fig. 7.1. Museum Bleulandinum, UMC Utrecht. This specimen is cataloged by the Museum Bleulandium, University Medical Centre, as "Child's Head with Turkish Cap."

Fig. 7.2. Amsterdam Museum.

Chapter Eight

Page 82: Lucas Kilian (German, 1579–1637), after Johann Remmelin (German, 1583–1632), *Visio catoptri microcosmici tertia* (detail). See cat. no. 51.

Fig. 8.1. bpk-Bildagentur / Weimar Kunstsammlungen zu Weimar Schlossmuseum / Olaf Mokansky / Art Resource, NY.

Figs. 8.2, 8.5, 8.7. Library Special Collections for Medicine and the Sciences, Louise M. Darling Biomedical Library, UCLA.

Figs. 8.3a, 8.3b. © Raffaello Bencini / Bridgeman Images.

Fig. 8.4. Whipple Museum of the History of Science, University of Cambridge.

Fig. 8.6. Bundesarchiv, Bild 102-16746 / Photo: Georg Pahl.

Catalog

Page 95: Antonio Cattani (Italian, active 1777–ca. 1790), after Ercole Lelli (Italian, 1702–66), Écorché figure, seen from the front (detail), 1780. See cat. no. 36.

Cat. nos. 13, 41. Huntington Library, San Marino, California.

Cat. nos. 15, 16, 33, 39. Library Special Collections for Medicine and the Sciences, Louise M. Darling Biomedical Library, UCLA.

Cat. no. 15, fig. D. Rijksmuseum File#RP-T-1940-288 http://hdl.handle.net/10934/RM0001.COLLECT.310709.

Cat. no. 16, fig. E. National Trust Photo Library / Art Resource, NY

Cat. no. 30, fig. I. J. Paul Getty Museum.

Cat. no. 35, fig. J. © Vanni Archive / Art Resource, NY.

Cat. nos. 36, 37, 38, fig. K. Renée DeVoe Mertz.

Cat. nos. 36, 37, 38, fig. L. © Alma Mater Studiorum University of Bologna / Sistema Museale di Ateneo / Photo: Fulvio Simoni.

Cat. no. 42. Library Special Collections, Charles E. Young Research Library, UCLA.

Cat. no. 43, fig. M. © Bernard Terlay / Musée Granet, Ville d'Aix-en-Provence.

Cat. no. 46. © 2021 Robert Rauschenberg Foundation and Gemini G.E.L., LLC / Licensed by VAGA at Artists Rights Society (ARS), NY / Published by Gemini G.E.L., LLC / Digital Image © 2021 Museum Associates / LACMA / Licensed by Art Resource, NY.

Cat. no. 47. Photo by Andrea D'altoè Neonlauro / Courtesy of the Artist & Marian Goodman Gallery.

Cat. no. 48. Digital Image © 2022 Museum Associates / LACMA / Licensed by Art Resource, NY.

Cat. no. 48, fig. P. Hirshhorn Museum and Sculpture Garden Collection.

Cat. no. 49a, b. Courtesy of the Cynthia Maughan Estate.

Cat. no. 50. Gift of the Robert Mapplethorpe Foundation to the J. Paul Getty Trust and the Los Angeles County Museum of Art. © Robert Mapplethorpe Foundation. Used by permission.

Cat. no. 52, fig. Q. Universitätsbibliothek Heidelberg / *Tabulae neurologicae ad illustrandam historiam anatomicam cardiacorum nervorum, noni nervorum cerebri, glossopharyngaei, et pharyngaei ex octavo cerebri* / pl. 5.

Cat. no. 56. © James Allen, 2014. Library Special Collections for Medicine and the Sciences, Louise M. Darling Biomedical Library, UCLA.

index

Page numbers in *italics* refer to illustrations.

This publication is issued on the occasion of the exhibition *Flesh and Bones: The Art of Anatomy,* on view at the Getty Research Institute at the Getty Center, Los Angeles, from 22 February to 10 July 2022.

Getty Research Institute Publications Program
Mary E. Miller, *Director, Getty Research Institute*
Gail Feigenbaum, *Associate Director*

Published by the Getty Research Institute, Los Angeles
Getty Publications
1200 Getty Center Drive, Suite 500
Los Angeles, California 90049-1682
getty.edu/publications

Lauren Edson, *Editor*
Kurt Hauser, *Designer*
Michelle Woo Deemer, *Production*
Karen Ehrmann and Pam Moffat, *Image and Rights Acquisition*
Tina Henderson, *Typesetter*

Distributed in the United States and Canada by the University of Chicago Press

Distributed outside the United States and Canada by Yale University Press, London

Type composed in Adobe Univers Condensed and Adobe Warnock Pro

Separations by iocolor, Seattle

Printed in China by Artron Art Group

Library of Congress Cataloging-in-Publication Data
Names: Kornell, Monique, author. | Travers, Erin, contributor. | Gensler, Thisbe, contributor. | Takahatake, Naoko, 1977- contributor. | Getty Research Institute, host institution, issuing body.
Title: Flesh and bones : the art of anatomy / by Monique Kornell.
Description: Los Angeles : Getty Research Institute, [2022] | Issued on the occasion of an exhibition on view at the Getty Research Institute at the Getty Center, Los Angeles, from 22 February to 10 July 2022. | Includes bibliographical references and index. | Summary: "Focused on the intersection of science and art, this book explores themes of anatomy from the Renaissance to modern times"— Provided by publisher.
Identifiers: LCCN 2021035906 (print) | LCCN 2021035907 (ebook) | ISBN 9781606067697 (hardback) | ISBN 9781606067703 (pdf)
Subjects: LCSH: Anatomy, Artistic—Exhibitions. | Medical illustration—History—Exhibitions. | Human figure in art—Exhibitions. | Medicine and art—Exhibitions.
Classification: LCC NC760 .F59 2022 (print) | LCC NC760 (ebook) | DDC 704.9/42—dc23
LC record available at https://lccn.loc.gov/2021035906
LC ebook record available at https://lccn.loc.gov/2021035907

Front cover: Antonio Cattani (Italian, active 1777–ca. 1790), after Ercole Lelli (Italian, 1702–66), Écorché figure, seen from the back (detail), 1781. See cat. no. 37.

Back cover: Lucas Kilian (German, 1579–1637), after Johann Remmelin (German, 1583–1632), *Third Vision from Mirrors of the Microcosm* (detail). See cat. no. 51.

Endpapers: Jean Bosq (French, active 1801–40), after Jean-Galbert Salvage (French, 1770–1813), *Muscles of the Arm and of the Hand* (detail). See fig. 8.8.

Frontispiece: After Jan Steven van Calcar (North Netherlandish, ca. 1515–ca. 1546), Profile skeleton (detail). See fig. 2.8.

Title page: Arnauld Éloi Gautier Dagoty (French, 1741–before 1780), after Jan Wandelaar (Dutch, 1690–1759), Sixth muscle figure (detail). See fig. 1.9.

Page iv: (*Left*) Domenico Bonaveri (Italian, 1653–1731), after Jan Steven van Calcar (North Netherlandish, ca. 1515–ca. 1546), Lamenting or praying skeleton (detail). See cat. no. 2. (*Right*) After Nicolas Beatrizet (French, 1507 or 1515–ca. 1565), after Gaspar Becerra (Spanish, ca. 1520–ca. 1570), after Jan Steven van Calcar (North Netherlandish, ca. 1515–ca. 1546), Walking muscle figure in profile with two views of the anatomy of the eye (detail). See cat. no. 9.

Page v: (*Left*) Pacifico Battistelli (Italian, active late 1830s–1840), after Costantino Squanquerillo (Italian, active late 1830s–1840), *The Superficial Muscles of the Neck* (detail), 1837. See fig. 2.12. (*Right*) Carlo Cesi (Italian, ca. 1622–82), Foreshortened écorché with a raised leg (detail). See cat. no 20.

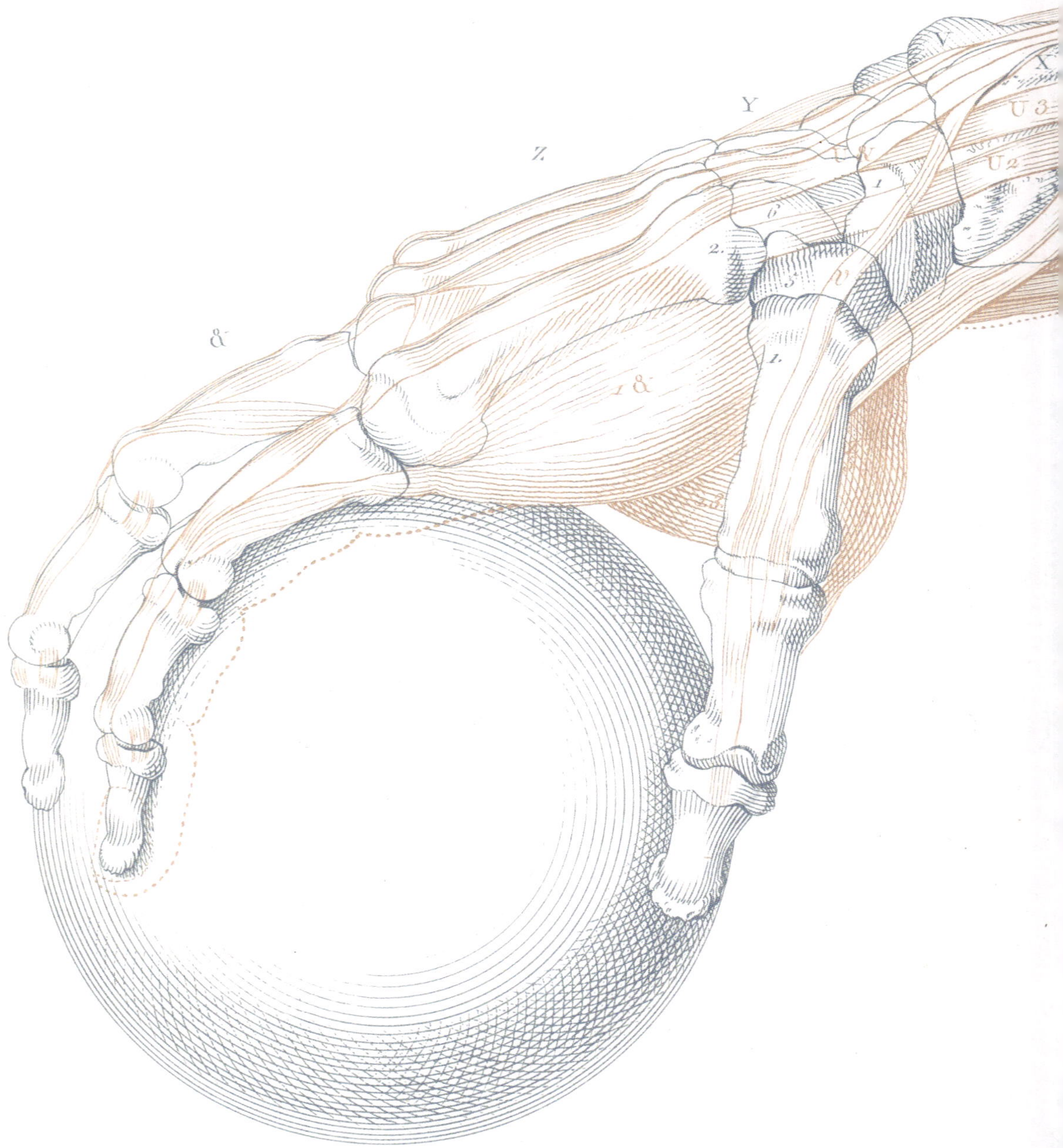

MUSCLES DU BRAS

Xx *Biceps.*

VV *Brachial.*

Z 4 *Emi*

1.3. *Add*

du tendon